The Future of Post-Human Data Analysis

The Future of Post-Human Data Analysis

A Preface to a New Theory of Quantitative and Qualitative Research

Peter Baofu

Viva Books

New Delhi | Mumbai | Chennai | Kolkata | Bengaluru | Hyderabad | Kochi | Guwahati

First Indian Edition 2012

VIVA BOOKS PRIVATE LIMITED

- 4737/23 Ansari Road, Daryaganj, New Delhi 110 002
  E-mail: vivadelhi@vivagroupindia.net, Tel. 42242200
- 76, Service Industries, Shirvane, Sector 1, Nerul, Navi Mumbai 400 706
  E-mail: vivamumbai@vivagroupindia.net, Tel. 27721273, 27721274
- Jamals Fazal Chambers, 26 Greams Road, Chennai 600 006
  E-mail: vivachennai@vivagroupindia.net, Tel. 28290304, 28294241
- B-103, Jindal Towers, 21/1A/3 Darga Road, Kolkata 700 017
  E-mail: vivakolkata@vivagroupindia.net, Tel. 22836381, 22816713
- 7, Sovereign Park Apartments, 56-58, K. R. Road, Basavanagudi, Bangalore 560 004
  E-mail: vivabangalore@vivagroupindia.net, Tel. 26607409, 26607410
- 101-102 Mughal Marc Apartments, 3-4-637 to 641, Narayanguda, Hyderabad 500 029
  E-mail: vivahyderabad@vivagroupindia.net, Tel. 27564481, 27564482
- Beevi Towers, First Floor, SRM Road, Kaloor, Kochi 682 018, Kerala
  E-mail: vivakochi@vivagroupindia.net, Tel: 0484-2403055, 2403056
- 232, GNB Road, Beside UCO Bank, Silpukhuri, Guwahati 781 003
  E-mail: vivaguwahati@vivagroupindia.net, Tel: 0361-2666386

www.vivagroupindia.com

Published by arrangement with
Cambridge International Science Publishing
7, Meadow Walk, Great Abington
Cambridge CB1 6AZ, UK

ISBN : 978-81-309-1994-2

Published by Vinod Vasishtha for Viva Books Private Limited, 4737/23, Ansari Road, Daryaganj, New Delhi-110 002. Printed and bound by Syndicate Binders, Noida.

To Those beyond Quantitative and Qualitative Research in Future History

BOOKS ALSO BY PETER BAOFU

Category I: Social Sciences and Related Fields

- 43. *The Future of Post-Human Gambling* (2011)
- 42. *The Future of Post-Human Education* (2011)
- 41. *The Future of Post-Human War and Peace* (2010)
- 40. *The Future of Post-Human Law* (2010)
- 39. *The Future of Post-Human Organization* (2009)
- 38. *The Future of Post-Human Mass Media* (2009)
- 37. *The Rise of Authoritarian Liberal Democracy* (2007)
- 36. *Beyond the World of Titans, and the Remaking of World Order* (2007)
- 35. *Beyond Capitalism to Post-Capitalism* (2005)
- 34. Volume 1: *Beyond Democracy to Post-Democracy* (2004)
- 33. Volume 2: *Beyond Democracy to Post-Democracy* (2004)
- 32. *The Future of Capitalism and Democracy* (2002)

Category II: Formal Sciences and Related Fields

- 31. *The Future of Post-Human Computing* (2011)
- 30. *The Future of Post-Human Formal Science* (2010)
- 29. *The Future of Post-Human Language* (2009)
- 28. *The Future of Post-Human Geometry* (2009)
- 27. *The Future of Post-Human Mathematical Logic* (2008)
- 26. *The Future of Information Architecture* (2008)

Category III: Natural Sciences and Related Fields

- 25. *The Future of Post-Human Chemistry* (2011)
- 24. *The Future of Post-Human Acoustics* (2011)
- 23. *The Future of Post-Human Geology* (2010)
- 22. *Beyond Cosmology to Post-Cosmology* (2010)
- 21. *The Future of Post-Human Personality* (2010)
- 20. *The Future of Post-Human Sexuality* (2010)
- 19. *The Future of Post-Human Urban Planning* (2009)
- 18. *The Future of Post-Human Engineering* (2009)
- 17. *The Future of Post-Human Unconsciousness* (2008)
- 16. *The Future of Complexity* (2007)
- 15. *Beyond Nature and Nurture* (2006)
- 14. *The Future of Post-Human Space-Time* (2006)
- 13. *The Future of Post-Human Consciousness* (2004)

Category IV: The Humanities and Related Fields

- 12. *The Future of Post-Human Literature* (2011)
- 11. *The Future of Post-Human Humor* (2011)
- 10. *Beyond Ethics to Post-Ethics* (2011)
- 9. *The Future of Post-Human Religion* (2010)
- 8. *The Future of Post-Human Chess* (2010)
- 7. *The Future of Post-Human Martial Arts* (2009)
- 6. *The Future of Post-Human Creative Thinking* (2009)
- 5. *The Future of Post-Human Knowledge* (2008)
- 4. *The Future of Aesthetic Experience* (2007)
- 3. *Beyond Civilization to Post-Civilization* (2006)
- 2. Volume 1: *The Future of Human Civilization* (2000)
- 1. Volume 2: *The Future of Human Civilization* (2000)

CONTENTS

Part Three: Qualitative Research

Part Four: Conclusion

TABLES

Category III: Visions on the Mind

Category IV: Visions on History

Category VII. Visions on Society (Socio-Political)

Category VIII. Visions on Society (Socio-Economic)

Category IX: Visions on Culture

FOREWORD

The Information Age has brought on a tsunami of information that threatens to overwhelm the human brain. Luckily, humankind has also come upon data analysis, whereby tons of information can be sifted, sorted and made understandable, so that it is useful.

Dr. Peter Baofu has rendered a great service to present and future researchers by taking a critical look at the techniques used in data analysis. By surveying the field, he highlights how some techniques may be more appropriate than others in certain contexts; what benefits flow from certain techniques, and what shortcomings exist in them.

It looks like data analysis is here to stay, for without it there can be no progress. Dr. Baofu's original contribution and his vision of the future great transformations of data analysis propose to make it more valuable.

Sylvan Von Burg
School of Business
George Washington University

ACKNOWLEDGMENTS

This book is written to challenge the conventional wisdom on data analysis and to replace it with an alternative novel way of thinking on the future fate of data analysis (in relation to quantitative and qualitative research).

Unsurprisingly, this book receives no external funding nor help from any formal organization or institution, because of its political incorrectness—as this is something that I often stressed in all my previous books.

My only reward is that joy of conceiving something original in intellectual history.

In any event, I greatly appreciate the foreword by Sylvan von Burg at George Washington University.

And as always, I bear the sole responsibility for the ideas presented in this book.

ABBREVIATIONS

ALD = Peter Baofu. 2007. *The Rise of Authoritarian Liberal Democracy: A Preface to a New Theory of Comparative Political Systems*. Cambridge, England: Cambridge Scholars Publishing, Ltd.

BCIV = Peter Baofu. 2006. *Beyond Civilization to Post-Civilization: Conceiving a Better Model of Life Settlement to Supersede Civilization*. NY: Peter Lang Publishing, Inc.

BCPC = Peter Baofu. 2005. *Beyond Capitalism to Post-Capitalism: Conceiving a Better Model of Wealth Acquisition to Supersede Capitalism*. NY: The Edwin Mellen Press.

BCOS = Peter Baofu. 2010. *Beyond Cosmology to Post-Cosmology: A Preface to a New Theory of Different Worlds*. Cambridge, England: Cambridge International Science Publishing, Ltd.

BDPD1 = Peter Baofu. 2004. Volume 1. *Beyond Democracy to Post-Democracy: Conceiving a Better Model of Governance to Supersede Democracy*. NY: The Edwin Mellen Press.

BDPD2 = Peter Baofu. 2004. Volume 2. *Beyond Democracy to Post-Democracy: Conceiving a Better Model of Governance to Supersede Democracy*. NY: The Edwin Mellen Press.

BEPE = Peter Baofu. 2011. *Beyond Ethics to Post-Ethics: A Preface to a New Theory of Morality and Immorality*. Charlotte, NC: Infomration Age Publishing.

BNN = Peter Baofu. 2006. *Beyond Nature and Nurture: Conceivng a Better Way to Understand Genes and Memes*. Cambridge, England: Cambridge Scholars Publishing, Ltd.

BWT = Peter Baofu. 2007.. *Beyond the World of Titans, and the Renaking of World Order: A Preface to a New Logic of Empire-Building*. Cambridge, England: Cambridge Scholars Publishing, Ltd.

FAE = Peter Baofu. 2007. *The Future of Aesthetic Experience: Conceiving a Better Way to Understand Beauty, Ugliness and the Rest*. Cambridge, England: Cambridge Scholars Publishing, Ltd.

FC = Peter Baofu. 2007. *The Future of Complexity: Conceiving a Better Way to Understand Order and Chaos.* London, United Kingdom: World Scientific Publishing Co.

FCD = Peter Baofu. 2002. *The Future of Capitalism and Democracy.* MD: The University Press of America.

FHC1 = Peter Baofu. 2000. Volume 1. *The Future of Human Civilization.* NY: The Edwin Mellen Press.

FHC2 = Peter Baofu. 2000. Volume 2. *The Future of Human Civilization.* NY: The Edwin Mellen Press.

FIA = Peter Baofu. 2008. *The Future of Information Architecture: Conceiving a Better Way to Understand Taxonomy, Network, and Intelligence.* Oxford, England: Chandos Publishing (Oxford) Limited.

FPHA = Peter Baofu. 2011. *The Future of Post-Human Acoustics: A Preface to a New Theory of Sound and Silence.* Cambridge, England: Cambridge International Science Publishing, Ltd.

FPHC = Peter Baofu. 2004. *The Future of Post-Human Consciousness.* NY: The Edwin Mellen Press.

FPHCHEM = Peter Baofu. 2011. *The Future of Post-Human Chemistry: A Preface to a New Theory of Substances and their Changes.* England: Cambridge Scholars Publishing, Ltd.

FPHCHESS = Peter Baofu. 2010. *The Future of Post-Human Chess: A Preface to a New Theory of Tactics and Strategy.* Cambridge, England: Cambridge International Science Publishing, Ltd.

FPHCOM = Peter Baofu. 2011. *The Future of Post-Human Computing: A Preface to a New Theory of Hardware and Software.* Cambridge, England: Cambridge International Science Publishing, Ltd.

FPHCT = Peter Baofu. 2009. *The Future of Post-Human Creative Thinking: A Preface to a New Theory of Invention and Innovation.* Cambridge, England: Cambridge Scholars Publishing, Ltd.

FPHDA = Peter Baofu. 2011. *The Future of Post-Human Data Analysis: A Preface to a New Theory of Quantitative and Qualitative Research.* Cambridge, England: Cambridge International Science Publishing, Ltd.

FPHE = Peter Baofu. 2009. *The Future of Post-Human Engineering: A Preface to a New Theory of Technology.* Cambridge, England: Cambridge Scholars Publishing, Ltd.

FPHEDU = Peter Baofu. 2011. *The Future of Post-Human Education: A Preface to a New Theory of Teaching and Learning.* Cambridge, England: Cambridge International Science Publishing, Ltd.

FPHFS = Peter Baofu. 2010. *The Future of Post-Human Formal Science: A Preface to a New Theory of Abstraction and Application.* Cambridge, England: Cambridge Scholars Publishing, Ltd.

FPHG = Peter Baofu. 2009. *The Future of Post-Human Geometry: A Preface to a New Theory of Infinity, Symmetry, and Dimensionality.* Cambridge, England: Cambridge Scholars Publishing, Ltd.

FPHGAM = Peter Baofu. 2011. *The Future of Post-Human Gambling: A Preface to a New Theory of Risk and Caution.* Cambridge, England: Cambridge International Science Publishing, Ltd.

FPHGEOL = Peter Baofu. 2010. *The Future of Post-Human Geology: A Preface to a New Theory of Statics and Dynamics.* Cambridge, England: Cambridge International Science Publishing, Ltd.

FPHH = Peter Baofu. 2011. *The Future of Post-Human Humor: A Preface to a New Theory of Joking and Laughing.* Cambridge, England: Cambridge International Science Publishing, Ltd.

FPHK = Peter Baofu. 2008. *The Future of Post-Human Knowledge: A Preface to a New Theory of Methodology and Ontology.* Oxford, England: Chandos Publishing (Oxford) Limited.

FPHL = Peter Baofu. 2009. *The Future of Post-Human Language: A Preface to a New Theory of Structure, Context, and Learning.* Cambridge, England: Cambridge Scholars Publishing, Ltd.

FPHLAW = Peter Baofu. 2010. *The Future of Post-Human Law: A Preface to a New Theory of Necessity, Contingency, and Justice.* Cambridge, England: Cambridge Scholars Publishing, Ltd.

FPHLIT = Peter Baofu. 2011. *The Future of Post-Human Literature: A Preface to a New Theory of Fiction and Non-Fiction.* Cambridge, England: Cambridge International Science Publishing, Ltd.

FPHMA = Peter Baofu. 2009. *The Future of Post-Human Martial Arts: A Preface to a New Theory of the Body and Spirit of Warriors.* Cambridge, England: Cambridge Scholars Publishing, Ltd.

FPHML = Peter Baofu. 2008. *The Future of Post-Human Mathematical Logic: A Preface to a New Theory of Rationality.* Cambridge, England: Cambridge Scholars Publishing, Ltd.

FPHMM = Peter Baofu. 2009. *The Future of Post-Human Mass Media: A Preface to a New Theory of Technology.* Cambridge, England: Cambridge Scholars Publishing, Ltd.

FPHO = Peter Baofu. 2009. *The Future of Post-Human Organization: A Preface to a New Theory of Communication, Decision-Making, and Leadership.*.Cambridge, England: Cambridge Scholars Publishing, Ltd.

FPHP = Peter Baofu. 2010. *The Future of Post-Human Personality: A Preface to a New Theory of Normality and Abnormality.* Cambridge, England: Cambridge International Science Publishing, Ltd.

FPHR = Peter Baofu. 2010. *The Future of Post-Human Religion: A Preface to a New Theory of Spirituality.* Cambridge, England: Cambridge International Science Publishing, Ltd.

FPHS = Peter Baofu. 2010. *The Future of Post-Human Sexuality: A Preface to a New Theory of the Body and Spirit of Love-Makers.* Cambridge, England: Cambridge Scholars Publishing, Ltd.

FPHST = Peter Baofu. 2006. *The Future of Post-Human Space-Time: Conceivng a Better Way to Understand Space and Time.* New York: Peter Lang Publishing, Inc.

FPHU = Peter Baofu. 2008. *The Future of Post-Human Unconsciousness: A Preface to a New Theory of Anomalous Experience.* Cambridge, England: Cambridge Scholars Publishing, Ltd.

FPHUP = Peter Baofu. 2009. *The Future of Post-Human Urban Planning: A Preface to a New Theory of Density, Void, and Sustainability.* Cambridge, England: Cambridge Scholars Publishing, Ltd.

FPHWP = Peter Baofu. 2010. *The Future of Post-Human War and Peace: A Preface to a New Theory of Aggression and Pacificity.* Cambridge, England: Cambridge Scholars Publishing, Ltd.

• PART ONE •

Introduction

CHAPTER 1
INTRODUCTION—THE SIGNIFICANCE OF DATA ANALYSIS

> The law is this: that each of our leading conceptions, each branch of our knowledge, passes successively through three different theoretical conditions: the Theological, or fictitious; the Metaphysical, or abstract; and the Scientific, or positive.
> —Auguste Comte (1853)

The Blank Check of Data Analysis

Is data analysis (based on modern scientific methodology) really so promising in helping us understand reality better that, as Auguste Comte (1853) once famously wrote, "[t]he law is this: that each of our leading conceptions, each branch of our knowledge, passes successively through three different theoretical conditions: the Theological, or fictitious; the Metaphysical, or abstract; and the Scientific, or positive"?

However, this blank check of data analysis (based on modern scientific methodology) can be contrasted with an opposing view in our time, as Thomas Kuhn equally made a name of himself in arguing that the progress of scientific methodology over time does not lead to a truer understanding of reality but only reflects "paradigm shifts" for different worldviews. (WK 2011)

Should we then side with the view of Comte or that of Kuhn—or maybe half of each? Contrary to these conflicting views (and other ones as will be discussed in the book), data analysis (in relation to both quantitative and qualitative research) is neither possible nor desirable to the extent that the respective ideologues (on different sides) would like us to believe.

This challenge to these diverse views about data analysis does not mean, however, that data analysis is useless, or that those fields of study (related to data analysis) like methodology, epistemology, philosophy of science, psychology, linguistics, mathematics, logic, sociology, anthropology, and so on should be ignored too. Of course, neither of these extreme views is reasonable.

Instead, this book provides an alternative (better) way to understand the future of data analysis, especially in the dialectic context of quantitative and qualitative research—while learning from different approaches in the literature but without favoring any one of them (nor integrating them, since they are not necessarily compatible with each other).

In other words, this book offers a new theory (that is, *the complex theory of data analysis*) to go beyond the existing approaches in an original way.

If successful, this seminal project is to fundamentally change the way that we think about data analysis, from the combined perspectives of the mind, nature, society, and culture, with enormous implications for the human future and what I originally called its "post-human" fate.

Quantitative and Qualitative Research in Data Analysis

A good point of departure is to define the term "data analysis" at the outset, which refers to "a process of inspecting, cleaning, transforming, and modeling data with the goal of highlighting useful information, suggesting conclusions, and supporting decision making. Data analysis has multiple facets and approaches, encompassing diverse techniques under a variety of names, in different business, science, and social science domains." (WK 2011)

The term "data analysis" should not be confused with some related terms; for instance, "data modeling" is "sometimes used as a synonym" for "data analysis," whereas "data integration" is "a precursor to data analysis." (WK 2011)

In any event, data analysis is done with the focus or "research" question of a study in mind, and the word "data" in data analysis refers to information which can be distinguished in terms of two main types of data for research, namely, (a) "quantitative" data and (b) "qualitative" data. (WK 2011)

This then suggests that the study of data analysis requires an examination of both quantitative and qualitative research (with the two types of data aforecited), as described below (and summarized in *Table 1.1*).

Quantitative vs. Qualitative Data

The terms "quantitative" and "qualitative" need to be defined at the outset, as addressed below, in that order.

Quantitative Data

On the one hand, the term "quantitative" means the converse of qualitative data, in that "items are more precisely described as data in terms of quantity and in which numerical values are used"—and "results can be displayed using tables, charts, histograms and graphs." (WK 2011c & 2011e)

Good examples of quantitative data are shown below, for illustration: (WK 2011e & 2011f)

- "Counts"
  –Ex: "there are 643 dots on the ceiling"
  –Ex: "there are 735 pieces of bubble gum"
  –Ex: "there are 8 planets in the solar system"
- "Measurements"
  –Ex: "the length of this table is 1.892m"
  –Ex: "the temperature at 12:00 was 18.9 degrees Celsius"
- "Types"
  –Ex: "extensive quantity" (e.g., "energy, volume and mass")
  –Ex: "intensive quantity" (e.g., "density and pressure")

Of course, there can be other examples, so the list above is solely illustrative (not exhaustive).

Qualitative Data

On the other hand, the term "qualitative" means "a non-numerical data collection or explanation based on the attributes of the graph or source of data. For example, if you are asked to explain in qualitative terms a thermal image displayed in multiple colours, then you would explain the colour differences rather than the heat's numerical value." (WK 2011b)

Examples of qualitative data are shown below, for illustration: (WK 2011g)

- Ex: "the results of open-ended interviews"
- Ex: "notes of direct observation"
- Ex: "written documents (answers to questionnaires, diaries, program records, etc.)"

And the research techniques which researchers use to collect these qualitative data can be many, and hereafter are a few examples, for illustration: (WK 2011g)

- Ex: "participant observation"
- Ex: "direct observation"
- Ex: "unstructured interviewing"
- Ex: "case studies"
- Ex: "content analysis"
- Ex: "protocol analysis"
- Ex: "focus groups"

Needless to say, there can be other examples, so the list above is solely illustrative (not exhaustive).

Quantitative vs. Qualitative Research

With these examples of quantitative and qualitative data for research in mind—there are some fundamental distinctions between quantitative and qualitative research, as explained below (and summarized in *Table 1.1*).. (WK 2011b)

- Purposefulness—receives more attention in qualitative research, because "cases can be selected purposefully, according to whether or not they typify certain characteristics or contextual locations." But in quantitative research, the factor of purposefulness is minimized.
- Researcher's Role—receives "greater critical attention" in qualitative research, because "the possibility of the researcher taking a 'neutral' or transcendental position is seen as more problematic in practical and/or philosophical terms. Hence qualitative researchers are often exhorted to reflect on their role in the research process and make this clear in the analysis." But in quantitative research, neutrality is emphasized.

- Meaning—receives a greater attention in qualitative research, because of its interest in vagueness, ambivalence, and subjectivity embedded in "language, signs and meaning." But in quantitative research, preciseness and clarity are emphasized.
- Context and the Whole—are more important in qualitative research, because "qualitative research approaches analysis holistically and contextually, rather than being reductionistic and isolationist." But in quantitative research, reductionism is more tolerated.
- Exploration—is more pervasive, especially in the social sciences for qualitative research, when "qualitative methods are used for exploration (i.e., hypothesis-generating) or for explaining puzzling quantitative results. Quantitative methods, by contrast, are used to test hypotheses. This is because establishing content validity—do measures measure what a researcher thinks they measure?—is seen as one of the strengths of qualitative research. Some consider quantitative methods to provide more representative, reliable and precise measures through focused hypotheses, measurement tools and applied mathematics. By contrast, qualitative data is usually difficult to graph or display in mathematical terms."
- Policy and Program Evaluation—are more appropriate for "qualitative research since it can answer certain important questions more efficiently and effectively than quantitative approaches. This is particularly the case for understanding how and why certain outcomes were achieved (not just what was achieved) but also for answering important questions about relevance, unintended effects and impact of programs such as: Were expectations reasonable? Did processes operate as expected? Were key players able to carry out their duties? Did the program cause any unintended effects?"
- Diversity and Flexibility—are more valued in qualitative research, which "allow[s]…more diversity in responses as well as the capacity to adapt to new developments or issues during the research process itself. While qualitative research can be expensive and time-consuming to conduct, many fields of research employ qualitative techniques that have been specifically developed to provide more succinct, cost-efficient and timely results. Rapid Rural Appraisal is one formalized example of these adaptations but there are many others."

Of course, there can be other distinctions between quantitative and qualitative research, so the examples above are solely illustrative (not exhaustive), nor are the distinctions as listed absolute in comparison.

In any event, the point here is not to favor either the qualitative method or the quantitative one but to understand their differences. After all, "the use of one or other type of method has become a matter of controversy and even ideology, with particular schools of thought within each discipline favouring one type of method and pouring scorn on to the other. Advocates of quantitative methods argue that only by using such methods can the social sciences become truly scientific; advocates of qualitative methods argue that quantitative methods tend to obscure the reality of the social phenomena under study because they underestimate or neglect the non-measurable factors, which may be the most important." (WK 2011d)

Stages of Data Analysis

However, both quantitative and qualitative research have something fundamental in common, and a good illustration concerns data analysis as a process.

As defined above, data analysis is "a process of inspecting, cleaning, transforming, and modeling data with the goal of highlighting useful information, suggesting conclusions, and supporting decision making." (WK 2011a)

The process of data analysis can be understood in terms of two stages, namely, (a) initial data analysis and (b) main data analysis, as explained below.

Initial Data Analysis

Initial data analysis phase differs from the main analysis phase in that "during initial data analysis, one refrains from any analysis that are aimed at answering the original research question. The initial data analysis phase is guided by the following four questions," as shown below, for illustration: (WK 2011a)

- "Checks on Data Cleaning"—are important. For instance, "have decisions influenced the distribution of the variables? The distribution of the variables before data cleaning is compared to the distribution of the variables after data cleaning to see whether data cleaning has had unwanted effects on the data."
- "Analysis of Missing Observations"—are needed. For instance, "are there many missing values, and are the values missing at random? The missing observations in the data are analyzed to see whether more than 25% of the values are missing, whether they are missing at

random (MAR), and whether some form of imputation (statistics) is needed."

- "Analysis of Extreme Observations"—are necessary, since "outlying observations in the data are analyzed to see if they seem to disturb the distribution."
- "Comparison and Correction of Differences in Coding Schemes"—are also required, in that "variables are compared with coding schemes of variables external to the data set, and possibly corrected if coding schemes are not comparable."

Surely, there can be other questions, so the examples above are solely illustrative (not exhaustive).

Main Data Analysis

But in the main analysis phase, "analyses aimed at answering the research question are performed as well as any other relevant analysis needed to write the first draft of the research report." (WK 2011a; H. Adèr 2008)

Depending on the particular analysis of a researcher's preference, here are some possibilities to choose from, as shown below: (WK 2011a; H. Adèr 2008)

- "Exploratory and Confirmatory Approaches"—can be used. "Usually the approach is decided before data is collected. In an exploratory analysis no clear hypothesis is stated before analysing the data, and the data is searched for models that describe the data well. In a confirmatory analysis, clear hypotheses about the data are tested."
- "Stability of Results"—is also examined. For instance, "it is important to obtain some indication about how generalizable the results are. While this is hard to check, one can look at the stability of the results. Are the results reliable and reproducible? There are two main ways of doing this," for example,
  - –"crossvalidation" (i.e., "by splitting the data in multiple parts we can check if analyzes (like a fitted model) based on one part of the data generalize to another part of the data as well")
  - –"sensitivity analysis" (i.e.., "a procedure to study the behavior of a system or model when global parameters are...varied")
- "Statistical Methods"—can be used. "A very brief list of four of the more popular methods is" shown below:
  - –"General linear model: A widely used model on which various statistical methods are based (e.g., t test, ANOVA, ANCOVA,

MANOVA). Usable for assessing the effect of several predictors on one or more continuous dependent variables"

–"Generalized linear model: An extension of the general linear model for discrete dependent variables"

–"Structural equation modelling: Usable for assessing latent structures from measured manifest variables"

–"Item response theory: Models for (mostly) assessing one latent variable from several binary measured variables (e.g., an exam)"

Again, there can be other possibilities, so the examples above are solely illustrative (not exhaustive).

The Varieties of Data Analysis

These different possibilities about data analysis as a process (and the differences between quanitative and qualitative methods) also say much about the varieties of data analysis.

The next step then is to provide a classification of data analysis into some main sub-fields—as shown below (and summarized in *Table 1.1*).

It should be stressed, however, that the following sub-fields are not exhaustive (but solely illustrative), as some other sub-fields will be addressed in the rest of the book.

Data Mining

The first sub-field of data analysis to be introduced here is called "data mining." (WK 2011)

By definition, "data mining" refers to "the process of extracting patterns from large data sets by combining methods from statistics and artificial intelligence with database management." (WK 2011h; B. Palace 1996)

This sub-field of data analysis becomes more popular, especially "with recent tremendous technical advances in processing power, storage capacity, and inter-connectivity of computer technology" and "is seen as an increasingly important tool by modern business to transform unprecedented quantities of digital data into business intelligence giving an informational advantage." (WK 2011h)

Of course, "the process of extracting patterns from large data sets" has a long history (even before the existence of data mining), since "[e]arly methods of identifying patterns in data include Bayes' theorem (1700s) and regression analysis (1800s)," for instance. (WK 2011h)

However, the problem of extracting patterns from data sets becomes more challengning, since "data sets have grown in size and complexity," so "direct hands-on data analysis has increasingly been augmented with indirect, automatic data processing. This has been aided by other discoveries in computer science, such as neural networks, clustering, genetic algorithms (1950s), decision trees (1960s) and support vector machines (1990s)." (WK 2011h)

In this historical perspective, "data mining is the process of applying these methods to data with the intention of uncovering hidden patterns." (WK 2011h; M. Kantardzic 2003)

Statistical Applications

The second sub-field of data analysis to be introduced here is called "statistical applications" (or "statistics" in short).

By definition, "statistics" refers to "the science of the collection, organization, and interpretation of data. It deals with all aspects of this, including the planning of data collection in terms of the design of surveys and experiments." (WK 2011i; Y. Dodge 2003)

Statistics can be divided into three applications for summary, namely, (a) descriptive statistics, (b) exploratory data analysis, and (c) confirmatory data analysis, as shown below.

Descriptive Statistics

Descritpive statistics "describe the main features of a collection of data quantitatively," and good examples include the following: (WK 2011j; P. Mann 1995)

- Ex: "distribution" (e.g., "the frequency of individual or ranges of values for a variable")
- Ex: "central tendency" (e.g., "the mean, the median, and the mode")
- Ex: "dispersion" (e.g., "the range and the standard deviation")

In short, descriptive statistics are to summarize data.

However, "descriptive statistics" should not be confused with "inferential statistics," because "descriptive statistics aim to summarize a data set, rather than use the data to learn about the population that the data are thought to represent. This generally means that descriptive statistics, unlike inferential statistics, are not developed on the basis of probability theory." (WK 2011j; Y. Dodge 2003)

Exploratory Data Analysis (EPA)

Exploratory data analysis (EPA), unlike descriptive statistics, does not summarize data but is to analyse data "for the purpose of formulating hypotheses worth testing, complementing the tools of conventional statistics for testing hypotheses." (WK 2011k)

Some of the major objectives of exploratory data analysis are shown below: (WK 2011k)

- "Suggest hypotheses about the causes of observed phenomena"
- "Assess assumptions on which statistical inference will be based"
- "Support the selection of appropriate statistical tools and techniques"
- "Provide a basis for further data collection through surveys or experiments"

John Tukey contrasts exploratory data analysis with "Confirmatory Data Analysis, the term used for the set of ideas about hypothesis testing, p-values, confidence intervals etc. which formed the key tools in the arsenal of practising statisticians at the time." (WK 2011k)

Confirmatory Data Analysis (CPA)

Confirmatory data analysis (CPA) is a good contrast with exploratory data analysis (EPA), since it focuses on "statistical hypothesis testing," that is, using data to confirm hypotheses, instead of using data to suggest hypotheses to test (as in exploratory data analysis).

In other words, "[i]n a confirmatory analysis clear hypotheses about the data are tested," but "[i]n an exploratory analysis no clear hypothesis is stated before analysing the data, and the data is searched for models that describe the data well." (WK 2011a)

In general, confirmatory data analysis involves the following steps: (WK 2011dd; H. Adèr 2008)

- "The first step is to state the relevant null and alternative hypotheses."
- "The second step is to consider the statistical assumptions being made about the sample in doing the test; for example, assumptions about the statistical independence or about the form of the distributions of the observations. This is equally important as invalid assumptions will mean that the results of the test are invalid."
- "Decide which test is appropriate, and stating the relevant test statistic T."

- "Derive the distribution of the test statistic under the null hypothesis from the assumptions. In standard cases this will be a well-known result. For example the test statistics may follow a Student's t distribution or a normal distribution."
- "The distribution of the test statistic partitions the possible values of T into those for which the null-hypothesis is rejected, the so called critical region, and those for which it is not."
- "Compute from the observations the observed value tobs of the test statistic T."
- "Decide to either fail to reject the null hypothesis or reject it in favor of the alternative. The decision rule is to reject the null hypothesis H_0 if the observed value t_{obs} is in the critical region, and to accept or 'fail to reject' the hypothesis otherwise."

Surely, there can be other steps, but the list above is illustrative enough.

Predictive Analytics

The third sub-field of data-analysis to be introduced here is called "predictive analytics."

By definition, "predictive analytics" refers to "an area of statistical analysis that deals with extracting information from data and using it to predict future trends and behavior patterns." (WK 2011l)

In so doing, "predictive analytics relies on capturing relationships between explanatory variables and the predicted variables from past occurrences, and exploiting it to predict future outcomes. It is important to note, however, that the accuracy and usability of results will depend greatly on the level of data analysis and the quality of assumptions." (WK 2011l)

For illustration, two main groups of predictive analytics are "regression techniques" and "machine learning techniques," as described below: (WK 2011l)

- Ex: "regression techniques"—the concern here is to "establish…a mathematical equation as a model to represent the interactions between the different variables in consideration. Depending on the situation, there is a wide variety of models that can be applied while performing predictive analytics."
  –for continuous variables (e.g., linear regression model)
  –for discrete variables (e.g., logistic regression)

- Ex: "machine learning techniques"—the concern here is "to develop techniques to enable computers to learn. Today,...it includes a number of advanced statistical methods for regression and classification....In certain applications it is sufficient to directly predict the dependent variable without focusing on the underlying relationships between variables. In other cases, the underlying relationships can be very complex and the mathematical form of the dependencies unknown. For such cases, machine learning techniques emulate human cognition and learn from training examples to predict future events" (like "neural networks" and "support vector machines").

Of course, there can be other examples, so the list above is solely illustrative (not exhaustive).

Text Analytics

And the fourth sub-field of data analysis to be introduced here is called "text analytics."

By definition, "text analytics" refers to "information retrieval, lexical analysis to study word frequency distributions, pattern recognition, tagging/annotation, information extraction, data mining techniques including link and association analysis, visualization, and predictive analytics. The overarching goal is, essentially, to turn text into data for analysis via application of natural language processing (NLP) and analytical methods." (WK 2011x)

In order to distinguish "text analytics" from "data mining"—an earlier term ("text mining") for text analytics is not used much nowadays. For instance, "Ronen Feldman modified a 2000 description of 'text mining' in 2004 to describe 'text analytics.' The latter term is now used more frequently in business settings while 'text mining' is used in some of the earliest application areas, dating to the 1980s, notably life-sciences research and government intelligence." (WK 2011x; R. Feldman 2004)

In any event, a typical text-analytics includes the following tasks, for illustration: (WK 2011x)

- "Information Retrieval"—"is a preparatory step: collecting or identifying...textual materials, on the Web or held in a file system, database, or content management system, for analysis."
- "Named Entity Recognition"—"is the use of gazetteers or statistical techniques to identify named text features: people, organizations,

place names, stock ticker symbols, certain abbreviations, and so on. Disambiguation—the use of contextual clues—may be required to decide where, for instance, 'Ford' refers to a former U.S. president, a vehicle manufacturer, a movie star (Glenn or Harrison?) or some other entity."

- "Recognition of Pattern Identified Entities"—involves the use of some "features such as telephone numbers, e-mail addresses, quantities (with units)" which "can be discerned via regular expression or other pattern matches."
- "Coreference"—is the "identification of noun phrases and other terms that refer to the same object."
- "Relationship, Fact, and Event Extraction"—"is identification of associations among entities and other information in text."
- "Sentiment Analysis"—"involves discerning subjective (as opposed to factual) material and extracting various forms of attitudinal information: sentiment, opinion, mood, and emotion. Text analytics techniques are helpful in analyzing sentiment at the entity, concept, or topic level and in distinguishing opinion holder and opinion object."

Surely, there can be other examples, so the list above is solely illustrative (not exhaustive).

The Theoretical Debate

These different sub-fields of data analysis then lead us to consider a more important question concerning the theoretical debate on the very nature of data analysis.

Although there can be many different theoretical approaches in the debate—there are five main ones to be summarized hereafter, and they can be called, in the absence of better words, namely, (a) the *mental* argument, (b) the *natural* argument, (c) the *societal* argument, (d) the *cultural* argument, and (e) the *complex* argument (which is my original contribution to the debate), to be elabroated hereafter, respectively.

Lest misunderstanding occurs, three clarifications should be made here, as this is something that I regularly emphasize in my previous books, whenever a theoretical debate is summarized for a topic, *almost verbatim*.

Firstly, there can be other arguments besides the first four here, that is, the "mental," the "natural," the "societal," and the "cultural" arguments (as indicated above). The advantage to select these four here has to do with their illustrative representation of the diverse theories in the literature (which are deemed sufficient for the purpose at hand in this book).

Secondly, the examples (theories) given to illustrate each argument are also supposed to be illustrative, not exhaustive. After all, the arguments will be addressed again in the rest of the book, with other examples (theories) to be introduced.

And thirdly, the four arguments are not necessarily mutually exclusive. For instance, those who advocate the "mental" argument can also consider the "cultural" argument, although they do not focus on the latter.

And the reverse is also true, in that those who make the "cultural" argument can also consider the "mental" argument, although they do not focus on the latter.

In other words, their disagreement is often one in degree, not in kind.

The Mental Argument

With these three clarifications in mind—the first major theoretical approach is called, in the absence of better words, the *mental* argument, which focuses, relatively speaking of course, more on the biological and psychological aspects of data analysis.

Thus, the word "mental" in the title of the argument is suggestive, because it refers to the focus on the biological and psychological aspects of data analysis so understood.

But the mental argument has different issues of concern, and different defenders of the argument offer their own distinctive stands on the issues, depending on their own interests.

For illustration, consider two major versions of the mental argument, namely, (a) discourse on cognitive biases, and (b) phenomenology—to be summarized below, respectively.

Discourse on Cognitive Biases

The first version of the mental argument has to do with "cognitive bias."

The term "cognitive bias" here means "a pattern of deviation in judgment that occurs in particular situations. Implicit in the concept of a 'pattern of deviation' is a standard of comparison; this may be the judgment of people outside those particular situations, or may be a set of independently verifiable facts. The existence of some of these cognitive biases has been verified empirically in the field of psychology." (WK 2011m)

A good example of cognitive bias is "confirmation bias," which refers to "the tendency to search for or interpret information in a way that confirms one's preconceptions." (WK 2011m; M. Oswald 2004)

This is important to the field of data analysis, since one of the main reasons for the appeal of exploratory data analysis (EPA) to some researchers like John Tukey is his critique of confirmatory data analysis (CPA), in that "too much emphasis in statistics was placed on statistical hypothesis testing (confirmatory data analysis); more emphasis needed to be placed on using data to suggest hypotheses to test. In particular, he held that confusing the two types of analyses and employing them on the same set of data can lead to systematic bias owing to the issues inherent in testing hypotheses suggested by the data." (WK 2011k)

However, for the critics, the discourse on "cognitive biases" has some major problems.

For instance, one major criticism of the discourse on "cognitive biases" is that there is no consensus about their causes. Over the decades, there have been different theories to explain them, and good examples are "bounded rationality" (by H. Simon and others), "attribution theory" (by Fritz Heider, Harold Kelley, Edward E. Jones, Lee Ross, Bernard Weiner, and others), "cognitive dissonance" (by L. Festinger and others), "heuristics" (by A. Tversky, Paul Slovic, Melissa Finucane, Ellen Peters, Donald G. MacGregor, and others), "adaptive bias" (by Martie G. Haselton, David M. Buss, and others), "misinterpretations or misuse of statistics" (by many different scholars), and so on. (WK 2011o, 2011p, 2011q, 2011r, 2011s, 2011t & 2011u)

Phenomenology

The second version of the mental argument has to do with "phenomenology."

The term "phenomenology" refers to "a broad philosophical movement emphasizing the study of conscious experience. It was founded in the early years of the 20th century by Edmund Husserl [1970], expanded together with a circle of his followers at the universities of Göttingen and Munich in Germany…." (WK 2011v)

For Husserl, "phenomenology…is primarily concerned with the systematic reflection on and analysis of the structures of consciousness, and the phenomena which appear in acts of consciousness." (WK 2011v)

More fundamentally, "phenomenology is a method of philosophical inquiry that rejects the rationalist bias that has dominated Western thought since Plato in favor of a method of reflective attentiveness that discloses the individual's 'lived experience.'" (WK 2011v)

Rooted in the skeptic tradition called "epoché," Husserl's phenomenology "entails the suspension of judgment while relying on the intuitive grasp of knowledge, free of presuppositions and intellectualizing. Sometimes depicted as the 'science of experience,' the phenomenological method is rooted in intentionality, Husserl's theory of consciousness (developed from Brentano). Intentionality represents an alternative to the representational theory of consciousness which holds that reality cannot be grasped directly because it is available only through perceptions of reality which are representations of it in the mind." (WK 2011v)

This is important to the field of data analysis, since phenomenology questions the objective reality that data analysis tries to interpret and instead proposes a methodological alternative which is more based on a subjective, intuitive grasp of reality, "free of presuppositions and intellectualizing." (WK 2011v)

However, for the critics, phenomenology has some major problems.

One major criticism of phenomenology is that it is reductionistic to the subjective, intuitive basis of the subject. For instance, Graham Harman criticised phenomenology "to be deficient in that it constantly subordinates the independent life of objects to our (human) access to them," and proposed a "radical break with the traditional use of terms such as intentionality" in phenomenology. (WK 2011v) And "Martin Heidegger believed that Husserl's approach overlooked basic structural features of both the subject and object of experience (what he called their 'being')" and proposed the need "to encompass our understanding and experience of Being itself." (WK 2011v)

The Natural Argument

The second theoretical approach is called, in the absence of better words, the *natural* argument, which focuses, relatively speaking of course, more on the natural aspects of data analysis, especially in relation to the state of nature.

Therefore, the word "natural" in the title of the argument is suggestive, because it refers to the focus on the natural aspects of data analysis so understood.

Like the mental argument—the natural argument has different issues of concern, and different defenders of the argument offer their own distinctive stands on the issues, depending on their own interests.

For illustration, consider two major versions of the natural argument, namely, (a) positivism, and (b) evolutionary theory—to be summarized below, in that order.

Positivism

The first version of the natural argument has to do with "positivism."

The term "positivism" refers to "a set of epistemological perspectives and philosophies of science which hold that the scientific method is the best approach to uncovering the processes by which both physical and human events occur," especially as it "was developed in the early 19th century by the philosopher and founding sociologist, Auguste Comte." (WK 2011w; SGC 2011)

Positivism has different principles, and five of them are shown below, for illustration: (WK 2011w)

- "The unity of the scientific method…is the same across all sciences (social and natural)."
- "The goal of inquiry is to explain and predict."
- "Scientific knowledge is testable."
- "Science does not equal common sense."
- "[S]cience should be as value-neutral as possible, and the ultimate goal of science is to produce knowledge…."

However, for the critics, "positivism" has some major problems.

One major criticism of "positivism" is that it is too uncritical to the scientific method itself.

For instance, Max Weber argued that "sociology may be loosely described as a 'science' as it is able to identify causal relationships—especially among ideal types, or hypothetical simplifications of complex social phenomena. As a nonpositivist, however, one seeks relationships that are not as 'ahistorical, invariant, or generalizable' as those pursued by natural scientists." (WK 2011w; D. Ashley 2005)

And "the critical theorist Jürgen Habermas has critiqued pure instrumental rationality (in its relation to the cultural 'rationalisation' of the modern West) as meaning that scientific thinking becomes something akin to ideology itself. Positivism may be espoused by 'technocrats' who believe in the inevitability of social progress through science and technology." (WK 2011w; W. Outhwaite 1988; D. Schunk 2004.)

Evolutionary Theory

The second version of the natural argument has to do with "evolutionary theory."

The term "evolutionary theory" refers to "the scientific study of evolution [which] began in the mid-nineteenth century, when research into the fossil record and the diversity of living organisms convinced most scientists that species evolve. The mechanism driving these changes remained unclear until the theory of natural selection was independently proposed by Charles Darwin and Alfred Wallace in 1858. In the early 20th century, Darwinian theories of evolution were combined with genetics, palaeontology, and systematics, which culminated into a union of ideas known as the modern evolutionary synthesis. The synthesis became a major principle of biology as it provided a coherent and unifying explanation for the history and diversity of life on Earth." (WK 2011y; P. Bowler 2003; U. Kutschera 2004; IPII 2006; AAAS 2006)

Evolutioary theory has been used for data analysis. For instance, "[i]n computer science, simulations of evolution using evolutionary algorithms and artificial life started in the 1960s, and was extended with simulation of artificial selection....Practical applications also include automatic evolution of computer programs. Evolutionary algorithms are now used to solve multi-dimensional problems more efficiently than software produced by human designers, and also to optimise the design of systems." (WK 2011y; J. Koza 1992; M. Jamshidi 2003; A. Fraser 1958)

However, for the critics, "evolutinary theory" for data analysis has some major problems.

For instance, one major criticism of "evolutionary theory" for data analysis is that the use of "evolutionary algorithms" (EAs) is inherently limited due to "computational complexity." (WK 2011z)

As we now know, "[i]n most of real applications of EAs, computational complexity is a prohibiting factor. In fact, this computational complexity is due to fitness function evaluation," to the extent that "fitness approximation" is used instead. (WK 2011z)

The Societal Argument

The third theoretical approach is called, in the absence of better words, the *societal* argument, which focuses, relatively speaking of course, more on the societal aspects of data analysis.

In this sense, the word "societal" in the title of the argument is suggestive, because it refers to the societal aspects of data analysis so understood.

Like the mental and natural arguments—the societal argument has different issues of concern, and different defenders of the argument offer their own distinctive stands on the issues, depending on their own interests.

For illustration, consider two major versions of the societal argument, namely, (a) critical theory and (b) social constructivism—to be summarized in what follows, in that order.

Critical Theory

The first version of the societal argument has to do with the "critical theory."

The term "critical theory" refers to "an examination and critique of society and culture, drawing from knowledge across the social sciences and humanities. The term has two different meanings with different origins and histories: one originating in sociology and the other in literary criticism. This has led to the very literal use of 'critical theory' as an umbrella term to describe any theory founded upon critique." (WK 2011aa)

Jürgen Habermas (1968) in *Knowledge and Human Interests* made an epistemological distinction between these two branches of critical theory, as shown below: (WK 2011aa)

- Literary Critical Theory—"is ultimately a form of hermeneutics, i.e. knowledge via interpretation to understand the meaning of human texts and symbolic expressions—including the interpretation of texts which are themselves implicitly or explicitly the interpretation of other texts."
- Social Critical Theory—"is, in contrast, a form of self-reflective knowledge involving both understanding and theoretical explanation to reduce entrapment in systems of domination or dependence, obeying the emancipatory interest in expanding the scope of autonomy and reducing the scope of domination."

Critical theory is important to the field of data analysis, because it provides a qualitative approach which is essentially critical in orientation, by rejecting "the idea that a researcher's work is considered" as "objective" at all. (WK 2011aa)

However, for the critics, "critical theory" has some major problems.

Firstly, one major critique of critical theory in relation to the version in literary studies is that "literary critical theory" has no "normative" dimension, because "it is focused on interpretation and explanation rather than on social transformation." (WK 2011aa)

Secondly, another major critique of critical theory in relation to the version in the social sciences is that, while its emancipatory interest can be worthwhile, there is no consensual basis for everyone to agree upon, especially in regard to what "social transformation" is to be carried out in the human future. For instance, "who" is to say what is "good" or "right" for society to be transformed?

Social Constructivism

The second version of the societal argument has to do with the "social constructivism."

The term "social constructivism" is based on a constructivist epistemology which refers to "an epistemological perspective in philosophy about the nature of scientific knowledge. Constructivists maintain that scientific knowledge is constructed by scientists and not discovered from the world. Constructivists claim that the concepts of science are mental constructs proposed in order to explain our sensory experience. Another important tenet of Constructivist theory is that there is no single valid methodology in science, but rather a diversity of useful methods." (WK 2011bb)

In this sense, constructivist epistemology is "opposed to positivism, which is a philosophy that holds that the only authentic knowledge is that which is based on actual sense experience and what other individuals tell us is right and wrong." (WK 2011bb)

This is important to the field of data analysis, since constructivist epistemology treats the reality of any data analysis as constructed by the researchers in question, as something not objective about the world at all.

However, for the critics, "constructivist epistemology" has some major problems.

One major problem with "constructivist epistemology" is that "it either explicitly advocates or implicitly reduces to relativism. This is because it takes the concept of truth to be a socially 'constructed' (and thereby socially relative) one. This leads to the charge of self-refutation: if what is to be regarded as 'true' is relative to a particular social formation, then this very conception of truth must itself be only regarded as being 'true' in this society. In another social formation, it may well be false. If so, then social constructivism itself would be false in that social formation." (WK 2011bb)

The Cultural Argument

The fourth theoretical approach is called, in the absence of better words, the *cultural* argument, which focuses, relatively speaking of course, more on the cultural aspects of data analysis.

Therefore, the word "cultural" in the title of the argument is suggestive, because it refers to the focus on the cultural aspects of data analysis so understood.

Like the mental, natural, and societal arguments—the cultural argument has different issues of concern, and different defenders of the argument offer their own distinctive stands on the issues, depending on their own interests.

For illustration, consider two major versions of the cultural argument, namely, (a) postpositivism, and (b) hermeneutics —to be summarized in what follows, respectively.

Postpositivism

The first version of the cultural argument has to do with "postpositivism."

The term "postpostivism" refers to "a metatheoretical stance that critiques and amends positivism. Postpositivists believe that human knowledge is based not on unchallengeable, rock-solid foundations, but rather upon human conjectures. As human knowledge is thus unavoidably conjectural, the assertion of these conjectures is warranted, or more specifically, justified by a set of warrants, which can be modified or withdrawn in the light of further investigation." (WK 2011cc)

However, unlike constructivist epistemology—"postpositivism is not a form of relativism, and generally retains the idea of objective truth.... Postpositivism is an amendment to positivism that recognizes these and other critiques against logical positivism. It is not a rejection of the scientific method, but rather its reformation to meet these critiques. It preserves the basic assumptions of positivism: ontological realism, the possibility and desirability of objective truth, and the use of experimental methodology." (WK 2011cc)

However, for the critics, "postpositivism" has some major problems too.

For instance, one major problem with "postpositivism" is that its quest for "objective truth" remains elusive, even though its critique of postivisim is sensible.

Hermeneutics

And the second version of the cultural argument has to do with "hermeneutics."

The term "hermeneutics" refers to "the study of interpretation theory, and can be either the art of interpretation, or the theory and practice of interpretation. Traditional hermeneutics—which includes Biblical hermeneutics—refers to the study of the interpretation of written texts, especially texts in the areas of literature, religion and law. Contemporary, or modern, hermeneutics encompasses not only issues involving the written text, but everything in the interpretative process. This includes verbal and nonverbal forms of communication as well as prior aspects that affect communication, such as presuppositions, preunderstandings, the meaning and philosophy of language, and semiotics." (WK 2011ff; S. Ferguson 1988)

Hermeneutics is important for the field of data analysis, because, for instance, "hermeneutic consistency" requires "coherent explanation" in any analysis, "encompassing all forms of communication and expression; written, verbal, artistic, geo-political, physiological, sociological etc." (WK 2011ff)

However, for the critics, "hermeneutics" has some major problems.

For instance, one major criticism of "hermeneutics" is that "a hermeneutic approach is too relativist and that their own interpretations are based on common-sense evaluation." (WK 2011ff)

The Complex Argument

And fifthly, unlike the previous four arguments (viz., the mental, the natural, the societal, and the cultural)—the fifth one to be introduced here can be called, in the absence of better words again, the *complex* argument, which learns from all other theoretical approaches (which include the previous four arguments and also other examples as will be introduced in the rest of this book), without, however, favoring any of them, nor trying to integrate them (as they are not necessarily compatible with each other, as the introduction of the theoretical debate in the previous sub-sections has shown).

The Complex Theory of Data Analysis

My *complex* argument can be more precisely called *the complex theory of data analysis*, since it is to propose a new theory of quantitative and qualitative research in data analysis.

My *complex theory of data analysis* has four major distinct features.

Firstly, my theory makes good use of all theoretical approaches to data analysis in relation to quantitative and qualitative research, be they about the mental argument, the natural argument, the societal argument, or the cultural argument—especially from the most comprehensive combined perspectives of the mind, nature, society, and culture (as already dealt with in Chapter Two and Chapter Three).

Secondly, just like many other theories of mine in my previous books, my theory here does not heavily favor any specific theory over others in the literature, nor trying to integrate them (as they are not necessarily compatible with each other).

Thirdly, my theory treats the issue of data analysis in the distinctive dialectic context of no quantitative research without qualitative research, and vice versa.

And fourthly, it contains twenty-five major theses, namely, (a) the first thesis: the formalness-informalness principle, (b) the second thesis: the absoluteness-relativeness principle, (c) the third thesis: the partiality-totality principle, (d) the fourth thesis: the predictability-unpredictability principle, (e) the fifth thesis: the explicability-inexplicability principle, (f) the sixth thesis: the fiction-reality principle, (g) the seventh thesis: the finiteness-transfiniteness principle, (h) the eighth thesis: the preciseness-vagueness principle, (i) the ninth thesis: the simpleness-complicatedness principle, (j) the tenth thesis: the openness-hiddenness principle, (k) the eleventh thesis: the denseness-emptiness principle, (l) the twelfth thesis: the change-constancy principle, (m) the thirteenth thesis: the order-chaos principle, (n) the fourteenth thesis: the slowness-quickness principle, (o) the fifteenth thesis: the expansion-contraction principle, (p) the sixteenth thesis: the theory-praxis principle, (q) the seventeenth thesis: the convention-novelty principle, (r) the eighteenth thesis: the evolution-transformation principle, (s) the nineteenth thesis: the symmetry-asymmetry principle, (t) the twentieth thesis: the softness-hardness principle, (u) the twenty-first thesis: the seriousness-playfulness principle, (v) the twenty-second thesis: the regression-progression principle, (w) the twenty-third thesis: the sameness-difference principle, (x) the twenty-fourth thesis: the stability-reaction principle, and (y) the twenty-fifth thesis: the post-human complexity—to be addressed in Chapter Four.

Because of the last thesis on "the post-human complexity," the word "complex" in the title of my theory is suggestive, since it is to stress the emergence of complexity in reality, both in the human and post-human future.

However, it is important to clarify at the outset that, lest any misunderstanding easily occurs, the word "complexity" here does not endorse complexity theory (which I already criticized and rejected in *FC*) but simply stresses the ever more complex world that future humans and their successors, post-humans, will face, in a way that current humans have never known.

With this clarification in mind—the seminal project here, if successful, will fundamentally change the way that we think about quantitative and qualitative research in data analysis, from the combined perspectives of the mind, nature, society, and culture, with enormous implications for the human future and what I originally called its "post-human" fate.

Theory and Meta-Theory

The *complex* argument, or *the complex theory of data analysis*, presupposes some kind of meta-theory, in special relation to methodology and ontology—just as all theories, either explicitly or implicitly, require their own versions of meta-theory.

Meta-Theory

Theory and meta-theory, broadly speaking, can be further classified into different categories, as shown below:

- On Theory
  - —Ex: in relation to the mind
  - —Ex: in relation to nature
  - —Ex: in relation to society
  - —Ex: in relation to culture
- On Meta-Theory
  - —Ex: in relation to methodology
  - —Ex: in relation to ontology

Each category, in its turn, can then be further sub-divided into different levels of analysis—which will be elaborated in the next two sections (and summarized in *Table 4.43*).

With this reminder in mind, I already proposed (in my previous books) my distinctive contribution to meta-theory, namely, methodology (like "theory of methodological holism" and "contrastive theory of rationality," for instance) and ontology (like "theory of existential dialectics" and "critical-dialectic theory of formal science," for instance).

The next two sections are to introduce my meta-theory—especially, existential dialectics and sophisticated methodological holism, respectively—to be repeated, *often verbatim* (with only a few updated revisions), from my previous works, as this is something that I often do in each new book of mine as an introduction to some background information for the convenience of the readers.

In the process, I shall also introduce my numerous new theories (or to be exact, 53 of them in 44 books so far, as of May, 2011) in relation to the four categories of the mind, nature, society, and culture in the context of theory (as also summarized in *Table 4.43*).

A Unified Theory of Everything

In the end, my distinctive contribution to both meta-theory and theory serves as a foundation to unify all domains of knowledge for an overarching understanding of all forms of reality (by way of some ontological principles at the meta-theoretical level and the comprehensive perspectives of the mind, nature, society, and culture at the theoretical level—as will be explained in the next two sections).

This is so, without commiting the sins of reductionism and reverse-reductionism (as often seen in many holistic approaches, with the fad of systems approach as a most recent notorious example, as already critically explained in *The Future of Post-Human Formal Science*, or *FPHFS*), since my distinctive contribution makes good use of different schools of thought without heavily favoring any of them nor trying to integrate them (as they are not necessarily compatible with each other), so as to include (or adjust for) heterogeneity, conflict, subjectivity, and complexity, for example.

In this sense, the word "unified" does not have to mean an integrative approach, in a narrow sense, nor a systems approach, in a broad one, both of which I have critically rebuked (in my previous books), while learning from them.

This overarching project (consisting of numerous theories of mine, as will be summarized in the next section), thus, constitutes my grand unifed theory of everything in the world and beyond.

The Logic of Existential Dialectics

To start, the ontology of existential dialectics can be summarized in this section—*often verbatim* (with only a few updated revisions), from my previous works, as this is something that I regularly do in each new book of mine as an introduction to some background information for the convenience of the readers.

After all, this book, like all others of mine, is in conversation with all previous ones of mine, for the final aim to converge all of them into a larger grand project on the future of intelligent life, both here on earth and there in deep space unto multiverses.

The summary can be organized in relation to (a) the conception of existential dialectics (or its ontics), (b) the syntax of existential dialectics (or its ontomethodology and ontologic), (c) the semantics of existential dialectics (or its ontosemantics), and (d) the pragmatics of existential dialectics (or its ontopragmatics), respectively hereafter—again *often verbatim*.

This means that, for those readers who had read some of my previous books, the summary below is really not much new (*almost verbatim*, with only some updated revisions).

But for those who have never read my previous books, it is a good review of them.

The Conception of Existential Dialectics (or Its Ontics)

This ontology was originally designated as "existential dialectics" in *Beyond Capitalism to Post-Capitalism* (herein abbreviated as *BCPC*), although it was already analyzed in other books of mine like the 2 volumes of *The Future of Human Civilization* (*FHC*), *The Future of Capitalism and Democracy* (*FCD*), *The Future of Post-Human Consciousness* (*FPHC*), and the 2 volumes of *Beyond Democracy to Post-Democracy* (*BDPD*).

Lest any confusion occurs, it is important to stress at the outset that the word 'existential' in "existential dialectics" has nothing to do with Existentialism, which I rebuked in *FHC*, *FCD*, and also *FPHC*. Instead, the word here merely refers to the existence of intelligent life (both primitive and advanced) in a broad sense.

Some Basic Concepts

The conception of existential dialectics (or its ontics) makes use of different concepts (like "sets," "elements," "relations," "operations," "functions," "truth values," "axioms," "postulates," and "principles"—as shown in *Table 4.3*), which are important for the understanding of any logic of ontology.

That clarified—existential dialectics, as a language of ontology, can be analyzed in different ways that a language, as an analogy, is often analyzed, namely, in relation to phonology ("the study of patterns of a language's sounds"), phonetics ("the study of the physical aspects of sounds of human language"), morphonology ("the study of the internal structure of words"), syntax ("the study of how words combine to form grammatical sentences"), semantics ("the study of the meaning of words [lexical semantics] and fixed word combinations [phraseology]"), and pragmatics ("the study of how utterances are used…literally, figuratively, or otherwise…in communicative acts"). (WK 2007)

But since my theory of existential dialectics makes use of the English language for communication here, it does not propose a new way to make the patterns of a language's sounds (as in phonology), to study the physical aspects of sounds (as in phonetics), or even to strive for a new organization of the internal structure of words for the English language (as in morphonology).

For this reason, the language of existential dialectics to be analyzed hereafter is not concerned with the phonological, phonetic, and morphonological aspects of the English language. Instead, the analysis will explore the syntax, semantics, and pragmatics of existential dialectics as a language of ontology to improve our understanding of reality.

The Ontic Dispute

But first, there is an ontic dispute to be addressed, which is related to what I called "the dilemma of ontology" as introduced earlier in *Sec. 1.4* of *The Future of Post-Human Knowledge* (*FPHK*).

A good point of departure concerns what I want to call *the ontic dispute*, on how to understand the very nature of ontology.

Traditionally, scholars in the field used to study the conception of ontology in terms of its entities (e.g., material objects, abstract numbers), properties (e.g., duration, plurality), and relationships (e.g., causation).

This is a good starting point, but it suffers from different problems. Consider three examples of bias below, which are quite common in the literature, and the recent version (of treating concepts and principles in ontology on the basis of "plurality," "dynamism," "duration," "interaction," "life," "consciousness," and "volition") by Reginald Firehammer (2005) only constitutes a latest addition to this old (impoverishing) tradition.

Firstly, the classification of entities, properties, and relationships is too material-centric, in often focusing more on the existence of material entities than on non-material ones.

Secondly, the classification is too anthropo-centric, in often favoring the factor of human agency in depicting reality.

And lastly, the classification is too confused among the three categories relating to each other, in often lumping them together without adequate consideration of their intricated interrelationships (both in relation to each and in relation to all).

A good alternative is precisely what I want to offer in this debate, that is, a new classification of ontological entities, properties, and their relationships in terms of *method*, *structure*, *process*, *agency*, and *outcome* instead. Let me explicate what these words mean below.

In relation to "method"—it concerns how ontology is to be studied. In this sense, the word 'method' here should not be confused with the different usage of the same term 'method' as a technique of investigation in the context of methodology (as already introduced in *Sec. 1.2* of *FPHK*).

In relation to "structure"—it concerns what the general patterns of entities, properties, and relationships are.

In relation to "process"—it concerns how these general patterns change over time.

In relation to "agency"—it concerns how agency affects any change of these general patterns over time (without putting, however, too much emphasis on the role of human agency).

And in relation to "outcome"—it concerns what the outputs of this change over time are.

The Syntax of Existential Dialectics (or Its Ontomethodology and Ontologic)

The syntax of existential dialectics, analogously speaking, refers here to the structure of an ontology which can be studied by way of combining ideas into complex relationships like ontological principles to depict reality in the world—and can be analyzed in seven sub-sections, namely, (a) the selection criteria, (b) further clarifications on classification, (c) ontomethodology, (d) ontologic, (e) the principles as short cuts, (f) the principles as family resemblances, and (g) the dialectic constraints imposed by the principles, respectively hereafter.

The Selection Criteria

To start, not any pair of relationship can be chosen as an ontological principle in existential dialectics; otherwise, there could be billions of them under the sky.

Five illustrations below suffice to clarify this point.

(a) Firstly, an ontological principle cannot *overlap* with other principles, in that it is too closely related with other ones.

Of course, one cannot totally rule out any relationship between two entities, especially in a complex world of everything being linked to everything else—but the comparison here is relative, not absolute.

In this sense, the flexibility-inflexibility pair cannot be an ontological principle, since it has something closely in common with other ones (like the simpleness-complicatedness principle, for instance).

As an illustration, something which is flexible in interpretation is more likely to allow multiple viewpoints and thus makes the overall picture more complicated than otherwise (simple).

(b) Secondly, an ontological principle cannot be *redundant* in relation to other principles, in that it somehow duplicates other principles.

In this sense, for instance, the directness-indirectness pair cannot be an ontological principle, because it somehow duplicates other ontological principles (like the preciseness-vagueness principle, for instance).

As an illustration, something which is direct means to be right to the point and does not need to go to an unnecessarily long loop—so it is less vague, in being clearer about the thing in question.

(c) Thirdly, an ontological principle cannot be *derived* from other ones—as if it is a child-parent relationship.

In this sense, the convenience-inconvenience pair cannot be an ontological principle, because it can be derived from other ones (like the slowness-quickness principle, for instance).

For instance, something which is convenient already implies that it can be obtained quickly (rather than slowly)—all things considered.

(d) Fourthly, an ontological principle cannot be *trivial* (or *parochial*)—as if it is only one component competing with other ones for inclusion in a set.

In this sense, the consistency-inconsistency pair cannot be an ontological principle, because its validity is limited (or parochial), as it can be easily challenged and replaced by opposing major (not minor) theories (e.g., the correspondence theory of truth, the pragmatic theory of truth, the semantic theory of truth, the disclosedness theory of truth, and whatnot), not just in accordance to the consistency (or coherence) theory of truth (for consistency).

Thus, the consistency argument is only one among different opposing arguments in relation to the logic of what constitutes "truth."

(e) And fifthly, an ontological principle cannot be *one-sided*, that is, part of one side (category) of a dialectic pair.

In this sense, the extroversion-introversion pair cannot be an ontological principle, since it is part of the side (category) of "normality" in the normality-abnormaltiy pair (because both extroversion and introversion are two different "normal" personality traits in psychology).

In addition, the normality-abnormality pair is already related to the symmetry-asymmetry principle, the density-emptiness principle, the convention-novelty principle, and the same-difference principle (as will be explained in the section on "the principles as family resemblances" later).

Although these examples are not exhaustive, they illustrative the selectiveness of any ontological principle to be existentially dialectic.

Further Clarifications on Classification

Lest any misunderstanding carelessly occurs, some further clarifications on classification should be made here (and summarized in *Table 4.8*).

(a) Firstly, the total number of ontological principles is unknown, to be discovered later, as our knowledge of the world becomes more advanced.

Consequently, the principles as introduced in my books are not exhaustive, with new ones being added, whenever more of them are discovered in later research.

At least, future generations can pick up where I leave off and continue the discovery.

In this sense, my work should be treated as a pioneering effort for the development of a systematic, comprehensive analysis of a new general ontology for the future of knowledge.

(b) Secondly, the principles are not rigidly classified, as they can be reclassified in a different way.

For instance, the preciseness-vagueness principle is classified under the category of "structure" but can be reclassified under the category of "method," although in so doing, it has a different meaning in the context of method.

The same logic applies to the same-difference principle under the category of "outcome," which can be reclassified under the category of "structure," although in so doing, once more, it has a different meaning in the context of structure.

(c) Thirdly, the principles are to be understood together, not that each prinicple is to be treated independently of others. After all, the principles are mutually constraining, in that they work together as a whole.

(d) Fourthly, the principles are relevant to all subject matters, but some principles are, relatively speaking of course, more useful to some subject matters than others under certain conditions—as implied in the symmetry-asymmetry principle.

(e) And fifthly, the principles constitute only two levels of analysis, this time, at the ontological and methodological levels—while other levels of analysis (from the perspectives of the mind, nature, society, and culture) are also needed, in order to understand reality in its totality (as explained in the section on "sophisticated methodological holism," so as to avoid the dual dangers of reductionism and reverse-reductionism).

Ontomethodology

With these selection criteria and clarifications in mind—the syntax of existential dialectics can be divided into two main smaller sections for analysis, since they are related to each other in the context of some fundamental principles, that is, (a) ontomethodology and (b) ontologic, to be addressed hereafter, respectively.

In the context of ontomethodology, some ontological principles are proposed here in relation to the nature of methodology, which however is tied up with the general issue of ontology in the context of knowledge.

In this sense, ontomethodology (in the context of these ontological principles in the category of method) are related with ontologic in general (in the context of ontological principles in general).

Some good instances of ontological principles involving ontomethodology include the formalness-informalness principle, the absoluteness-relativeness principle, the partiality-totality principle, the predictability-unpredictability principle, the explicability-inexplicability principle, and the fiction-reality principle, which will be introduced in conjunction with ontological principles in the ontologic of existential dialectics in general hereafter.

Ontologic

With this clarification of the relationship between ontomethodology and ontologic in mind (which is comparable to the relationship between methodology and ontology as described in *Sec. 1.3* of *FPHK*)—the first three principles in existential dialectics were introduced in *BCPC*. And more principles were later added to the logic in my subsequent books (like *FC*, *FAE*, and others), as explained below.

More specifically, in *BCPC*, I proposed three principles (based on previous works of mine, not just something out of the blue) for the ontological logic, namely, (a) the regression-progression principle on the direction of history, (b) the symmetry-asymmetry principle on the relationships among existents, and (c) the change-constancy principle on the alteration of things—or in a more elegant term, the dynamics of space-time (in *FC*).

In *The Future of Complexity* (*FC*), I added three more principles, on top of the three aforementioned, namely, (d) the partiality-totality principle on the relationships between parts and whole, (e) the order-chaos principle on the pattern of things, and (f) the predictability-unpredictability principle on the occurrence of things—as part of the ontological logic of existential dialectics.

In *The Future of Aesthetic Experience* (or simply *FAE*), another principle is added, namely, (g) the evolution-transformation principle (on the multiple kinds of agency).

Also in *FAE*, I then reclassified the 7 principles into four categories, namely, (i) in relation to method, for the partiality-totality principle and the predictability-unpredictability principle, (ii) in relation to process, for the change-constancy principle and the order-chaos principle, (iii) in relation to agency, for the symmetry-asymmetry principle and the evolution-transformation principle, and (iv) in relation to outcome, for the regression-progression principle.

In *The Rise of Authoritarian Liberal Democracy* (*ALD*), I added two more principles, namely, (h) the softness-hardness principle on the force of agency (which is to be classified under the category about agency) and (i) the same-difference principle on the metamorphosis of change (which is to be classified under the category about outcome).

In *The Future of Information Architecture* (*FIA*), I introduce three additional principles, that is, (j) the simpleness-complicadness principle on the interconnection among things, (k) the preciseness-vagueness principle on the refinement of things, and (l) the slowness-quickness principle on the speed of change—with the first two in relation to structure and the third in relation to process.

In *The Future of Post-Human Unconsciousness* (*FPHU*), I added three more principles, that is, (m) the openness-hidden principle on the detection of things, in relation to structure, (n) the explicability-inexplicability principle on the underlying mechanisms of things, in relation to method, and (o) the expansion-contraction principle on the growth of things, in relation to process.

In *The Future of Post-Human Knowledge* (*FPHK*), I added one more principle, that is, (p) the absoluteness-relativeness principle on the multiplicity of entities, in relation to method.

In *The Future of Post-Human Mathematical Logic* (*FPHML*), I added one more principle, that is, (q) the formalness-informalness principle on the formal requirements of systems, in relation to method.

In *The Future of Post-Human Engineering* (*FPHE*), I added one more principle, that is, (r) the theory-praxis principle on the duality of knowledge, in relation to agency.

In *The Future of Post-Human Creative Thinking* (*FPHCT*), I added one more principle, that is, (s) the convention-novelty principle on the nature of creative thinking, in relation to agency.

In *The Future of Post-Human Geometry* (*FPHG*) here, I added one more principle, that is, (t) the finiteness-transfiniteness principle on the nature of numbers, in relation to structure.

In *The Future of Post-Human Urban Planning* (*FPHUP*), I added one more principle, that is, (u) the denseness-emptiness principle on the distribution of entities in space, in relation to structure.

In *The Future of Post-Human Humor* (*FPHH*), I added one more principle, that is, (v) the seriousness-playfulness principle on the extent of seriousness, in relation to agency.

In *The Future of Post-Human Literature* (*FPHLIT*), I added one more principle, that is, (w) the fiction-reality principle on the different faces of reality, in relation to method.

And in *The Future of Post-Human Chemistry* (*FPHCHEM*), I added one more principle, that is, (x) the stability-reaction principle on the feedback mechanism for further changes, in relation to outcome.

With this update, there are twenty-four principles so far in existential dialectics, that is:

- six principles in the category of "method"—viz., the formalness-informalness principle, the absoluteness-relativeness principle, the partiality-totality principle, the predictability-unpredictability principle, the explicability-inexplicability principle, and the fiction-reality principle
- five principles in the category of "structure"—viz., the finiteness-transfiniteness principle, the simpleness-complicatedness principle, the preciseness-vagueness principle, the openness-hiddenness principle, and the denseness-emptiness principle
- four principles in the category of "process"—viz., the change-constancy principle, the slowness-quickness principle, the order-chaos principle, and the expansion-contraction principle
- six principles in the category of "agency"—viz., the theory-praxis principle, the convention-novelty principle, the symmetry-asymmetry principle, the evolution-transformation principle, the softness-hardness principle, and the seriousness-playfulness principle

- three principles in the category of "outcome"—viz., the regression-progression principle, the same-difference principle, and the stability-reaction principle

There can be more than twenty-four principles in existential dialectics, but they are what I have discovered so far (as summarized in *Table 4.4*).

The Principles as Short Cuts

Yet, these principles should be treated with caution, lest misunderstanding occurs, since they do not constitute rigid dualities (or dichotomies).

The reason is that each pair in an ontological principle consists of two opposites, which are, however, merely short cuts both for multiple variations and degrees, as well as for different interactions with multiple other entities unlike them.

In this light, each pair can end up having hundred (or even thousand, if not more) different versions, which interact with hundred (or even thousand, if not more) other entities—as summarized in *Table 4.5*.

There are two clarifications here.

Firstly, this conception of shortcuts is not mutually exclusive nor absolute, in that the opposites can come in all shapes and sizes, with different degrees. For instance, by analogy, just as there are different degrees of the two colors "white" and "black"—there are likewise different degrees of the opposites in each ontological pair, to the extent that there can be multiple entities (not only two) interacting with each other in each pair.

And secondly, this conception of shortcuts do not ignore other possible entities in interacting with the two opposites (with their different versions) in each pair. For instance, by analogy, there are not only the two colors "white" and "black" as opposites, since there are other colors too besides them like "yellow," "green," "purple," or else. The same logic can be applied to each ontological pair, in that they also interact with other entities, not with only two of them (with their different versions).

Therefore, with these two clarifications in mind—each pair in an ontological principle serves only as an abbreviation for something more complicated and, therefore, although it contains two opposites, it should not be confused as a duality (dualism).

Existential dialectics rejects any dualism (or dichotomy) as too rigid and instead allows the multiplicity of entities, to the extent that between the two opposites in each pair exist many other alternatives to choose from. They are named in that short form for aesthetic elegance only, instead of listing all possible entities between the two opposites in the title (which would be too cumbersome and sound extremely awkward).

To be dialectic is to go beyond any rigid dichotomy and transcend into something different altogether in the long haul. One may be tempted to call the dialectic logic here with a different name like existential "multilectics" (instead of "dialectics"), but this naming is incorrect (or even misleading), for the two reasons aforestated.

After all, all the possible entities between any two opposites in an ontological pair are still about the two opposites, although each ontological pair come in all shapes and sizes (with different degrees) and do not exist by themselves but also interact with other entities unlike them. The virtue of revealing an ontological pair is to show how they relate within themselves (in multiple versions) and also interact with others unlike them (also in multiple versions).

This will be clear in a later sub-section titled "the dialectic constraints imposed by the principles."

The Principles as Family Resemblances

In addition, each ontological principle is generic, with some other comparable ontological pairs to be put in the same family (like a *family resemblance*)—as summarized in *Table 4.6*.

Ludwig Wittgenstein (1953) once suggested the idea of "family resemblance" in explaining different games classified under the same family called *games*.

Why should, for instance, playing football and chess as playing "games," when it is well understood that football is not the same as chess? The answer is that, although each game is different and has different rules, many of them (though not all) share, more or less, some commonalites (e.g., scoring as necessary for winning).

And this is so, even though some games share more than some others in any given selection of critieria, and no two games are exactly identical. So, Wittgenstein's point here is that there is no essential core which is common to all games, and the best that one can look for is some characteristics which are common to many (but not all) games. (A. Biletzki 2006)

(a) Appropriate Family Resemblances

In the context of existential dialectics, I want to distinguish two types of family resemblances, and, in the absence of better words, let me call them (a) *appropriate* family resemblances and (b) *inappropriate* family resemblances.

In regard to appropriate family resemblances—good examples include some of the pairs as already explained in an earlier section on selection criteria, like the flexibility-inflexibility pair, which can be put in the family resemblance of the simpleness-complicatedness principle, although the two pairs are not exactly identical. Likewise, the directness-indirectness pair can be put in the family resemblance of the preciseness-vagueness principle, although, again, the two pairs are not exactly identical.

In this sense, which specific pair in a family should be used to designate the name of the family can be at times a bit arbitrary, but with good reason.

For illustration, in the context of method, the partiality-totality principle can take the different form like individualisticness-holisticness, just as the explicability-inexplicability principle can take the different form like underlyingness-regularness—although each two pairs are not exactly identical and have slightly different meanings and usages.

A more comprehensive listing of this family resemblance for all other principles is shown in *Table 4.6* on the syntax of existential dialectics.

In fact, this section on family resemblances overlaps with the semantics of existential dialectics, which will be introduced later in the next section (and summarized in *Table 4.11*).

(b) Inappropriate Family Resemblances

However, there are some family resemblances which are not appropriate, and let me distinugish them by calling them, in the absence of better words, (b1) *inadequate* family resemblances, (b2) *compound* family resemblances, (b3) *one-sided* family resemblances, and (b4) *vague* family resemblances (as summarized in *Table 4.10*).

(b1) In regard to "inadequate" family resemblances—there are some pairs which are not adequate, because they fail the test of selection criteria (as already pointed out in an earlier section), especially in relation to the criterion of not being *trivial* (or *parochial*)—as if it is only one component competing with other ones for inclusion in a set.

For instance, the *consistency-inconsistency* pair is inadequate, because it is only one component competing with other ones for inclusion in the larger set of the formalness-informalenss principle (like the *completeness-incompleteness* pair and the *soundness-unsoundness* pair, as discussed in my book titled *The Future of Post-Human Mathematical Logic*, or *FPHML* in short).

The same is true for the *cognition-noncognition* pair, which is related with both the completeness-incompleteness pair and the soundness-unsoundness pair in the larger set of the formalness-informalenss principle.

(b2) In regard to "compound" (or non-atomic) family resemblances—there are some pairs which are not appropriate for a different reason, because they fail the test of selection criteria (as already pointed out in an earlier section), especially in relation to the criterion of not *overlapping* with other principles, in that it is too closely related with other ones.

For instance, the *normality-abnormality* pair is compound (and thus inappropriate), because it overlaps with several other principles, like the symmetry-asymmetry principle, the density-emptiness principle, the convention-novelty principle, and the same-difference principle, for instance (as already discussed in *FPHP*).

The normality-abnormality pair overlaps with the symmetry-asymmetry principle, because being "abnormal" (say, for those with mental illnesses) is often treated not equally by others who are "normal." So, the relationship reveals the symmetry-asymmetry principle.

The normality-abnormality pair also overlaps with the density-emptiness principle, because being "abnormal" (say, for those with mental illnesses) is often not as dense in the overall (not local) distribution of a population as those who are "normal." So, the relationship reveals the density-emptiness principle.

The normality-abnormality pair also overlaps with the convention-novelty principle, because being "abnormal" (say, for those with mental illnesses) often exhibits non-conventional (non-conformist) ideas and behaviors when contrasted with those who are "normal." So, the relationship reveals the convention-novelty principle.

The normality-abnormality pair also overlaps with the same-difference principle, because being "abnormal" (say, for those with mental illnesses) shares some similarities in outcome when compared with those who are "normal" (in that both contribute to the control and regulation of human thoughts and behaviors in society, albeit in different ways). Yet, being "abnormal" also has something different from being "normal," in that abnormal thoughts and behaviors follow a different pattern when contrasted with those which are normal. So, the relationship reveals the same-difference principle.

(b2.1) There is a special case of compound family resemblances (as summarized in *Table 4.10*), which can be called, in the absence of better words, *one-sided and compound* family resemblances.

For instance, the extroversion-introversion pair is a good example of *one-sided and compound* family resemblances, because, firstly, it is "one-sided," in being part of the side (category) of "normality" in the normality-abnormality pair—and, secondly, the normality-abnormality pair is in turn "compounded," because it is related to the symmetry-asymmetry principle, the density-emptiness principle, the convention-novelty principle, and the same-difference principle, as already explained above.

(b3) In regard to "one-sided" family resemblances—there are some pairs which are not appropriate for another different reason, because they fail the test of selection criteria (as already pointed out in an earlier section), especially in relation to the criterion of not being *one-sided*, that is, part of one side (category) of a dialectic pair.

For instance, the extroversion-introversion pair cannot be an ontological principle, since it is part of the side (category) of "normality" in the normality-abnormaltiy pair (because both extroversion and introversion are two different "normal" personality traits in psychology).

By the same logic, the vision-blindness pair and the sound-silence pair cannot be ontological principles, because the pairs are part of the side (category) of "asymmetry" in the symmetry-asymmetry principle (since being able to see and hear for some individuals and being unable to see or hear for others are two contrastive abilities in the world of animals and humans, for instance).

These illustrations of indequate, compound, and one-sided family resemblances show how difficult it is to come up with an ontological principle in existential dialectics out of the blue, because many of them violate the selection criteria in one way or the other. It is all too easy to join two opposing words and simply make them into a pair, but it is all too difficult to pass the test of the selection criteria to be qualified as an ontological principle in existential dialectics.

(b4) And in regard to "vague" family resemblances—there are some pairs which are not appropriate for still another different reason, because the terms in the pairs are vague in meaning, so that they are inappropriate family resemblances if defined in some particular ways, although they can be appropriate family resemblances if defined in other particular ways.

For instance, such pairs like shortness-longness, narrowness-broadness, smallness-largeness, smallness-greatness, microness-macroness, and littleness-muchness can be understood in at least three different meanings and thus have three different possibilities. The first possibility is that they are one-sided as part of finiteness (in the finiteness-transfiniteness principle), if both opposites are defined on a finite scale. The second possibility is that they are one-sided as part of transfiniteness (in the finiteness-transfiniteness principle) if both opposites are defined on a transfinite scale. And the third possibility is that they are appropriate family resemblances in relation to the finiteness-transfiniteness principle if one opposite is defined on a finite scale and the other opposite is defined on a transfinite scale.

Other examples of inappropriate family resemblances are shown in *Table 4.10* on "The Syntax of Existential Dialectics VII: Types of Inappropriate Family Resemblances."

The Dialectic Constraints Imposed by the Principles

The principles, as they constitute the syntax of existential dialectics, are dialectic in character, such that, when they are applied, they impose dialectic constraints on how reality is to be understood.

Consider, say, the symmetry-asymmetry principle as an illustration here, in order to summarize two main characters of the dialectic constraints in question (as summarized in *Table 4.7*).

Firstly, to be dialectic here is to go beyond the narrow dichotomies (and, for that matter, any rigid multi-dimensional classificatory scheme), be they about "self" vs. "world," "freedom" vs. "unfreedom," "barbarity" vs. "civilization," "individuality" vs. "communality," and so on.

One way to do so (to go beyond) is to consider them all in terms of co-existence (without heavily favoring one over the rest). For instance, my theory of "post-civilization" (to be summarized later in the section on the pragmatics of existential dialectics) is to go beyond barbarity and civilization in terms of understanding barbarity and civilization as being co-existent.

And the same logic can be said in relation to my theories of "post-democracy," "post-capitalsm," and others (also to be introduced later in the section on the pragmatics of existential dialectics), in regard to freedom vs. unfreedom, equality vs. inequality, communality vs. individuality, spirituality vs. materiality, and so on.

But to consider them all (in the dichotomies—and, for that matter, in any rigid multi-dimensional classifactory scheme) as co-existent is not the same as to imply that the opposing categories in any classificatory scheme are all equal, since, in accordance to the symmetry-asymmetry principle (as an illustration here), if they are equal in terms of being considered as co-existent, they are asymmetric in terms of being unequal in dialectic interaction (e.g., X can be more dominant than Y in case A, or Y is more dominant than X in case B), although in some special cases, they can be relatively equal (e.g., X and Y are relatively equal in case C).

For this reason, there are different versions of "post-democracy" and "post-capitalism" in my theories. As an illustration, in version I of the theory of post-democracy, freedom is more dominant than equality, whereas in version II of the theory of post-democracy, equality is more so than freedom.

But this "X more than Y" has to be understood in the context of dialectic logic (not in symbolic logic, as conventionally understood), in that both "X" and "Y" are important in post-democracy (in the context of dialectic logic), but in an asymmetry way.

By contrast, in symbolic logic, it often favors one over the other—be it in regard to privileging freedom over equality (in Fascism), favoring freedom relatively more than equality (in Liberal Democracy), or favoring equality relatively more than freedom (in Socialist Democracy). In regard to the latter two cases (about Liberal Democracy and Socialist Democracy), the difference between dialectic logic and symbolic logic can be one in degree, not in kind—in this sense, albeit not in other senses.

The same reasoning can be said about the relatonships between individuality and communality, between spirituality and materiality, and between formal legalism and informal legalism in the different versions of my theory of "post-capitalism."

Secondly, to be dialectic is to go beyond the narrow dichotomies (and, for that matter, any rigid multi-dimensional classificatory scheme) in another way, this time, in a transcendent way, that is, in exploring other possibilities or even other issues not considered within the narrow confines of narrow dichotomies (and, for that matter, any rigid multi-dimensional classificatory scheme).

As an analogy, in this second meaning, to go beyond the narrow dichotomy of "black" vs. "white" is not just to choose both "black" and "white" (as in the first meaning) but also to explore other options (e.g., "green," "purple," "blue," etc.—and, alternatively, "shade," "line," "curve," etc.).

By the same reasoning, to go beyond "democracy" is to transcend democracy (as in version III of my theory of "post-democracy") and to explore other possibilties of lifeforms (e.g., "floating consciousness," "hyper-spatial consciousness," etc., to live beyond the narrow obsession with freedom and equality).

This dialectic character of the principles in existential dialectics has important implications for the pragmatics of existential dialectics (as will be clear shortly, in the section on the pragmatics of existential dialectics).

The Semantics of Existential Dialectics (or Its Ontosemantics)

The syntax of existential dialectics so understood in terms of ontological principles only gives us the structure of ontology in the world, in an abstract (general) sense. These principles by themselves do not tell us the specific meanings in a given context.

In order to grasp the specific meanings of the principles in a given context, it is necessary to study the semantics of existential dialectics (as summarized in *Table 4.11*).

And the analysis of the ontological principles as family resemblances in the previous section on syntax is only a starting point (and thus overlaps a bit with the semantics of existential dialectics here).

For this reason, I have often gone to great lengths in my previous books on different subjects to explain the specific meanings of the principles when applied in different contexts.

Consider two examples below for illustration, which are not exhaustive.

In *FPHST*, for instance, I used the first three principles (i.e., the change-constancy principle, the regression-progression principle, and the symmetry-asymmetry principle) to propose "the perspectival theory of space-time," for a better way to understand space and time—especially, though not exclusively, in relation to future post-human history (as summarized in *Table 1.3*, *Table 1.4*, *Table 1.5*, and *Table 1.6*).

In so doing, I had to introduce concepts and theories specific to the field of physics and other related fields (e.g., "absolute space" and "absolute time" in "classical mechanics" and "relative space-time" in "the theory of relativity").

And in *BNN*, I also exploited the three principles to propose the "transcendent" approach to the study of genes and memes as a new way to understand the interaction between nature and nurture.

In so doing, I had to explore concepts and theories in the world of evolutionary biology (e.g., "mutation," "variation," "adaptation," "selection," and "inheritance" in classical evolutionary theory) and neural biology (e.g., "chromosome," "gene," "DNA," "RNA," "protein," "neuron," "neural network," and "behavior").

Other examples, besides these two here, are given in other books of mine on different topics.

Therefore, this semantic dimension of existential dialectics is important, since, in each of my books when the ontological principles are used, each principle becomes relevant in a different meaning when contextualized within a given case study in question.

In this book, for instance, the principles are relevant in relation to a different subject matter (as summarized in the concluding chapter), and they can be shown in all shapes and sizes (since they are also to be treated as "short cuts," as indicated before), without, however, being reduced to "specific ontology" (as opposed to "general ontology"), as explained in *Table 4.9*.

In the technical jargons of semantics, these exercises perform two essential functions of translation (or interpretation), namely, (a) "intensive" interpretation (about the properties or relations of the principles in existential dialectics), and (b) "extensive" interpretation (about the classes or cases to which the properties and relations of the principles in existential dialectics are applied). (WK 2011oo)

But there are limits of translation (or interpretation), and an excellent example is that "there is a problem with the 'objectivity' of translation, as it is relative to different cultural (or sub-cultural) contexts. For instance, one of the still unresolved problems in linguistics is: 'Is there an objective gauge for the quality of translation?'" (*FPHCOM* & WK 2011pp)

This is one serious problem about the logic of translation and interpretation that I already addressed in *Sec. 3.5.1* of *FPHCOM* (and also, for that matter, in *FHC*, in the context of hermeneutics).

The Pragmatics of Existential Dialectics (or Its Ontopragmatics)

Even the semantics of existential dialectics is not enough to understand reality, since semantics by itself does not tell us the nature of intentionality and interests in relation to the matrix of power formations and conflicting interests when existential dialectics as an ontology is applied.

Surely, one cannot completely separate the pragmatics of existential dialectics from its semantics (and for that matter, its syntax), but the point here is to give each a more elaborated analysis.

For this reason, the pragmatics of existential dialectics needs to be studied too, in special relation to the inclusion of power formation and conflicting interests in the application of the principles of existential dialectics (as summarized in *Table 4.12*).

It is thus no wonder that all my previous books have shown in different ways how and why these ontological principles and their theoretical applications can reveal the future world to come in a way that humans have never known, in a totally different battleground for competing human (and later post-human) interests by myriad groups fighting for their own versions of hegemony.

Perhaps a summary of my previous works in relation to existential dialectics and their contributions to my new theories as proposed over the years is deemed revealing here, for the further understanding of the pragmatics of existential dialectics (in close relation to the semantics of existential dialectics too).

That stressed—the following summary in four sub-sections is something that I often do in my previous books too, as a way to introduce the intellectual background of my ideas.

So, for those readers who had read my previous works, the following sub-sections serve as a reminder, as they are repeated hereafter *often verbatim* (with only some updated changes, with the rest kept intact).

These sub-sections are, namely, (a) the two-way street connecting theory with meta-theory, (b) the theoretical application of existential dialectics, (c) direct and indirect applications of existential dialectics, and (d) multiple applications of existential dialectics—to be summaried herefafter, respectively.

The Two-Way Street Connecting Theory and Meta-Theory

However, lest any misunderstanding occurs, it is important to remember that the pragmatics of existential dialectics is not a one-way street (that is, using the ontological principles for theoretical insights in praxis) but a two-way one, namely, (a) from meta-theory to theory, and (b) from theory to meta-theory.

(a) On one side of the street, the ontological principles can inspire some theoretical insights in praxis, that is, in relation to some specific fields.

(b) On the other (opposing) side of the street, however, the study of a subject matter in the specific fields in turn reveals some more hitherto unknown ontological principles to be discovered and identified.

For this reason, three new principles were added in *FC*, one in *FAE*, two in *ALD*, and three in *FIA*, on top of the original three in *BCPC*, for example (as there are more)—after some research on the specific subject matters.

Both theory and meta-theory enrich each other in all my works over the years.

The Theoretical Application of Existential Dialectics

This two-way street connecting theory and meta-theory has contributed to my original construction of many theories in numerous books of mine on numerous fields in all domains of knowledge, which can be classified in terms of specific "meta-theories" (about epistemology, methodology, logic, ontology, etc.) and specific "theories" (about the mind, nature, culture, society, and the rest—as will be further explained later in the sub-section on "the holistic organization of an inquiry" in the section about "sophisticated methodological holism").

The final aim of my numerous new specific theories (on both theory and meta-theory) is to converge all of them (as proposed in my numerous books) into a grand project about the future of intelligent life, both here on earth and in deep space.

In this sense, all these theories of mine serve as a foundation to unify all domains of knowledge for an overarching understanding of all forms of reality (by way of some ontological principles and the comprehensive perspectives of the mind, nature, society, and culture), while allowing for heterogeneity, conflict, subjectivity, and complexity, for example.

With this grand project in mind—the summary of my specific theories on meta-theory are given in this section (on "existential dialectics") and the next (on "sophisticated methodological holism"), whereas the specific theories on theory will be summarized in the following categories of (a) society, (b) culture, (c) the mind, (d) nature, and (e) the rest, as shown below, in that order.

(a) Society

The category of "society" here can be divided into four sub-categories, i.e., (a1) social organizations, (a2) social institutions, (a3) social structure, and (a4) social systems (or trends).

(a1) Social Organizations

In relation to social organizations, I proposed a new thesis on the ambivalent nature of virtual organizations in the future, known as "the theory of e-civic alienation" in *FCD* (and summarized in *Table 4.19*).

And in *FPHO*, another theory called "the combinational theory of organization" is worked out, although it was first proposed in Ch.6 of *ALD*.

(a2) Social Institutions

There are different social institutions, be they about (a2.1) politics, (a2.2) economics, (a2.3) military affairs, (a2.4) mass media, (a2.5) law, (a2.6) education, and (a.7) gambling, for instance.

(a2.1) In relation to politics, I offered different visions on (a2.1.1) political theory, (a2.1.2) comparative politics, and (a2.1.3) international relations, to cite three main ones.

(a2.1.1) In the context of political theory, my vision of a path-breaking political system in future times concerns the different forms of "post-democracy" to supersede democracy unto the post-human age.

My vision of post-democracy was called "the theory of post-democracy" (as summarized in *Table 4.22*, *Table 4.23*, and *Table 4.24*), whereas *Table 4.25* explains the distinctions among democracy, non-democracy, and post-democracy.

And *Table 4.26* stresses the multiple causes of the emergence of post-democracy, while *Table 4.27* clarifies some possible misunderstandings in regard to post-capitalism and post-democracy.

(a2.1.2) In the context of comparative politics, I also showed, in *ALD*, how and why authoritarianism can be an advanced form of liberal democracy (just as it has its own counterpart in non-liberal democracy), or what I originally called "authoritarian liberal democracy" (as summarized in *Table 4.21*), before it is to be superceded by "post-democracy."

(a2.1.3) In the context of international relations (both at the institutional and structural levels), I provided, in *BWT*, "the theory of cyclical progression of empire-building," for a better way to understand the logic of empire-building on earth and beyond—with the rise of what I originally envisioned as "hyper-empires" (like "the Chinese Union" and "the Indian Union"), "meso-empires" (like "the European Union" and "the North American Union") and "micro-empires" (like "the Middle Eastern Union" and "the Latin American Union") in the coming "post-post-Cold War era" as a visionary illustration, as summarized in *Table 4.20.*

In addition, in *FPHWP*, I proposed "the multifaceted theory of war and peace" to understand war and peace, in a small sense—or aggression and pacificity, in a broad sense.

(a2.2) In relation to economics, I also offered another vision, this time to understand the future forms of economic systems, that is, the different forms of "post-capitalism" to supersede capitalism unto the post-human age.

My vision of post-capitalism was referred to (in *FCD*, *BDPD* and *BCPC*) as "the theory of post-capitalism" (as summarized in *Table 4.28*, *Table 4.29*, *Table 4.30*, *Table 4.31*, and *Table 4.32*). And *Table 4.33* clarifies the differences among capitalism, non-capitalism, and post-capitalism, whereas *Table 4.34* shows multiple causes of the emergence of post-capitalism.

This vision on "post-capitalism" constitutes the third theoretical application of existential dialectics, this time, in that there is no wealth without poverty (or the wealth/poverty dialectics).

(a2.3) In relation to military affairs, I also proposed "the multifaceted theory of war and peace" (in *FCD*, although it was first so labeled in *FPHK*) for a new way to understand the causes of war—and, for that matter, the rise and fall of different forms of warfare over time.

This theory was further expanded in *FPHWP*, where I suggested a new way to understand war and peace, in a small sense—or aggression and pacificity, in a broad sense.

(a2.4) In relation to mass media, I also proposed "the totalistic theory of communication" in *The Future of Post-Human Mass Media* (or in short, *FPHMM*).

(a2.5) In relation to law, I also proposed "the reconstruction theory of law" in *The Future of Post-Human Law* (or in short, *FPHLAW*).

(a2.6) In relation to education, I proposed "the heterodox theory of education" in *The Future of Post-Human Education* (or in short, *FPHEDU*).

(a2.7) And in relation to gambling, I proposed "the detached theory of gambling" in *The Future of Post-Human Gambling* (or in short, *FPHGAM*).

(a3) Social Structure

In relation to social structure, I first proposed two additional theories, that is, "the theory of the cyclical progression of hegemony" in *FCD* (and also in *FPHC* and *BDPD*), and "the theory of the cyclical progression of empire-building" (in *BWT*, as already indicated above)—with important implications for the symmetry-asymmetry princple in existential dialectics, just to cite one example.

Then, in *BNN*, I proposed another theory, namely, "the theory of contrastive advantages" to understand the persistence of social structure by gender, race, ethnicity, or any other groupings.

(a4) Social Systems (or Trends)

In relation to social systems, there are several sub-categories for illustration, which can be about (a4.1) urbanization, (a4.2) technology, (a4.3) demography, (a4.4) the environment, and (a4.5) social trends as a whole (which also include the current phenomenon of globalization), and (a4.6) migration.

(a4.1) In relation to urbanization, I proposed "the contingent theory of urban planning" in *FPHUP* for a new way to understand density and void.

(a4.2) In relation to technology, I worked out "the ambivalent theory of technology" in *FPHE* for a new way to understand the nature of technology.

Some other books (like *FPHMM*, *FAE*, *FCD*, and *FHC*) also deal with technology, and a good example concerns "the synthetic theory of information architecture" in *FIA* on taxonomy, network, and intelligence, in conjunction with the rise of technology.

(a4.3) In relation to demography, I proposed, first in *FHC* (and later in other books like *FCD*, *FPHC*, *FPHU*, etc.), "the theory of post-humanity," in relation to the "post-human" age at some distant point of "after-postmodernity," long after human extinction, to be eventually superseded by "post-humans" of various forms.

Some good candidates of the post-human lifeforms include, for instance, "thinking robots," "thinking machines," "cyborgs," "genetically altered superior beings," "floating consciousness," "hyper-spatial consciousness," "unfolding unconsciousness," "the hyper-sexual body," and "the hyper-martial body." This post-human vision of mine was first originally worked out in *FHC* and further elaborated in both *FCD*, *FPHC*, *FPHU*, *FPHS*, and *FPHMA*, for instance.

Lest any confusion occurs, I need to emphasize two clarifications here about the term "post-human" as a neologism in my works.

Firstly, the word "post-human" here should *not* be confused with another term which looks similar but has a totally different meaning in the literature of postmodernism, namely, "post-humanism"—which constitutes a critique of "humanism" as tradionally understood (especially, though not exclusively, in relation to the idea of progress in science and reason in the Enlightenment project). (WK 2008)

And secondly, the word "post-human" here should also *not* be confused with a similar term which is used to champion the ideology of technology for the future co-existence between humans and cyborgs in "trans-humanism." (WK 2008a)

Instead, my term "post-human" in relation to "posthuman-ism" also rejects "transhumanism" (especially, though not exclusively, in relation to the promises of technology) and refers to something else altogether, that is, the future extinction of humans and its post-human successors in deep space and beyond unto multiverses.

Later, in *BEPE*, I went further and provided a more comprehensive critique of "transhumanism," together with other issues.

With these two clarifications in mind (as summarized in *Table 1.31*)—in the end, a most fundamental question about intelligent life now has an answer, in that, if asked, "What is the future of human civilization?"—my answer in *FCD* (89) is thus: "As addressed in Ch.7 of *FHC*, a later epoch of the age of after-postmodernity (that is, at some point further away from after-postmodernity) will begin, as what I called the 'post-human' history (with the term 'post-human' originally used in my doctoral dissertation at M.I.T., which was finished in November 1995, under the title *After Postmodernity*, still available at M.I.T. library, and was later revised and published as *FHC*). The post-human history will be such that humans are nothing in the end, other than what culture, society, and nature (with some luck) have shaped them into, to be eventually superseded by post-humans (e.g., cyborgs, thinking machines, genetically altered superior beings, and others), if humans are not destroyed long before then."

(a4.4) In relation to the environment, I also addressed the issue of sustainability, especially in *FHC*, *FCD*, and *FPHUP*. For instance, in *FPHUP*, I specifically proposed a new theory, "the contingent theory of urban planning," on the need to go beyond the contemporoary obsession with sustainability, in the context of not only the environment but also other issues.

This is also related to the issue of "geology" in (d5).

(a4.5) In relation to social trends as a whole, I proposed in Ch.9 of *FCD* (367-8), that "civilizational history will continue into the following cyclical progression of expansion, before it is to be superseded (solely as a high probability, since humans might be destroyed sooner either by themselves or in a gigantic natural calamity) by posthumans at some distant point in after-postmodernity (which I already discussed in *FHC*)" unto multiverses (different constellations of universes):

Local → Regional → Global →Solar →
Galactic → Clustery…→ Multiversal

The current phenomeon of "globalization" lies in the third category on the line above and was extensively analyzed in *FCD*, in close connection with "globalism" at the cultural level—in the context of "postmodernity" (which will be introduced later in the section on history).

In *BDPD*, this thesis of mine was referred to as "the theory of the cyclical progression of system integration and fragmentation," at the systemic level—in close relation to another thesis analyzed in Ch.9 of *FCD*, which was called, in the absence of better words, "the multifaceted theory of war and peace" in *FPHK* (as already indicated above).

In *FC*, I also proposed "the dialectic theory of complexity" to account for the factor of uncertainty, which can have a major role to play in the world.

(a4.6) And in relation to migration, I proposed "the resettlement theory of geology" in *FPHGEOL*—which overlaps with the issue of "geology" in (d.5).

(b) Culture

The category of "culture" here can also be divided into different sub-categories like (b1) aesthetics, (b2) religion, (b3) popular (mass) culture, (b4) logic, (b5) civilization as a whole, (b6) ethics, (b7) epistemology (methodology), (b8) metaphysics (ontology), (b9) ideology, and (b10) language.

(b1) Aesthetics

In relation to aesthetics, I proposed "the transformative theory of aesthetic experience" in *FAE* about some great transformations of aesthetic experience in future history.

And in *FPHLIT*, I further proposed "the comparative-impartial theory of literature" for a new way to understand fiction and non-fiction.

(b2) Religion

In relation to religion, I proposed "the comparative theory of religion" in *FHC* (and also in *FCD*, although it was first so labeled in *FPHK*).

And in *FPHR*, I provided a solid foundation of this theory with a better name, that is, "the comparative-substitutive theory of religion" for a new way to understand spirituality.

(b3) Popular (Mass) Culture

In relation to popular (mass) culture, there are some sub-categories for illustration, which can be, for example, (b3.1) mass culture in general and (b3.2) sports (like chess).

(b3.1) In relation to mass culture in general, I proposed, in *FHC*, a new way to understand mass culture in terms of some fundamental dilemmas, which is called, in the absence of better words, "the dualistic theory of mass culture" in *FPHO*.

(b3.2) In relation to sports (like chess), I proposed, in *FPHCHESS*, "the mediative-variative theory of chess" for a new way to understand tactics and strategy (in different traditions).

(b4) Logic

In relation to logic, I also worked out, in *FPHML*, "the contrastive theory of rationality" for a new way to understand rationality and knowledge (in the context of mathematical logic).

(b5) Civilization as a Whole

In relation to civilization as a whole, I also argued, in *FHC*, no freedom without unfreedom (as shown in *Table 4.13*) and no equality without inequality (as shown in *Table 4.14*), especially in relation to the seven dimensions of life existence first worked out in *FHC* (i.e., the technological, the everyday, the true, the holy, the sublime/beautiful, the good, and the just).

Even more interestingly, in *BDPD* and later in *BCIV*, I further revealed a theoretical refinement of existential dialectics, this time, in arguing that there is no civilization without barbarity, with human civilization to be eventually superseded by what I originally analyzed as "post-human post-civilization" (which should not be confused with "post-human civilization"), in the context of the freedom/unfreedom and equality/inequality dialectics.

In the final analysis, civilization cannot live without barbarity and has to learn to co-exist with it in ever new ways. Preposteorus as this may seem to many contemporaries—it is no more imperative to preserve civilization than necessary to destroy barbarity, and the ideal of civilization is essentially bankrupt, to be eventually superseded by "post-civilization" (as summarized in *Tables 4.35-4.43*).

This freedom/unfreedom and equality/inequality dialectics holds in pre-moderny, modernity, postmodernity, and what I orignally called "after-postmodernity" in *FHC* and *FCD*—especially, though not exclusively, in the context of the emergence of "post-humans."

There are two specific theories about "after-postmodernity" here. The first theory about "after-postmodernity" is about the trinity of pre-modernity, modernity, postmodernity, and after-postmodernity at the cultural level—and was called, in *BDPD*, "the theory of the trinity of modernity to its after-postmodern counterpart," although it was first analyzed in *FHC* and later in *FCD* and also *BCIV* (as summarized in *Table 1.34*, *Table 1.35*, *Table 1.36*, and *Table 1.37* and will be elaborated further in the later sub-section on "the rest" about future history).

The second theory about "after-postmodernity" is about pre-modernity, modernity, postmodernity, and after-postmodernity at the historical level. For instance, a conclusion in *FHC*, which is shocking indeed for many contemporaries, is none other than that "[t]he post-human history will therefore mark the end of human history as we know it and, for that matter, the end of human dominance and, practically speaking, the end of humans as well. The entire history of human civilization, from its beginning to the end, can be summarized by four words, linked by three arrows (as already discussed in *FHC*)":

Pre-Modernity → Modernity →
Post-Modernity → After-Postmodernity

In *BDPD*, this thesis of mine was specifically called "the theory of the evolution from pre-modernity to after-postmodernity," at the historical level—as shown in *Table 1.34*, *Table 1.35*, *Table 1.36*, and *Table 1.37* (and will be elaborated further in the later sub-section on "the rest" about future history).

Therefore, "[t]he end of humanity in the coming human extinction is the beginning of post-humanity. To say an untimely farewell to humanity is to foretell the future welcome of post-humanity." (P. Baofu 2002: 89) This thesis of mine was known in *BDPD* as "the theory of post-humanity," at the systemic level—as already indicated in (a4.3) above.

(b6) Ethics

In relation to ethics, I proposed, in *BEPE*, "the theory of post-ethics" for a new way to understand morality and immorality.

(b7) Epistemology (Methodology)

In relation to epistemology (methodology), I propose "the complex theory of data analysis" in *FPHDA* for a different way to understand quantitative and qualitative research.

In addition, I already proposed "the critical-dialectic theory of formal science" in *FPHFS* and "the theory of methodological holism" in several books (like Ch.1 *FCD*, Ch.1of *FPHC*, *Sec.2.1* and *Sec.2.5* of *BCPC*, *FC*, *FPHK*, etc.) for a different way to do methodology.

(b8) Metaphysics (Ontology)

I also proposed "the theory of existential dialectics" in several books (like *FHC*, *FCD*, *FPHC*, *BDPD*, *FC*, *FAE*, *ALD, FIA, FPHK*, etc,) and "the contrastive theory of rationality" in *FPHML* for a different way to understand ontology.

(b9) Ideology

The issue about ideology was already discussed under the categories of "social institutions" and of "civilization as a whole," especially in relation to my theories of "post-democracy," "post-capitalism," "post-civilization," and so on.

(b10) Language

And I already proposed "the multilogical theory of learning" in *FPHL* for a new way to understand the nature of language structure and context .

(c) The Mind

The important vision about "post-humans" (as summarized earlier) brings us to the category of "the mind" here, which, as is often understood in everyday's usage, refers to the "brain" and the "body" (together with their interactions) and has different levels of consciousness which can be both "spiritual" and non-spiritual.

Therefore, the "mind" already presupposes the "brain," "body," and "spirit" (or "soul"), so that one does not fall into the intellectual trap of the "mind-body dualism" as debunked in modern (and especially, postmodern) philosophy (which I already went to great lengths to explain in *The Future of Human Civilization*).

That said—the category of "the mind" can be divided into two sub-categories, namely, in relation to (c1) biology and (c2) psychology.

(c1) Biology

In relation to biology, there are two concerns here, namely, (c1.1) genes and intelligence and (c1.2) the human body.

(c1.1) In relation to genes and intelligence, I worked out "the theory of contrastive advantages" (which was originally proposed in *FCD* and further worked out in *BNN*, as summarized in *Table 1.24*), to show the nature-nurture interactions of multiple levels in action (e.g., the biological, the psychological, the structural, the systemic, the cultural, and others) for humans and post-humans on earth and beyond, to the extent that different groups, be they on the basis of race, gender, ethnicity, class, age, or else, are not equal, on average, and have not been, nor will they be, even if there can be exceptions.

(c1.2) In relation to the human body, for instance, in *FPHMA*, I suggested "the expansive-contractive theory of martial arts," in which two great future transformations of the martial body were envisioned, that is, both "with the hyper-martial body" and "without the body" (as summarized in *Table 1.32*)—although this issue can also be classified under the category of "culture."

And in *FPHS*, I proposed "the theory of virtual sexuality," in which three great future transformations of the sexual body were envisioned, that is, "virtual pleasure," "the hyper-sexual body," and "the post-sexual floating mind" (as summarized in *Table 1.33*).

(c2) Psychology

In relation to psypchology (which also overlaps with biology too, however), I also proposed some greater transformations of the mind to come in the future, especially though not exclusively in the post-human age.

These greater transformations of the mind in psychology can be further sub-divided into different categories, namely, (c2.1) consciousness, (c2.2) unconsciousness (together with subconsciousness), (c2.3) creativity, (c2.4) learning, (c2.5) aggression and pacificity, (c2.6) personality, (c2.7) humor, and (c2.8) artificial intelligence.

(c2.1) In relation to consciousness, I analyzed, in *FPHC*, the nature of consciousness in relation to the conceptual dimensions of consciousness (as shown in *Table 1.12*), the theoretical levels of consciousness (as shown in *Table 1.13*), the thematic issues of consciousness (as shown in *Table 1.14*), the different dimensions of human existence (viz., having-ness, belonging-ness, and being-ness) in relation to consciousness (as shown in *Table 1.15*, *Table 1.16*, *Table 1.17*, and *Table 1.18*).

In *BCPC*, the nature of consciousness was further explored in relation to cognition, emotion, and behavior, which can be in consciousness and other mental states (e.g., subconsciousness and unconsciousness), in the context of existential dialectics, as shown in *Table 1.19*, *Table 1.20*, *Table 1.21*, *Table 1.22,* and *Table 1.23*.

In fact, in *FCD*, I already proposed three different theories in relation to the limits of cognition, emotion, and behavior, namely, "the theory of cognitive partiality," "the theory of emotional non-neutrality," and "the theory of behavioral alteration."

Then, I argued that the existence of human consciousness will be superseded one day with "floating consciousness" (as first proposed in *FCD*) and "hyper-spatial consciousness" (as first proposed in *FPHC*) as a climax of evolution in consciousness, after the future extinction of human consciousness (as shown below):

Primordial consciousness → Human consciousness →
Post-human consciousness (with floating consciousness and hyper-spatial consciousness as a climax in the evolution of consciousness)

In *BDPD*, these latest theses of mine were called "the theory of floating consciousness" and "the theory of hyper-spatial consciousness," both at the cosmological and psychological levels (as summarized in *Table 1.26* and *Table 1.25*).

(c2.2) In relation to unconsciousness, for instance, in *FPHU*, the unconsious mind will undergo tremendous change, to the rise of what I originally called "unfolding unconsciousess" in relation to anomalous experience, as a climax of evolution in unconsciousness, in conjunction with the conscious mind (both human now and post-human later), especially at the psychological level, as shown below:

Primordial unconsciousness → Human unconsciousness →
Post-human unconsciousness (with unfolding unconsciousness as a climax in the evolution of unconsciousness)

In *FPHU*, this thesis of mine is called "the theory of unfolding unconsciousness" in a small sense, or "the unfolding theory of anomalous experience," in a larger one, as shown in *Table 1.27* and *Table 1.28*.

(c2.3) In relation to creativity, for instance, in *FPHCT*, I also proposed "the comprehensive theory of creative thinking" to explore the role of creative thinking, in relation to technology and other factors, together with my original proposal of important creative techniques and traits (as shown in *Table 1.29*) and its double nature of desirability and undesirability (as shown in *Table 1.30*).

(c2.4) In relation to learning, for instance, in *FPHL*, I worked out "the multilogical theory of learning" for a new way to understand the nature of structure and context in language, in relation to learning and its future development in the context of the mind.

(c2.5) In relation to aggression (and pacificity), I proposed "the multifaceted theory of war and peace" in *FPHWP* (as mentioned above).

(c2.6) In relation to personality, I worked out "the contrarian theory of personality" in *FPHP* for a new way to understand normality and abnormality.

(c2.7) In relation to humor, I worked out "the metamorphic theory of humor" in *FPHH* for a new way to understand joking and laughing.

(c2.8) And in relation to artificial intelligence, I proposed "the supersession theory of computing" in *FPHCOM* for a new way to understand hardware, software, and the mind.

In addition, I also propose "the complex theory of data analysis" in *FPHDA* for a different way to understand quantitative and qualitative research in the context of the evolution of new minds.

(d) Nature

The category of "nature" here refers to the state of nature, which can be divided into some sub-categories like (d1) physics, (d2) cosmology, (d3) biology, (d4) geography, (d5) geology, (d6) acoustics, and (d7) chemistry.

But since biology in (d3) overlaps with biology in (c1)—only the rest of the categories are addressed hereafter.

(d1) Physics

In relation to physics, I also proposed "the perspectival theory of space-time" in *FPHST* (based on *FPHC*) for a new way to understand space and time, as shown in *Table 1.3*, *Table 1.4*, *Table 1.5*, and *Table 1.7*.

In *FPHG* (based on *FPHC* and *FPHST*), I proposed "the selective theory of geometry" for a new way to understand geometry, with implications for the understanding of space-time in the context of infinity, symmetry, and dimensionality for future lifeforms that our world has never known (as shown in *Table 1.6*, *Table 1.7*, *Table 1.8; Table 1.9*, *Table 1.10,* and *Table 1.11*).

And in *FC*, I proposed "the dialectic theory of complexity" for a new way to understand order and chaos in the state of nature and beyond.

(d2) Cosmology

In relation to cosmology, my theories on physics (as indicated above) already have important implications for understanding the nature and the future of the universe (which was also addressed in *FHC*, among other works aforcited).

I also proposed "the theory of hyper-spatial consciousness" in *FPHC* (and also in *FPHG*) for a new way to understand the future evolution of the mind to enter into different dimensions of space-time that our current world has never known.

In addition, in *BCOS*, I proposed "the theory of post-cosmology" for a new way to understand the beginnings and ends of the cosmos and beyond.

(d4) Geography

In relation to geography, I proposed "the theory of the geopower of nature" in *ALD* for a different way to understand the relationships between geography and other entities (like the mind, culture, and society).

(d5) Geology

In relation to geology, I proposed "the resettlement theory of geology" in *FPHGEOL* for a different way to understand the relationships between statics and dynamics.

(d6) Acoustics

In relation to acoustics, I proposed "the multilateral theory of acoustics" in *FPHA* for a different way to understand the relationships between sound and silence.

(d7) Chemistry

And in relation to chemistry, I proposed "the creational theory of chemistry" in *FPHCHEM* for a different way to understand substances and their changes.

(e) The Rest

The category of "the rest" here refers to what cannot be strictly classified in any of the previous categories (i.e., society, culture, the mind, and nature). A good illustration of "the rest" refers to that which is, namely, (e1) historical.

(e1) Historical

Precisely here, I proposed "the theory of the evolution from pre-modernity to after-postmodernity" (as first worked out in *FHC* and later in *FCD* and *FPHC*), which overlaps—as a reminder from (b5)—with the issue concerning "civilization as a whole" in (b5).

For instance, in both *FCD* and *FPHC*, I worked out the structure of "post-human civilization" in terms of the trinity of after-postmodernity (i.e., "free-spirited after-postmodernity," "post-capitalist after-postmodernity," and "hegemonic after-postmodernity").

Both conceptually and theoretically, the trinity of after-postmodernity is a sequential extension to the trinity of modernity (i.e., "free-spirited modernity," "capitalist modernity,"and "hegemonic modernity") and the trinity of postmodernity (i.e., "free-spirited postmodernity," "capitalist postmodernity," and "hegemonic postmodernity") as first proposed in *FHC*.

And the trinity of pre-modernity (i.e., "pre-free-spirited pre-modernity," "pre-capitalist pre-modernity" and "hegemonic pre-modernity") was later conceived in *BCIV* to complete the historical set from pre-modernity to after-postmodernity.

In *BDPD*, this thesis about the trinity of pre-modernity, modernity, postmodernity, and after-postmodernity was collectively known as "the theory of the trinity of modernity to its after-postmodern counterpart," at the cultural level—as already so mentioned in (b5) earlier, and summarized in *Table 1.34*, *Table 1.35*, *Table 1.36*, and *Table 1.37*.

At the structural level, all these trinities are subject to the existential constraints (e.g., the freedom/unfreedom and equality/inequality dialectics in the context of "the cyclical progression of hegemony"), be the historical epoch in pre-modernity, modernity, postmodernity, or after-postmodernity in future times. Each of the historical epochs has its ever new ways of coming to terms with the ever new (different) mixtures of freedom/unfreedom and equality/inequality.

This is importantly so, not because, as is falsely assumed in conventional wisdom, one certain way is superior (or better) than another in terms of achieving more freedom and less unfreedom, or more equality with less inequality.

On the contrary, indeed, in each of the historical epochs, each increase of unfreedom greets each freedom achieved, just as each increase of inequality welcomes each equality achieved, albeit in ever new (different) ways. In *BDPD*, this thesis of mine was labeled as "the theory of the cyclical progression of hegemony," at the structural level, though it was first analyzed in *FCD*.

In *BDPD*, more theoretical applications of existential dialectics were further examined, in relation to five main features, in the context of the duality of oppression, namely, (a) that each freedom/equality achieved is also each unfreedom/inequality created, (b) that the subsequent oppressiveness is dualistic, both by the Same against the Others and itself and by the Others against the Same and themselves, (c) that both oppression and self-oppression can be achieved by way of downgrading differences (between the Same and the Others) and of accentuating them, (d) that the relationships are relatively asymmetric among them but relatively symmetric within them, even when the Same can be relatively asymmetric towards itself in self-oppression, and the Others can be likewise towards themselves, and (e) that symmetry and asymmetry change over time, with ever new players, new causes, and new forms, be the locality here on Earth or in deep space unto multiverses—as summarized in *Table.4.13*, *Table 4.14*, *Table 4.15*, and *Table 4.16*.

The same logic also holds both in relation to wealth and poverty (as addressed in *BCPC* and summarized in *Table 4.17* on the wealth/poverty dialectics) and in relation to civilization and barbarity (as addressed in *BCIV* and summarized in *Table 4.18* on the civilization/barbarity dialectics).

In *BDPD*, this thesis on existential dialectics was labeled as "the theory of existential dialectics."

Direct and Indirect Applications of Existential Dialectics

A different way to appreciate the usefulness of existential dialectics is by way of the analysis of its (a) direct and (b) indirect applications.

(a) In direct applications, on the one hand, the logic of existential dialectics can shed some theoretical insights on diverse phenomena in the world, and good instances are the usage of the principles of existential dialectics for the theoretical insights on the freedom/unfreedom dialectics, the equality/inequality dialectics, and the wealth/poverty dialectics—as introduced earlier in (b5) and (e) in the previous sub-section.

For instance, my books like *FPHST*, *BNN*, *FAE*, *FC*, *FIA*, *FPHU*, *FPHK*, *FPHML*, *FPHE*, *FPHMM*, *FPHCT*, *FPHG*, *FPHUP*, *FPHL*, *FPHO*, *FPHMA*, *FPHS*, *FPHLAW*, *FPHFS*, *FPHWP*, *FPHP*, *BEPE*, *BCOS*, *FPHGEOL*, *FPHCHESS*, *FPHR*, *FPHEDU*, *FPHH*, *FPHA*, *FPHGAM*, *FPHCOM*, *FPHLIT*, and *FPHCHEM* use the principles to reveal some theoretical insights on the perspectives of space and time (as in *FPHST*), of nature and nurture (as in *BNN*), of beauty and ugliness (as in *FAE*), of order and chaos (as in *FC*), of taxonomy and network (as in *FIA*), of the unconscious mind (as in *FPHU*), of the nature of knowledge (as in *FPHK*), of mathematical logic (as in *FPHML*), of engineering (as in *FPHE*), of mass media (as in *FPHMM*), of creative thinking (as in *FPHCT*), of geometry (as in *FPHG*), of urban planning (as in *FPHUP*), of language (as in *FPHL*), of organization (as in *FPHO*), of martial arts (as in *FPHMA*), of sexuality (as in *FPHS*), of law (as in *FPHLAW*), of formal science (as in *FPHFS*), of aggression and pacificity (as in *FPHWP*), of normality and abnormality (as in *FPHP*). of morality and immorality (as in *BEPE*), of the beginnings and ends of the universe (as in *BOCS*), of statics and dynamics (as in *FPHGEOL*), of tactics and strategy (as in *FPHCHESS*), of secularness and sacredness (as in *FPHR*), of teaching and learning (as in *FPHEDU*), of joking and laughing (as in *FPHH*), of sound and silence (as in *FPHA*), of risk and caution (as in *FPHGAM*), of hardware and software (as in *FPHCOM*), of fiction and non-fiction (as in *FPHLIT*), and of substances and their changes (as in *FPHCHEM*).

(b) In indirect applications, on the other hand, the theoretical insights can further be used to reveal other phenomena directly from them (viz., the theoretical insights) and therefore indirectly from the principles themselves.

A good illustration is of course the use of the theoretical insights on the freedom/unfreedom and equality/inequality dialectics for the understanding of the civilization/barbarity dialectics.

This distinction between direct and indirect applications may sound a bit academic, since even in indirect applications, the phenomena under study can still be directly related back to the principles themselves.

In the previous example, as an illustration, the civilization/barbarity dialectics can be directly related to the principles of existential dialectics without the intermediate role of the freedom/unfreedom and equality/inequality dialectics.

Multiple Levels of Application

There is another issue to be clarified, however. In other words, the theoretical insights can be applied to multiple levels of analysis—even though, in a given example, it may refer to one level or a few only.

For instance, in the example concerning the freedom/unfreedom dialectics, it can be used at the structural level (e.g., in relation to the theory of the cyclical progression of hegemony), but it can also be exploited for other levels (e.g., the theory of post-capitalism at the institutional level).

All these levels of application should not be misleadingly construed, as I stress this before, as a one-way street (that is, to use the ontological principles for theoretical application at multiple levels of specific analysis) but a two-way one, in which theoretical insights in praxis, when studied in more specific contexts, can in turn refine the nature of existential dialectics (for example, with the addition of new principles).

Sophisticated Methodological Holism

The summary of my original meta-theory on methodology is provided below, again, to be repeated *often verbatim* from my previous works—as this is something I regularly do in each new book to introduce my past works, which are in conversation with this current one.

This means that, for those readers who had read some of my previous books, the summary below is really not much new (*almost verbatim*, with only some updated revisions).

That clarified—this original contribution of mine to the study of methodology takes the form of an approach known as "the theory of methodological holism" or "methodological holism" in short, as already worked out in *FPHC* (2004).

I have made good use of this methodological approach of mine for all of my previous works.

The Confusion of Holism with Systems Approach

Yet, it is imperative to stress at the outset, as I often emphasized this in all my previous books, that my approach of "methodological holism" does not oppose or exclude "methodological individualism" (as some readers may be tempted to assume, as is conventionally understood) but actually includes it.

The confusion here has to do with equating holism with systems approach in general (be it about systems theory, chaos theory, complexity theory, or else), which I debunked, while learning something from it, as already elaborated in *The Future of Complexity* (2007)—and *The Future of Post-Human Formal Science* (2010) is another addition to this critique of the intellectual fad of systems approach.

In other words, my meta-theory of holism, or sophisticated methodological holism in short, makes a distinction between doing holistic work and doing systems work, for two reasons.

The first reason is that holism, in my theory, does not imply systems approach, since it rejects systems approach because of the myriad problems that the literature on systems approach has been unable to resolve, while learning something from it.

And the second reason is that my theory is related to the other theory of mine, that is, existential dialectics, which does not exclude methodological individualism (but includes it, because it simply goes beyond both reductionism and reverse-reductionism in a dialectic way).

For these reasons (and others too, as summarized in *Table 4.1*), my version of methodological holism is *sophisticated*—not *vulgar* as sometimes used by inapt scholars employing the same term.

The Ontological Constraints

Sophisticated methodological holism is subject to some ontological constraints, and some good examples include "the formalness-informalness principle," "the absoluteness-relativeness principle," "the partiality-totality principle," "the predictability-unpredictability principle," "the explicability-inexplicability principle," and "the fiction-reallity principle" in existential dialectics.

Of course, there can be more than six ontological principles in relation to method, but the examples here are sufficient to illustrate the point in question.

With this caveat in mind—the first two were already analyzed in previous books of mine like *FC* and *FAE*, whereas the third one was addressed in *FPHU*, the fourth and fifth ones were accounted for in *FPHK* and *FPHML*, respectively, and the last one was addressed in *FPHLIT*.

For now, it suffices to show that sophisticted methodological holism, because of these ontological constraints on methodology, targets two major sins of methodology, namely, what I call, in the absence of better words, (a) *reductionism* and (b) *reverse-reductionism*—both of which come in all shapes and sizes, to be summarized below (and also shown in *Table 4.2*).

Against the Varieties of Reductionism

There are four versions of reductionism to be summarized here, which sophisticated methodological holism rejects, namely, (a1) conceptual, (a2) theoretical, (a3) methodological, and (a4) ontological forms of reductionism.

(a1) In conceptual reductionism, a good case in point concerns myriad dualities like mind vs. body, self vs. world, democracy vs non-democracy, and the like (as already addressed in *FHC*, *FPHC*, and *BDPD*, for instance).

(a2) In theoretical reductionism, an illuminating case study is best exemplified by what I originally called "the foundation fallacy" in *FPHST*, in any attempt to naively understand space-time from the physical perspective as the foundation and, consequently, to dangerously dismiss (or belittle) other perspectives.

In *FAE*, I elaborated further these versions of reductionism in the literature on aesthetics (e.g., form vs. content, representation vs. expression, critics vs. artists, and externalism vs. internalism).

In *FIA*, I revealed other forms of reductionism in the literature on information architecture (e.g., the constructivist argument and the representational argument).

In *FPHU*, I showed the persistent legacy of reductionism, this time, in the literature on anomalous experience (e.g., the obsession with physics, chemistry, and biology for explaining anomalous experience).

In *FPHE*, I examined another case of reductionism in action, in the context of engineering (e.g., technical constraints vs. normative constraints).

In *FPHMM*, I elaborated one more version of reductionism, in relation to the three domains of communication (e.g., the competing views on sending, connecting, and receiving).

In *FPHCT*, I explored another version of reductionism, in relation to invention and innovation (e.g., the bio-psychological argument vs. the socio-cultural arguments).

In *FPHG*, I identified another version of reductionism, in relation to infinity, symmetry, and dimensionality (e.g., the Euclidean argument vs. the non-Euclidean arguments).

In *FPHUP*, I analyzed another version of reductionism, in relation to density and void (e.g., the engineering argument and the ecology argument).

In *PFHL*, I examined another version of reductionism, in relation to structure and context (e.g., the structuralist argument and the contextualist argument).

In *PFHO*, I explored another version of reductionism, in relation to communcation, decision-making, and leadership (e.g., the rational-system argument vs. the natural-system argument vs. the open-system argument).

In *PFHMA*, I revealed another version of reductionism, in relation to the martial body and spirit (e.g., the spiritual argument vs. the materialist argument vs. the defensive argument).

In *PFHS*, I analyzed another version of reductionism, in relation to the sexual body and spirit (e.g., the naturalist argument vs. the constructivist argument).

In *PFHLAW*, I examined another version of reductionism, in relation to law (e.g., the necessity argument vs. the contengency argument).

In *FPHWP*, I show another version of reductionism, in relation to war and peace (e.g., the aggressivist argument vs. the pacifist argument).

In *BEPE*, I analyzed another version of reductionism, in relation to morality and immorality (e.g., the objectivist argument vs. the non-objectivist argument vs. the skeptical argument).

In *BCOS*, I examined another version of reductionism, in relation to the contested beginnings and speculative ends of the universe (e.g., the scientific argument vs. the religious argument vs. the esoteric argument vs. the metaphysical argument).

In *FPHP*, I showed another version of reductionism, in relation to normality and abnormality (e.g., the natural argument vs. the social argument vs. the cultural argument vs. the mental argument).

In *FPHGEOL*, I scrutinized another version of reductionism, in relation to statics and dynamics (e.g., the catastrophe argument vs. the uniformity argument vs. the revision argument).

In *FPHCHESS*, I showed another version of reductionism, in relation to tactics and strategy (e.g., the natural argument vs. the social argument vs. the cultural argument vs. the mental argument).

In *FPHR*, I revealed another version of reductionism, in relation to secularness and sacredness (e.g., the critical argument vs. the skeptical argument vs. the theist argument).

In *FPHEDU*, I examined another version of reductionism, in relation to teaching and learning (e.g., the teacher-centered argument vs. the student-centered argument vs. the balanced argument).

In *FPHH*, I explored another version of reductionism, in relation to joking and laughing (e.g., the natural argument vs. the social argument vs. the cultural argument vs. the mental argument).

In *FPHA*, I revealed another version of reductionism, in relation to sound and silence (e.g., the natural argument vs. the social argument vs. the cultural argument vs. the mental argument).

In *FPHGAM*, I revealed another version of reductionism, in relation to risk and caution (e.g., the natural argument vs. the social argument vs. the cultural argument vs. the mental argument).

In *FPHCOM*, I showed another version of reductionism, in relation to hardware and software (e.g., the natural argument vs. the social argument vs. the cultural argument vs. the mental argument).

In *FPHLIT*, I proposed another version of reductionism, in relation to fiction and non-fiction (e.g., the natural argument vs. the social argument vs. the cultural argument vs. the mental argument).

In *FPHCHEM*, I proposed another version of reductionism, in relation to substances and their changes (e.g., the natural argument vs. the social argument vs. the cultural argument vs. the mental argument).

And in *FPHDA*, I propose another version of reductionism, in relation to quantitative and qualitative research (e.g., the natural argument vs. the social argument vs. the cultural argument vs. the mental argument).

(a3) In methodological reductionism, a good illustration can be the debate between different versions of qualitative and quantitative methods (as already analyzed in *FC* and also *FHC*).

In *FPHML*, I examined similar reductionism, this time, in the literature on mathematical logic (e.g., the obsession with consistency, soundness, and completeness).

And in *FPHFS*, I explored the problems of reductionism in the context of formal science (e.g., the analytical argument, as opposed to the synthetic argument).

(a4) In ontological reductionism, an excellent instance involves another debate, this time between emergentism and reductionism in complexity theory (as addressed in *FC*) and also in psychology (as elaborated in *FPHC*, in the context of Being and Belonging).

But to target against the varieties of reductionism constitutes only a side of the same coin, and the other side concerns the varieties of reverse-reductionism.

Against the Varieties of Reverse-Reductionism

The other side of the same coin is the reverse version of reductionism, which is what I want to call, in the absence of better words, *reverse-reductionism*.

My sophisticated version of methodological holism targets against the varieties of reverse-reductionism (just as it also rejects the ones of reductionism).

Perhaps nothing expresses better the popularity of reverse-reductionim than the "anything-goes" mentality in postmodernism of our time, as shown in the following four versions.

(b1) In conceptual reverse-reductionism, any concept of "art" (e.g., fine art, applied art, outsider art, junk art) is welcome in postmodernism (as already analyzed in Ch.4 of *FHC*—and also in *FAE*).

(b2) In theoretical reverse-reductionism, a variety of art and literary theories co-exist. Take the case of literary studies, as there are now Literary Structuralism, Marxist Literary Criticism, New Criticism, Phenomenology, Hermeneutics, Language-Game Literary Criticism, Feminist Literary Criticism, Reception Theory, Reader Response Criticism, Poststructuralism, Semiotics, Pyschoanalytic Literary Criticism, just to cite some well-known ones, with no one being said to be better than any others (as detailedly analyzed in Ch.4 of *FHC*). (S. Raman 1997)

In *BNN*, I even introduced "the compromise fallacy" as another good illustration of theoretical reverse-reductionism, in misleadingly treating both genetic and environmental approaches as equally valid.

(b3) In methodological reverse-reductionism, multiple methodologies are deemed as acceptable in postmodernism (e.g., doing art without praxis, doing art with praxis, and doing art by sublation), as analyzed in Ch.4 of *FHC*.

And in *FPHFS*, I exposed the problems of reverse-reductionism in the context of formal science in relation to systems theory (e.g., the synthetic argument).

(b4) In ontological reverse-reductionism, no privileged ontology is allowed, and the door is open practically for anything in postmodernism (e.g., the equal status of the ontology of Being vs. that of Becoming, as already addressed in Ch.4 of *FHC*—and also in *FPHC*).

In *FAE*, I also introduced another version of reverse-reductionism, that is, "the pluralist fallacy," in the context of understanding aesthetic experience, for instance—although this fallacy has been committed not exclusively in relation to the ontological level (but also at the conceptual, theoretical, and methodological ones).

These dual dangers against reverse-reductionism (in this sub-section) and reductionism (in the previous sub-section) are something that sophisticated methodological holism rejects. And in *FPHFS*, I addressed these two problems further.

This point about the dual dangers is important enough, since many scholars often favor one against the other, but it is vital to target both.

The Holistic Organization of an Inquiry

With these dual dangers against reductionism and reverse-reductionism in mind—sophisticated methodological holism suggests that an inquiry of any given phenomenon is more complete, if treated in the context of a comprehensive analysis at all relevant levels, which challengingly encompass all the fields of human knowledge, ranging from the natural sciences and formal sciences through the social sciences to the humanities.

This section is something that I had already stressed in all my previous books and repeat (*often verbatim*) hereafter. But, for those readers who had read my previous books, this serves as a reminder.

With this reminder in mind—there are multiple ways to engage in a holistic inquiry with all relevent levels of analysis. Over the years, I have proposed different ways to fulfill this holistic methodological requirement, as repeated in my previous books.

Hereafter is a summary of four major ways, namely, (a) by discipline, (b) by domain, (c) by subject, and (d) by meta-analysis—all of which fulfill the holistic requirement of sophisticated methodological holism on a given issue.

(a) In a holistic organization by discipline, a good illustration is the multiple levels of analysis, namely, (a1) the micro-physical, (a2) the chemical, (a3) the biological, (a4) the psychological, (a5) the organizational, (a6) the institutional, (a7) the structural, (a8) the systemic, (a9) the cultural, (a10) the macro-physical (cosmological), and (a11) other relevant levels, which are either a combination of all these levels or the practical applications with a combination of them.

This holistic organization by discipline is used in *FCD* and *BCPC*, for instance.

(b) In a holistic organization by domain, some good candidates include the classification of different perspectives of analysis in relation to (b1) nature, (b2) the mind, (b3) society, and (b4) culture, as already worked out in *BCIV*, *FPHST*, *FC*, *FAE*, *FPHU*, *FPHK*, *FPHE*, and *FPHFS*, just to cite some examples.

Culture in (b4), in this re-classification, is the same as culture in (a9) and can be further divided into different sub-categories, with good examples like (a9i) epistemology (methodology), (a9ii) aesthetics, (a9iii) ethics, (a9iv) metaphysics (ontology), and (a9v) religion. For clarification, epistemology (methodology) in (a9i) and metaphysics (ontology) in (a9iv) overlap with the category of "meta-theory."

Society in (b3) includes the organizational in (a5), the institutional in (a6), the structural in (a7), and the systemic in (a8).

The mind (b2) has more to do with the chemical in (a2), the biological in (a3), and the psychological in (a4), although it can overlap with (a1), (b3) and (b4), for instance.

It should be stressed that the "mind" here also includes the categories like the "brain," "body," and the "spirit" (or the "soul")—whenever needed in a given context, so that one does not fall into the intellectual trap of the "mind-body dualism" in the history of modern (and especially, postmodern) philosophy, which I already went to great lengths to explain in *The Future of Human Civilization*. After all, in everyday's usage, the "mind" already implies the "brain" and the "body" (together with their interactions) and has different levels of consciousness which can be both "spiritual" and non-spiritual.

And nature in (b1) refers to the micro-physical in (a1) and the macro-physical (cosmological) in (a10), although it can also overlap with (a2), (a3) and (a4).

(c) In a holistic organization by subject, some excellent examples concern the analysis of consciousness in *FPHC* (i.e., on Having, Belonging, and Being); of civilization in *FHC* (i.e., the True, the Holy, the Everyday, the Technological, the Beautiful/Sublime, the Good, and the Just); of international politics in *BWT* (i.e., Hyper-Empires, Meso-Empires, and Micro-Empires); of nature and nurture in *BNN* (i.e., genes and memes); of information architecture in *FIA* (i.e., taxonomy and network); of mathematical logic in *FPHML* (i.e., consistency, soundness, and completeness); of communication in *FPHMM* (i.e., sending, connecting, and receiving); of creative thinking in *FPHCT* (i.e., invention and innovation); of geometry in *FPHG* (i.e., infinity, symmetry, and dimensionality); of urban planning in *FPHUP* (i.e., density and void), of language in *FPHL* (i.e., structure and context), and of organization in *FPHO* (i.e., communication, decision-making, and leadership), in *FPHMA* (i.e., the martial body and the martial spirit), in *FPHS* (i.e., the sexual body and the sexual spirit), in *FPHLAW* (i.e., necessity and contingency), in *FPHWP* (i.e., aggression and pacificity), in *FPHP* (i.e., normality and abnormality), in *BEPE* (i.e., morality and immorality), in *BCOS* (i.e., the beginnings and ends of the universe), in *FPHGEOL* (i.e., statics and dynamics), in *FPHCHESS* (i.e., tactics and strategy), in *FPHR* (i.e., secularness and sacredness), in *FPHEDU* (i.e., teaching and learning), in *FPHH* (i.e., laughing and joking), in *FPHA* (i.e., sound and silence), in *FPHGAM* (i.e., risk and caution), in *FPHCOM* (i.e., hardware and software), in *FPHLIT* (i.e., fiction and non-fiction), and in *FPHCHEM* (i.e., substances and their changes)—by incorporating all the relevent levels of analysis as cited above in each.

(d) And in a holistic organization by meta-analysis, two good cases in point involve the works on post-capitalism in *BCPC* and on authoritarian liberal democracy in *ALD*, with the classification of analysis in theory and meta-theory.

The essential point to remember here, as I thus stress in all my previous books, is that the multiple levels of analysis can be reorganized in many different ways, insofar as none of the levels (if relevant to an inquiry in question) is ignored or dismissed, to avoid the danger of reductionism (and for that matter, the one of reverse-reductionism, in the opposite direction).

Three Clarifications

Three clarifications are needed here, however, to avoid any hasty misunderstanding.

Firstly, the entities in each classification are not mutually exclusive. For instance, in the holisitc organization by domain, nature in (b1) can also be linked to the chemical in (a2), the biological in (a3), and the systemic in (a8). And the mind (b3) can alternatively be related to the micro-physical in (a1) and the chemical in (a2), for example.

Secondly, there is also the important factor of luck (or uncertainty in conventional usage), but it is already implied or allowed in each of the organizations of an inquiry—especially in relation to the predictability-unpredictability principle and the order-chaos principle in existential dialectics. This topic about uncertainty was already addressed in *FHC*—and more extensively, in *FC*, in the context of order and chaos (and later, to an extent, in *FPHK*, *FPHFS*, and *FPHGAM*, for example).

And lastly, the comparison in a classification is not absolute, but relative, as there are often some shades of gray, not exactly black or white (figuratively speaking).

Some further clarifications and qualifications of sophisticated methodological holism are shown in *Table 4.2*.

Some Distinctive Usefulness

In the end, sophisticated methodological holism—when applied, especially though not exclusively, as illustrated in my numerous works—can enrich the understanding of reality in some distinctive ways. Here are four examples (as revealed in each of my books).

Firstly, it provides a comprehensive analysis of a subject matter, from which much can be learned about reality, both in relation to the perspectives of the mind, nature, society, and culture.

Secondly, it offers a new classification of the subject matter.

Thirdly, it suggests some visions of the future in relation to the subject matter in question.

And fourthly, it proposes some insights on meta-theory (e.g., methodology and ontology) in general—with the clear understanding, however, of the dilemma of specific vs. general ontology (as shown in *Table 4.9* on the syntax of existential dialectics in the context of the dilemma of ontology).

For this very reason, all of these ways are important, without reducing one into the analysis of another. (*FPHCT*)

Chapter Outline

With this summary of my two original meta-theories (on methodology and ontology, *almost verbatim* from my previous works) in mind—this book is organized in four main parts, namely, (a) Part I on the introduction to this book, (b) Part II on quantitative research, (c) Part III on qualitative research, and (d) Part IV on the conclusion of the book.

The book is thus organized in four chapters.

The introductory chapter, that is, Chapter One titled "Introduction—The Significance of Data Analysis," begins by questioning the blank check of data analysis and its opposing view, then introduces quantitative and qualitative research in data analysis, as well as the varieties of data analysis—together with the theoretical debate in data analysis, my complex theory of data analysis, the relationship between theory and meta-theory, existential dialectics, sophisticated methodological holism, and some further clarifications.

Chapter Two titled "Quantitative Research and its Inadequacy" examines the nature of quantitative research in data analysis from the four perspectives of the mind, nature, society, and culture, with a constructive critique of each.

Chapter Three titled "Qualitative Research and its Limitation" explores the nature of qualitative research in data analysis from the four perspectives of the mind, nature, society, and culture, with a deconstructive analysis of each.

The last chapter titled "Conclusion—The Future of Data Analysis" summarizes the analysis in the book in the dialectic context of quantitative and qualitative research, with twenty-five major theses, namely, (a) the first thesis: the formalness-informalness principle, (b) the second thesis: the absoluteness-relativeness principle, (c) the third thesis: the partiality-totality principle, (d) the fourth thesis: the predictability-unpredictability principle, (e) the fifth thesis: the explicability-inexplicability principle, (f) the sixth thesis: the fiction-reality principle, (g) the seventh thesis: the finiteness-transfiniteness principle, (h) the eighth thesis: the preciseness-vagueness principle, (i) the ninth thesis: the simpleness-complicatedness principle, (j) the tenth thesis: the openness-hiddenness principle, (k) the eleventh thesis: the denseness-emptiness principle, (l) the twelfth thesis: the change-constancy principle, (m) the thirteenth thesis: the order-chaos principle, (n) the fourteenth thesis: the slowness-quickness principle, (o) the fifteenth thesis: the expansion-contraction principle, (p) the sixteenth thesis: the theory-praxis principle, (q) the seventeenth thesis: the convention-novelty principle, (r) the eighteenth thesis: the evolution-transformation principle, (s) the nineteenth thesis: the symmetry-asymmetry principle, (t) the twentieth thesis: the softness-hardness principle, (u) the twenty-first thesis: the seriousness-playfulness principle, (v) the twenty-second thesis: the regression-progression principle, (w) the twenty-third thesis: the sameness-difference principle, (x) the twenty-fourth thesis: the stability-reaction principle, and (y) the twenty-fifth thesis: the post-human complexity—to be elaborated in the rest of the book, with a summary in the concluding chapter.

This seminal project is to fundamentally alter the way that we think about quantitative and qualitative research in data analysis, from the combined perspectives of the mind, nature, society, and culture, with enormous implications for the human future and what I originally called its "post-human" fate.

Some Clarifications

However, some additional (last) clarifications are needed.

These clarifications, lest any potential misunderstanding by the readers occurs, refer to (a) the conversation with my previous books, (b) the illustrative selection of case studies and examples, (c) detailed analysis versus overall analysis, (d) two distinctive features of using quotations, and (e) the use of neologisms.

These clarifications are something that I also regularly repeat in my previous books too, so these points are repeated hereafter, *often verbatim*—albeit in a different context this time.

In Conversation with My Previous Books

Firstly, my *creational theory chemistry* is constructed on the theoretical foundation of my previous books (i.e., *FHC*, *FCD*, *FPHC*, *BDPD*, *BCPC*, *BCIV*, *FPHST*, *BNN*, *BWT*, *FC*, *FAE*, *ALD*, *FIA*, *FPHU*, *FPHK*, *FPHML*, *FPHE*, *FPHMM*, *FPHCT*, *FPHG*, *FPHUP*, *FPHL*, *FPHO*, *FPHMA*, *FPHS*, *FPHLAW*, *FPHFS*, *FPHWP*, *FPHP*, *BCOS*, *BEPE*, *FPHGEOL*, *FPHCHESS*, *FPHR*, *FPHEDU*, *FPHH*, *FPHA*, *FPHGAM*, *FPHCOM*, *FPHLIT*, and *FPHCHEM*—as explicated in "The List of Abbreviations").

For this reason, many of my books have some familiar words in their titles like "the future of post-human..." or "beyond...to," as well as some tables for background information—as they constitute a series of books with the final aim to converge all of my original theories on numerous fields into a unified theory of everything (as already summarized in the section on theory and meta-theory in Chapter One).

Just consider the following two illustrations, namely, (a) in relation to other related theoretical debates and (b) in relation to different visions as worked out in my previous books.

(a) The first illustration is that the theoretical debate here also involves a few other related theoretical issues.

These other related theoretical issues have to do with some deeply contested theoretical debates, which, however, I had already gone to great lengths to analyze in my numerous (previous) books, as summarized below. It should also be stressed, lest misunderstanding occurs, that each of the theoretical debates (as listed below) does not exhaust the issues covered in the books, as each book covers more issues than what is listed below (for illustration only).

- On Complexity
  –Ex: *The Future of Complexity*
- On Computation
  –Ex: *The Future of Post-Human Computing*
  –Ex: *The Future of Complexity*
  –Ex: *The Future of Post-Human Formal Science*
- On Systems Theory
  –Ex: *The Future of Post-Human Formal Science*
  –Ex: 2 volumes, *The Future of Human Civilization*
- On Epistemology (Methodology)
  –Ex: *The Future of Post-Human Knowledge*
  –Ex: 2 volumes, *The Future of Human Civilization*
- On Rationality
  –Ex: *The Future of Post-Human Mathematical Logic*
  –Ex: *The Future of Post-Human Humor*
- On Philosophy of Science
  –Ex: 2 volumes, *The Future of Human Civilization*
  –Ex: *The Future of Post-Human Mathematical Logic*
  –Ex: *The Future of Post-Human Knowledge*
- On Language (Meaning)
  –Ex: *The Future of Post-Human Language*
- On Textual Analysis
  –Ex: *The Future of Post-Human Literature*
- On Communication
  –Ex: *The Future of Post-Human Mass Media*
  –Ex: *The Future of Post-Human Organization*
- On Risk and Caution
  –Ex: *The Future of Post-Human Gambling*
  –Ex: *The Future of Post-Human Chess*
- On Creativity
  –Ex: *The Future of Post-Human Creative Thinking*
  –Ex: *The Future of Aesthetic Experience*
- On Learning
  –Ex: *The Future of Post-Human Language*
  –Ex: *The Future of Post-Human Education*
- On Cognition
  –Ex: *The Future of Post-Human Unconsciousness*
  –Ex: *The Future of Post-Human Consciousness*

–Ex: *The Future of Capitalism and Democracy*
–Ex: *The Future of Post-Human Creative Thinking*
–Ex: *The Future of Post-Human Personality*

- On Intuition
–Ex: *The Future of Post-Human Unconsciousness*
–Ex: *The Future of Post-Human Consciousness*
–Ex: *Beyond Capitalism to Post-Capitalism*
–Ex: *The Future of Capitalism and Democracy*
- On Emotion
–Ex: *The Future of Post-Human Unconsciousness*
–Ex: *The Future of Post-Human Consciousness*
–Ex: *The Future of Aesthetic Experience*
–Ex: *Beyond Capitalism to Post-Capitalism*
–Ex: *The Future of Capitalism and Democracy*
–Ex: *The Future of Post-Human Personality*
- On Behavior
–Ex: *The Future of Post-Human Language*
–Ex: *The Future of Post-Human Unconsciousness*
–Ex: *The Future of Post-Human Consciousness*
–Ex: *Beyond Capitalism to Post-Capitalism*
–Ex: *The Future of Capitalism and Democracy*
–Ex: *The Future of Post-Human Sexuality*
–Ex: *The Future of Post-Human Personality*
–Ex: *The Future of Post-Human Martial Arts*
- On Being and Belonging
–Ex: *The Future of Post-Human Consciousness*
–Ex: 2 volumes, *The Future of Human Civilization*
- On Different Dimensions of Human Existence
–Ex: 2 volumes, *The Future of Human Civilization*
- On Modernity, Post-Modernity, and After-Postmodernity
–Ex: 2 volumes, *The Future of Human Civilization*
–Ex: *Beyond Civilization to Post-Civilization*
- On Pre-Modernity
–Ex: *Beyond Civilization to Post-Civilization*
- On Spirituality
–Ex: 2 volumes, *The Future of Human Civilization*
–Ex: *The Future of Capitalism and Democracy*
–Ex: *The Future of Post-Human Religion*
–Ex: *Beyond Civilization to Post-Civilization*
–Ex: *The Future of Post-Human Sexuality*
–Ex: *The Future of Post-Human Martial Arts*

–Ex: *The Future of Post-Human Unconsciousness*
–Ex: *The Future of Post-Human Consciousness*
–Ex: *Beyond Ethics to Post-Ethics*

- On No Morality Without Immoralitiy
  –Ex: 2 volumes, *The Future of Human Civilization*
  –Ex: *The Future of Capitalism and Democracy*
  –Ex: *Beyond Democracy to Post-Democracy*
  –Ex: *Beyond Ethics to Post-Ethics*
- On No Freedom Without Unfreedom
  –Ex: 2 volumes, *The Future of Human Civilization*
  –Ex: *The Future of Capitalism and Democracy*
  –Ex: *Beyond Capitalism to Post-Capitalism*
  –Ex: *Beyond Democracy to Post-Democracy*
  –Ex: *Beyond Civilization to Post-Civilization*

Therefore, those who are interested in these other theoretical issues (and debates) which are related to the current context can consult these other books of mine for more details.

(b) And the second illustration is that one of the major theses of this book concerns the future, be it human and later post-human, which is closely linked with the different visions that I had worked out in my previous books, in relation to the mind, nature, society, and culture.

Of course, for the convenience of the readers, a summary of some of these visions will be provided whenever necessary, as shown in the two sections on meta-theory above (*often verbatim*) and in different tables at the end of this chapter and the concluding chapter, for instance.

Yet, I still expect the readers to read my previous books directly for more details—as this book is written in conversation with my previous ones.

Case Studies and Examples

Secondly, the case studies and examples in the chapters are not exhaustive but solely illustrative. Nor are they necessarily mutually exclusive, as they can be reclassified in a different way. And exceptions are allowed.

One is tempted to assume, however, that, since the case studies and their examples are not exhaustive, more case studies and examples are needed. But the problem here is that just adding more of them does not necessarily change the conclusions to be drawn and may even lead to redundancy.

After all, the case studies and examples used in this book are chosen with care and deemed sufficient for the purpose at hand, even though they are solely illustrative.

Detailed Analysis versus Overall Synthesis

Thirdly, there is a distinction between detailed analysis and overall synthesis, in that the former is, especially though not exclusively, more for the specialized viewpoint of a specific discipline, whereas the latter is, especially though not exclusively, more for the broad horizon of an interdisciplinary approach—relatively speaking, of course.

Or, by analogy, there is a distinction between depicting a "tree" (with tiny details) in a "forest" and viewing the entire "forest" itself (with the broader horizon).

This book, like all previous ones of mine, tries to strike a balance between the two forms of understanding, although this does not imply that the two forms should be exactly equally offered, in an exact 50-50 balance.

Instead, the book, more often than not, does not allow the tiny details to obscure the larger picture in a given case study, because the entire book is inter-disciplinary in nature and tries to look at the entire "forest," not being bogged down with the detailed analysis of an individual "tree" in the forest, by analogy.

In this sense, although detailed analysis can be provided whenever needed for a given case study, the important point to remember is the overall horizon with its broadness of scope.

Two Distinctive Features of Using Quotations

Fourthly, the way that quotes are used in this book (and others of mine) has two distinctive features, as explained below.

The first distinctive feature is that, at the end of each quotation, sometimes there are more than one reference—in which case the first reference is for the original citation, and the rest of the references are for the convenience of the reader for more information about the issue under quotation.

And the second feature is that quotations are used as often as possible, so as to let others speak for themselves, without the nuisance of putting my own words into their mouths. Even more importantly, this allows me to use the original text as an evidence for critical analysis at the end of each section.

The Use of Neologisms

And finally, I use different neologisms in my books, mostly to introduce my original concepts and theories, and good instances include those here or elsewhere in my previous books (e.g., "the complex theory of data analysis," "post-democracy," "hyper-spatial consciousness," "post-capitalism," and whatnot). Needless to say, they are used here solely for our current intellectual convenience, as they will be renamed differently in different ways in future history.

As I thus wrote in *FCD* (508-9), "all these terms 'post-capitalism,' 'post-democracy'...and other ones as introduced in...[the] project (e.g.,...'posthuman elitists,' and 'posthuman counter-elitists,' just to cite a few of them) are more for our current intellectual convenience than to the liking of future humans and post-humans, who will surely invent more tasteful neologisms to call their own eras, entities, and everything else, for that matter. But the didactic point here is to use the terms to foretell what the future might be like, not that its eras and entities must be called so exactly and permanently. After all, William Shakespeare (1995: Act II, Scene II, Line 47) well said long ago: 'What is in a name? That which we call a rose by any other name would smell as sweet.'"

As I had also stressed time and again before, each of the neologisms can be re-written as a different "X," only to be re-named differently by the powers that be in different eras of future history.

With these clarifications for thought in mind—I now proceed to Chapter Two on quantitative research and its inadequacy.

Table 1.1. Different Faces of Data Analysis (Part I)

• **Quantitative vs. Qualitive Research**

—*Purposefulness*: receives more attention in qualitative research, because "cases can be selected purposefully, according to whether or not they typify certain characteristics or contextual locations." But in quantitative research, the factor of purposefulness is minimized.

—*Researcher's Role*: receives "greater critical attention" in qualitative research, because "the possibility of the researcher taking a 'neutral' or transcendental position is seen as more problematic in practical and/or philosophical terms. Hence qualitative researchers are often exhorted to reflect on their role in the research process and make this clear in the analysis." But in quantitative research, neutrality is emphasized.

—*Meaning*: receives a greater attention in qualitative research, because of its interest in vagueness, ambivalence, and subjectivity embedded in "language, signs and meaning." But in quantitative research, preciseness and clarity are emphasized.

—*Context and the Whole*: are more important in qualitative research, because "qualitative research approaches analysis holistically and contextually, rather than being reductionistic and isolationist." But in quantitative research, reductionism is more tolerated.

—*Exploration*: is more pervasive, especially in the social sciences for qualitative research, when "qualitative methods are used for exploration (i.e., hypothesis-generating) or for explaining puzzling quantitative results. Quantitative methods, by contrast, are used to test hypotheses. This is because establishing content validity—do measures measure what a researcher thinks they measure?—is seen as one of the strengths of qualitative research. Some consider quantitative methods to provide more representative, reliable and precise measures through focused hypotheses, measurement tools and applied mathematics. By contrast, qualitative data is usually difficult to graph or display in mathematical terms."

(continued on next page)

Table 1.1. Different Faces of Data Analysis (Part II)

- **Quantitative vs. Qualitive Research (*cont'd*)**
  —*Policy and Program Evaluation*: are more appropriate for "qualitative research since it can answer certain important questions more efficiently and effectively than quantitative approaches. This is particularly the case for understanding how and why certain outcomes were achieved (not just what was achieved) but also for answering important questions about relevance, unintended effects and impact of programs such as: Were expectations reasonable? Did processes operate as expected? Were key players able to carry out their duties? Did the program cause any unintended effects?"
  —*Diversity and Flexibility*: are more valued in qualitative research, which "allow[s]…more diversity in responses as well as the capacity to adapt to new developments or issues during the research process itself. While qualitative research can be expensive and time-consuming to conduct, many fields of research employ qualitative techniques that have been specifically developed to provide more succinct, cost-efficient and timely results. Rapid Rural Appraisal is one formalized example of these adaptations but there are many others."

- **Sub-Fields of Data Analysis**
  —Ex: data mining
  —Ex: statistical applications
  —Ex: predictive analytics
  —Ex: text analytics

Notes: The categories are solely illustrative (not exhaustive), and the comparison is relative (not absolute), nor are they necessarily mutually exclusive. And some can be easily re-classified elsewhere. As generalities, they allow exceptions.
Source: From *Sec. 1.2* & *Sec. 1.3* of *FPHDA*. See book for citations.

Table 1.2. Six Great Future Transformation of Data Analysis

- **New Sub-Fields**
  —Ex: in the post-huamn era, the post-humans will further transform data analysis by different new studies of data analysis in a way that the human world has never known.

- **New Subjectivities**
  —Ex: in the post-huamn era, the post-humans will further transform data analysis by different new subjectivities (like emotions, intuition, imagination, heuristics, experience, and so on) for data analysis in a way that the human world has never known.

- **New Actions**
  —Ex: in the post-huamn era, the post-humans will further transform data analysis by different new actions in praxis (like personal experience, social change, and so on) in a way that the human world has never known.

- **New Narratives**
  —Ex: in the post-huamn era, the post-humans will further transform data analysis by different new narratives (like different techniques of interpretation) in a way that the human world has never known.

- **New Technologies**
  —Ex: in the post-huamn era, the post-humans will further transform data analysis by different new technologies for data analysis in a way that the human world has never known.

- **New Minds**
  —Ex: in the post-huamn era, the post-humans will further transform data analysis by different new minds to deal with complex problems in a way that the human world has never known.

Notes: The categories and examples are solely illustrative (not exhaustive), and the comparison is relative (not absolute), nor are they necessarily mutually exclusive. And some can be easily re-classified elsewhere. As generalities, they allow exceptions.
Source: From *Ch.4* of *FPHDA*. See text for citations.

Table 1.3. The Theoretical Debate on Space-Time (Part I)

- **Isaac Newton's Absolutist (Substantivist) Theory of Space-Time**
  —space and time are independent from each other. The structure of space-time is $E3 \times E1$ (with the structure of space, *P*, as the set of spatial locations in a three-dimensional Euclidean space, *E3*, and the structure of time as the set of temporal moments, *T*, in the one-dimensional real time, *E1*).
  —space and time are also independent from the effects of matter and events. The existence of space and time is possible even in a world absent of matter (and, for that matter, even in a world absent of events), as if they were material objects but with their total unchangingness thorough time.

- **Albert Einstein's Relativist Theory of Space-Time**
  —space and time are interchangeable (not absolute), just as matter and energy are equivalent (not independent) with the famous equation, $E = mc^2$ (as in the special theory of relativity in 1905).
  —space-time and matter-energy are also relative in a grand union (as in the general theory of relativity in 1915). Thus, each pair affects the other pair, as "matter 'tells' spacetime how to curve [in a non-Euclidean geometry] and curved spacetime 'tells' matter how to behave….Space contracts near mass and dilates away from it. Time dilates near mass and contracts away from it. Clocks positioned farther away from the mass of the earth run faster than clocks closer to the earth." (L. Shlain 1991: 328-330)

(continued on next page)

Table 1.3. The Theoretical Debate on Space-Time (Part II)

• **Peter Baofu's Perspectival Theory of Space-Time**

—space and time can be understood from multiple perspectives, be they in relation to culture, society, nature, and the mind, with each perspective revealing something about the nature of space-time and simultaneously delimiting its view. This is subject to "the regression-progression principle" in existential dialectics.

—each perspective of space and time exists in society and culture with good reasons, with some being more successful and hegemonic (dominant) than others. This is subject to "the symmetry-asymmetry principle" in existential dialectics.

—space and time will not last, to be eventually superseded (altered) by post-humans in different forms (e.g.,stretching/shrinking space-time, engineering more dimensions of space-time, and manipulating multiverses), be they here in this universe or in multiverses. Thus, even the physical existence of space-time cannot last forever, with ever more transformations in the process. This is subject to "the change-constancy principle" in existential dialectics.

—the conventional wisdom (especially by physicists) of treating the physical perspective of space and time as the foundation of all other perspectives (of space and time) and of regarding them as much less important is a form of reductionism,committing what I call *the foundation fallacy*, in misleadingly dismissing the multiple perspectives of space and time in relation to culture, society, nature, and the mind.

Notes: The examples in each category are solely illustrative (not exhaustive), and the comparison is relative (not absolute), nor are they necessarily mutually exclusive. Some can be easily re-classified elsewhere. As generalities, they allow exceptions.

Source: A summary of *Sec.1.1*, *Sec.1.2*, and *Sec.1.3* of *FPHST*—and, for that matter, the rest of *FPHST*

Table 1.4. Main Reasons for Altering Space-Time

- **The Need to Make New Energy-Matter**
  —Ex: manipulating molecular bonds for new materials
  —Ex: creating nanotechnologies on the atomic scale
  —Ex: engineering the atomic nucleus
  —Ex: restructuring most elementary particles
  —Ex: inventing new forms of matter and energy

- **The Need to Create New Space-Time**
  —Ex: creating "warp drive" (as in science fiction) for space travel
  —Ex: creating "pocket universes"

- **The Need to Conquer the Cosmos unto Multiverses**
  —Ex: spreading floating consciousness and hyper-spatial consciousness, besides other forms that humans have never known, in the cosmos and beyond unto multiverses for ultimate conquest

Notes: The examples in each category are solely illustrative (not exhaustive) nor necessarily mutually exclusive, and the comparison is relative (not absolute). As generalities, they allow exceptions. Also, it should be stressed that the three reasons are all related, in that they all contribute to the evolution of intelligent life in the cosmos unto multiverses in the most distant future beyond our current knowledge.
Sources: A summary of *Sec.6.2* of *FPHST*. See also *FHC*, *FCD*, *FPHC*, *FC*, *FPHU*, and *FPHG*, for example.

Table 1.5. The Technological Frontiers of the Micro-World

- **Type I-Minus**
  —Ex: building structures and mining

- **Type II-Minus**
  —Ex: playing with the genetic makeups of living things

- **Type III-Minus**
  —Ex: manipulating molecular bonds for new materials

- **Type IV-Minus**
  —Ex: creating nanotechnologies on the atomic scale

- **Type V-Minus**
  —Ex: engineering the atomic nucleus

- **Type VI-Minus**
  —Ex: restructuring most elementary particles

- **Type Ω-Minus**
  —Ex: altering the structure of space-time

Notes: As already indicated in *Sec.4.4.2.2* of *FPHC*, the problem with this micro-classification (from Barrow's work) is that the civilization types (with the exception of Type Ω-Minus, for example) are not quite distinct, since many of them can be achieved more or less in a civilization, to the extent that Type II-minus and Type III-minus, just to cite two plausible types, can be historically contemporaneous, relatively speaking, unlike the vast historical distance between, say, Type O and Type I (or Type I and Type II) civilizations. In other words, the micro-classification here is not very useful to understand civilization types but is revealing to see the technological frontiers of the micro-world.
Sources: A reconstruction from J.Barrow (1998:133), as originally shown in *Table 4.7* of *FPHC*. See *FPHC* for more info.

Table 1.6. Finity, Transfinity, and Infinity (Part I)

• **Infinity**

—The term "infinity," which is "symbolically represented with ∞," derives from the Latin *infinitas* (or "unboundedness"), to refer to "several distinct concepts—usually linked to the idea of 'without end'—which arise in philosophy, mathematics, and theology." (WK 2008c)

—In the context of mathematics, "infinity" is often "treated as if it were a number (i.e., it counts or measures things: 'an infinite number of terms') but it is a different type of 'number' from the real numbers. Infinity is related to limits,…large cardinals,…projective geometry, extended real numbers and the absolute Infinite," for instance. (WK 2008c)

• **Transfinity**

—To avoid confusion, my usage of the word "transfinite" (as used here in the ontological principle) is not the same as the one which was originally coined by Georg Cantor to refer to "cardinal numbers or ordinal numbers that are larger than all finite numbers, yet not necessarily absolutely infinite" (in the mind of God), as opposed to "relative" infinity (in the mind of Man). (WK 2008d) In other words, for Cantor, there are two kinds of infinity, "absolute" and "relative"—with "relative" infinity or "transfinity" (in the mind of Man) being dependent on "absolute" infinity (in the mind of God) for its existence.

—By contrast, my usage of the word "transfinite" here differs radically from the one used by Cantor (and other mathematicians) and is more limited, in light of the problems confronting any attempt to understand the idea of infinity, be it by intuition, imagination, and conception (as already shown in the prevous three sub-sections).

—With this clarification in mind, I allow numbers which can be many times larger—or smaller, for that matter—than the finite things that we encounter in daily life, but they do not have to be related to the idea of infinity at all (which may not exist).

(continued on next page)

Table 1.6. Finity, Transfinity, and Infinity (Part II)

• **Transfinity** ***(cont'd)***

—Of course, there may be some *borderline* cases, in which it is not clear whether the number in question is transfinite (in my usage) or simply a mathematical convenience. A good example of a borderline case is the Planck unit of length for "the smallest space possibly measured in nature," which is "less than billionths of trillionths of trillionths of an inch" (or something like 1.6×10^{-35} meters). (P. Baofu 2006a; N. McAleer 1987: 219; WKV 2008; D. Corbett 2008)

—The ideas concerning symmetry and dimensions in the examples above were addressed in Chapter Three and Chapter Four of *FPHG*. For now, it suffices to further add that it is mind-boggling to even conceive or imagine extreme numbers like the Planck unit of length (or others). Does it really exist at all? Or is it merely a mathematical convenience to describe a physical situation that the human mind does not understand properly?

—In any event, "transfinity" (in my usage) can be used as a further extension (of number) to the finite numbers in daily life—albeit with some borderline cases in mind.

• **Finity**

—It refers to numbers which are bounded (that is, with an end), especially (though not exclusively) in relation to things in everyday life.

Source: From *Sec. 1.4* and *Sec. 2.2.3* of *FPHG*

Table 1.7. Theoretical Speculations of Multiverses

• **"Baby Universes" (Ex: Andre Linde and others)**
—Ex: In a flat universe theory, "even if our part of it eventually collapses,…some spots in the cosmos would suddenly start inflating on their own, creating brand-new 'baby universes.'" (P. Baofu 2000: 623)

• **"Parallel Universes" (Ex: Stephen Hawking and others)**
—Ex: In quantum cosmology, there allows the existence of infinite numbers of parallel universes, with tunneling among them. (M. Kaku 1994: 256) Hawking later revised his views on this.

• **"Pocket Universes" (Ex: Alan Guth)**
—Ex: "As the pocket universes live out their lives and recollapse or dwindle away, new universes are generated to take their place....While life in our pocket universe will presumably die out, life in the universe as a whole will thrive for eternity." (A. Guth 1997: 248; P. Baofu 2002: 482)

• **"Brane Worlds" (Ex: Warren Siegel, Lisa Randall, and others)**
—Ex: Our universe is stuck on a membrane of space-time embedded in a larger cosmos, with different brane worlds connecting and/or colliding with each other.

Notes: These examples are solely illustrative (not exhaustive), and some of the items can be reclassified somewhere else. Nor are they always mutually exclusive. Since they are generalities, exceptions are expected.
Source: From *Table 4.8* of *FPHC*

Table 1.8. The Confusion between "Many Worlds" and "Multiverse" (Part I)

- **First Problem**

—The first problem in this second confusion concerns a debate between the "many-worlds" interpretation of quantum mechanics by Hugh Everett (1957) and others, and the Copenhagen interpretation by Niels Bohr and Werner Heisenberg on the effects of quantum states after interacting with an external environment (e.g., by a measuring observer).

—On the one hand, the Copenhagen interpretation of quantum mechanics is more indeterministic in its concern with the probabilities of the outcomes of a wave function after interacting with an external environment (e.g., by a measuring observer using light to detect, for example, "an electron in a particular region around the nucleus at a particular time." (WK 2008e) Each wave function for the state of a particle refers to "a mathematical representation used to calculate the probability for it to be found in a location, or state of motion." (WK 2008f) And the "wavefunction collapse" means that "the act of measurement causes the calculated set of probabilities to 'collapse' to the value defined by the measurement." (WK 2008f)

—On the other hand, the Everett's interrpetation is more deterministic, in rejecting "the objective reality of wavefunction collapse" and "instead explaining the subjective appearance of wavefunction collapse with the mechanism of quantum decoherence." (WK 2008g) The words "quantum decoherence" here refers to "the mechanism by which quantum systems interact with their environments to exhibit probabilistically additive behavior," in such a way that "the quantum nature of the system is simply 'leaked' into the environment." (WK 2008g) For example, "[a]s a result of an interaction, the wave functions of the system and the measuring device become entangled with each other. Decoherence happens when different portions of the system's wavefunction become entangled *in different ways* [italic added] with the measuring device...." (WK 2008g) Thus, for Everett, the appearance of "wavefunction collapse" can be explained by "quantum decoherence" for quantum states to branch out in different independent worlds after interacting with an external environment.

(continued on next page)

Table 1.8. The Confusion between "Many Worlds" and "Multiverse" (Part II)

• **First Problem *(cont'd)***
—Yet, a major criticism against Everett's interpretation is that these different independent worlds in Everett's sense "will never be accessible to us," because of the interference by the measuring observer in an external environment. (WK 2008g)
—For illustration, "once a measurement is done, the measured system becomes entangled with both the physicist who measured it and a huge number of other particles, some of which are photons flying away towards the other end of the universe; in order to prove that the wave function did not collapse one would have to bring all these particles back and measure them again, together with the system that was measured originally. This is completely impractical, but even if one can theoretically do this, it would destroy any evidence that the original measurement took place (including the physicist's memory)." (WK 2008g)
—But then, the critics ask, If these independent worlds in Everett's sense are inaccessible to us, how dc we know much of anything really about them?

• **Second Problem**
—What is relevant at the quantum mechanical level may not necessarily apply to the world at the physicsl level.
—In fact, the persistent inability among contemporoary physicists to unifiy the theory of relativity for large objects at the physical level and quantum mechanics for tiny objects at the quantum mechanical level should warn us about the danger of reductionism, by reducing the understanding of different universes ("parallel universes") at the physical level from a generalization of "many worlds" at the quantum mechanical level.

• **Third Problem**
—There is no empirical evidence to suggest that there are infinite (or extremely many) possible worlds in the cosmos as the many-worlds model by Everett would have us to believe.

Source: From *Sec.4.4.1* of *FPHG*

Table 1.9. Hyperspace and Its Challenge (Part I)

- **First Problem**
  - —Hyperspace for intra- and inter-universal travel is assumed to be possible, on the basis of four main assumptions; although there may be others, these are the often cited ones.
  - —The first assumption concerns the existence of "entry" and "exit" points in hyperspace, which is easier said than done to enter, go through, and then exit hyperspace.
  - —For instance, using hyperspace "requires complex calculations…of single or multiple hyperspace jumps and the control of the jump….Peturbations such as those experienced by ship in space from the gravitional field around an object such as a planet or even a star are exacerbated in hyperspatial travel, since mass in real space distorts hyperspace in an equal measure. 'Jumping' near to a gravitational mass is likely to make resulting exit from hyperspace to be highly uncertain, with the level of improbability *i* increasing with the square of the distance to the nearest gravitional 'well.'" (WK 2008h)
  - —Besides, "[o]nly spaceships equipped with a special force field can enter hyperspace, because exposure to hyperspace even for short period of time is hazardous to unprotected humans." (WK 2008h)

- **Second Problem**
  - —The second assumption concerns the faster-than-light speed. It is well known that Einstein argued for the speed of light as the ultimate limit, but nowadays phyiscists do allow the violation of this limit under certain conditions (like the hypothetical existence of tachyons, as will be explained further in a later sub-section on time travel).
  - —Even then, there is a challenge to explain in science fiction "why ships can travel faster than light in hyperspace," and two good arguments are that "hyperspace may be smaller than real space and therefore a star ship's propulsion seems to be greatly multiplied, or else the speed of light in hyperspace is not a barrier as it is in real space. Whatever the reasoning, the general effect is that ships traveling in hyperspace seem to have broken the speed of light, appearing at their destinations much more quickly and without the shift in time that the Theory of Relativity would suggest." (WK 2008h)

(continued on next page)

Table 1.9. Hyperspace and Its Challenge (Part II)

• **Third Problem**

—The third assumption concerns the speed of time (to be measured, say, by a clock), but this is elaborated in a different section on the related issue of "time travel" (as shown in a different table on time travel).

• **Fourth Problem**

—The fourth assumption concerns spatial shortcut, as it is possible, because of the curvature of space, in that "the manifold of ordinary three-dimensional space is curved in four or more 'higher' spacial dimensions (a 'hyperspace' in the geometric sense…). This curvature causes certain widely separated points in three-dimensional space to nonetheless be 'adjacent' to each other four-dimensionally. Creating an aperture in 4D space (a wormhole) between these locations can allow instantaneous transit between the two locations; a common comparison is that of a folded piece of paper, where a hole punched through two folded sections is more direct than a line drawn between them on the sheet. This idea probably arose out of certain popular descriptions of General Relativity and/or Riemannian manifolds, and may be the original form from which later concepts of hyperspace arose." (WK 2008h)

—But this does not mean that hyperpsace is always safe. In fact, "in some science fiction, the danger of hyperspace travel is due to the chance that the route through hyperspace may take a ship too close to a celestial body with a large gravitational field, such as a star. In such scenarios, if a starship passes too close to a large gravitational field while in hyperspace, the ship is forcibly pulled out of hyperspace and reverts to normal space. Therefore, certain hyperspace 'routes' may be mapped out that are safe, not passing too close to stars or other dangers." (WJ 2008h)

Source: From *Sec.4.4.1* of *FPHG*

Table 1.10. The Problems of Time Travel into the Future (Part I)

- **First Problem**
  —The first problem concerns "time dilation under the Theory of Special Relativity" (which separates space-time from mass-energy as special cases). Suppose you are on a moving train and an observer outside the train is watching you along the tracks. As your train moves along at the speed of light (or something close to it), "time, as measured by your watch, ticked along at a slower pace than time measured by the observer. Not only that, distance changed, too. For the observer, a one-foot ruler whizzing by on the train would have measured less than a foot." (NO 2000)
  —In fact, you may not even notice that this slower clicking of the clock; in fact, everything looks normal to you inside the train as it was before: "The weird thing is that, for you on the train, time wouldn't seem to be moving slower and your ruler wouldn't be shorter—all would appear normal." (NO 2008) By contrast, you instead think that "time on the rest of the Earth would appear to be ticking along slower and its rulers would be shorter." (NO 2000) So, both of you insist that the other clock (not his) is clicking more slowly—assuming, of course, an inertial frame of reference.
  —Now, change the story a bit, in that, this time, instead of an "inertial reference frame," acceleration is allowed, and you are on your way to board a spaceship. As an illustration, suppose "[y]ou board a spaceship and take off for deep space. The ship approaches the speed of light. Time for you seems to pass as it always has. It takes you about five seconds to tie your shoe. But to an observer on Earth (assuming he or she could watch you), you are moving at a snail's pace. It takes hours to tie your shoe." (NO 2000) In any event, "you continue on your journey. You slow down, stop, and accelerate back to Earth. You arrive home. You have aged two years during your flight. Two hundred years have passed on Earth. You have successfully travelled forward through time." (NO 2000) In this case, both the observer (if he is still alive) and you agree that you have aged less.

(continued on next page)

Table 1.10. The Problems of Time Travel into the Future (Part II)

• **First Problem** ***(cont'd)***

—In other words, when you are back to Earth, a lot of things have changed since, and you may find rather strange to be in a new environment which you hardly recognize, when compared with what you used to think as your city, or the Earth (two years ago, well, in accordance to your time of measurement). This means that you are in the future (two hundred years later), even though you are still relatively young (only two years older). But, can you go back? Well, no. The reason is that "[a]ccording to relativity, you can only move through time in one direction." (NO 2000)

• **Second Problem**

—The second problem concerns "time dilation under the Theory of General Relativity" (which unifies space-time with mass-energy in a general theory), the same result of time dilation occurs.

—For instance, when "one clock is deeper in a gravity well than the other," a result is that "the clock deeper in the well" will "tick…more slowly; this effect must be taken into account when calibrating the clocks on the satellites of the Global Positioning System, and it could lead to significant differences in rates of aging for observers at different distances from a black hole." (WK 2008i)

—More precisely, by way of another example, "it has been calculated that, under general relativity, a person could travel forward in time at a rate four times that of distant observers by residing inside a spherical shell with a diameter of 5 meters and the mass of Jupiter. For such a person, every one second of their 'personal' time would correspond to four seconds for distant observers. Of course, squeezing the mass of a large planet into such a structure is not expected to be within our technological capabilities in the near future." (WK 2008i; R. Gott 2002)

—The point here, however, is that time travel into the future, in this sense, is allowed, theoretically speaking, but in a very limited way, from the theory of relativity. But one can also ask, What if the theory of relativity were wrong? Well, in that case, there would be a need to find a better theory to explain time travel into the future.

Source: From *Sec.4.4.2* of *FPHG*

Table 1.11. The Problems of Time Travel into the Past (Part I)

- **First Problem**

—The first problem concerns the faster-than-light-speed, since, in accordance to the theory of special relativity, when an object goes faster than the speed of light, something weird can happen, in that, instead of having the clock clicking more slowly, it is moving backward, in relation to some inertial frame of reference—which then raisies the issue of whether time travel into the past is possible at all.

—But the problem here is about the violation of the law of casuality. In everyday language, an excellent illustration of causality violation is that, suppose you travel back into the past and then kill your parents, this then means that you could not have been born. But this is not true, since you are alive.

—Yet, "in the case of a hypothetical signal moving faster than light, there would always be some frames in which the signal was received before it was sent, so that the signal could be said to have moved backwards in time." The violation of casuality exists in this case.

—The first part of the problem is that, in accordance to the theory of special relativity, "it would take an infinite amount of energy to accelerate a slower-than-light object to the speed of light"—let alone the energy needed to propel an object to go faster than the speed of light. (WK 2008i) Although some suggest the possibility of "negative energy," it remains to be seen to what extent this is true.

—And the second part of the problem is that, "although relativity does not forbid the theoretical possibility of tachyons which move faster than light at all times, when analyzed using quantum field theory it seems that it would not actually be possible to use them to transmit information faster than light and there is no evidence for their existence." (WK 2008i; S. Chase 1993)

(continued on next page)

Table 1.11. The Problems of Time Travel into the Past (Part II)

• **Second Problem**

—The second problem concerns cosmic string and black holes, since it is the theory of general relativity which "extends the special theory to cover gravity, illustrating it in terms of curvature in spacetime caused by mass-energy and the flow of momentum." (WK 2008i)

—On the other hand, "[g]eneral relativity describes the universe under a system of field equations…and there exist solutions to these equations that permit what are called 'closed time-like curves,' and hence time travel into the past….The first of these was proposed by Kurt Gödel, a solution known as the Gödel metric, but his (and many others') example requires the universe to have physical characteristics that it does not appear to have." (WK 2008i; K. Thorne 1994)

—But then, the problem now becomes: "Whether general relativity forbids closed time-like curves for all realistic conditions is unknown." (WK 2008i)

• **Third Problem**

—The third problem concerns wormholes and related ones, and "[w]ormholes are a hypothetical warped spacetime which are also permitted by the Einstein field equations of general relativity, although it would be impossible to travel through a wormhole unless it was what is known as a traversable wormhole." (WK 2008i; M. Visser 1995) In other words, "a wormhole is a hypothetical topological feature of spacetime that is fundamentally a 'shortcut' through space and time. Spacetime can be viewed as a 2D surface, and when 'folded' over, a wormhole bridge can be formed. A wormhole has at least two mouths which are connected to a single throat or tube. If the wormhole is traversable, matter can 'travel' from one mouth to the other by passing through the throat." (WK 2008j)

—But the problem here is three-fold, as shown in what follows.

(continued on next page)

Table 1.11. The Problems of Time Travel into the Past (Part III)

- **Third Problem *(cont'd)***
  —The first part of the problem is that "there is no observational evidence for wormholes." (WK 2008j)
  —The second part of the problem is that it is not clear how exactly wormholes (and black holes, for that matter) could allow time travel into the past at all.
  —And the third part of the problem is that if using wormholes (and black holes, for that matter) to travel into the past requires something to go faster than light too, then this raises another issue concerning the two problems as previously indicated in the sub-section on the faster-than-light-speed.

- **Fourth Problem**
  —The fourth problem is that some physicists have argued that, even should time travel into the past be possible, by way of these three methods, the person still would not return to the same past history that he exactly experienced before, but in a different one. (WK 2008i)
  —In other words, "strictly speaking," time travel into the past is not "really" possible, in light of these criticisms, so there is no need to worry that one could travel into the past and kill one's parent, as a way to violate the law of casuality.

Source: From *Sec.4.4.2* of *FPHG*

Table 1.12. The Conceptual Dimensions of Consciousness (and Other Mental States)

- **On Heredity and Time**
  —Heredity and the Environment
  —The Past and the Present

- **On Layers of Mental States and Abnormality**
  —Consciousness, Unconsciousness, and Preconsciousness
  —Normality and Paranormality

- **On Organicity and Motivation**
  —Mechanicity and Organicity
  —Primary Motivations and Multiple Motivations

- **On Other (Mostly Epistemic) Considerations**
  —Synthesis and Analysis
  —Situation and the Subject
  —Process and Outcome
  —Reasoning and Other Modes of Thinking
  —Meta-Conceptual Nominalism and Realism

Notes:: These examples are solely illustrative (not exhaustive), and some of the items can be reclassified somewhere else. Nor are they always mutually exclusive. Since they are generalities, exceptions are expected.
Sources: From *FPHC*. A re-construction, but with my own contribution, originally from G.Lindzey & C.Hall, *Introduction to Theories of Personality* (NY: John Wiley & Sons, 1985).

Table 1.13. The Theoretical Levels of Consciousness (and Other Mental States) (Part I)

- **At the Micro-Physical Theoretical Level**
  —Quantum-Mechanics
  —Electromagnetism

- **At the Chemical Theoretical Level**
  —Biochemistry

- **At the Biological Theoretical Level**
  —Evolutionary Biology
  —Neuroscience
  —Artificial Intelligence

- **At the Psychological Theoretical Level**
  —Psychodynamic Psychology
    - Psychoanalytical Psychology
    - Analytical Psychology
    - Socially Oriented Psychology

  —Experimental (Behavioral) Psychology
    - Operant Reinforcement Theory
    - Stimulus-Response Theory
    - Social Learning Theory

  —Cognitive (Gestalt) Psychology
    - Humanist Psychology
    - Existential Psychology
    - Field Theory

  —Psychometric Psychology
    - Idiographics
    - Constitutional Psychology

  —Social Psychology
    - Symbolic Interactive Theory
    - Social Exchange Theory

(continued on next page)

Table 1.13. The Theoretical Levels of Consciousness (and Other Mental States) (Part II)

- **At the Organizational Theoretical Level**
  —Managerial-Bureaucratic Theory
  —Oligarchic Theory
  —Network Theory

- **At the Institutional Theoretical Level**
  —Functionalist Theory
  —Anomic Theory

- **At the Structural Theoretical Level**
  —Conflict Theory
    - Marxian Theory
    - Critical Theory
    - Weberian Theory

  —Games Theory (in Formal Theory)
  —Feminist Theory
    - Feminist Concerns
    - Feminist Hopes

- **At the Systemic Theoretical Level**
  —Equilibrium Theory
  —System Theory
  —Chaos Theory

- **At the Cosmological Theoretical Level**
  —Superluminal Model
  —The Theory of Floating Consciousness

(continued on next page)

Table 1.13. The Theoretical Levels of Consciousness (and Other Mental States) (Part III)

- **At the Cultural Theoretical Level**
  - —Substantive Theories
    - Structuralist Theory
    - Post-Structuralist Theory (in Postmodernism)
  - —Meta-Theories
    - Epistemic Objectivism vs. Epistemic Historicism
    - Epistemic Subjectivism vs. Epistemic Non-Subjectivism
      - –Phenomenology
      - –Ethnomethodology
      - –Hermeneutics
    - Epistemic Relativism vs. Epistemic Absolutism
    - Epistemic Reductionism vs. Epistemic Emergencism

- **At Other Levels**
  - —Structuration Theory
  - —Reflexive Socioanalysis

Notes: These examples are solely illustrative (not exhaustive), and some of the items can be reclassified somewhere else. Nor are they always mutually exclusive. Since they are generalities, exceptions are expected.
Source: From many different sources as indicated in *FPHC*

Table 1.14. The Thematic Issues of Consciousness (and Other Mental States)

- **The Factor of History**
  —Pre-Modernity
  —Modernity
  —Postmodernity
  —After-Postmodernity

- **The Importance of Needs**
  —Having (e.g., the everyday, the technological)
  —Belonging (e.g., the just and the good)
  —Being (e.g., the true, the holy, the sublime/beautiful, and the good)

Notes: These examples are solely illustrative (not exhaustive). Nor are they always mutually exclusive. Since they are generalities, exceptions are expected. Also, some categories overlap with others; for instance, the good is also related with being-ness, just as the everyday can be connected with belong-ness.
Sources: From *FPHC*, based on *FHC* and *FCD*

Table 1.15. Having, Belonging, and Being in Consciousness

- **Having (e.g., About the Everyday and Technological)**
  —Ex: Physiological needs (e.g., thirst, hunger, sex)
  —Ex: Safety (e.g., stability, freedom from fear and anxiety)

- **Belonging (e.g., About the Good and Just)**
  —Ex: Social interaction
  —Ex: Friendship, acquaintance
  —Ex: Love, family
  —Ex: Self-respect, respect from others

- **Being (e.g., About the True, Holy, Beautiful/Sublime, and Good)**
  —Ex: Understanding of reality about self and world
  —Ex: Spiritual quest for holiness, beauty, sublimity, and goodness

Notes: These examples are solely illustrative (not exhaustive), and some of the items can be reclassified somewhere else. Nor are they always mutually exclusive. Since they are generalities, exceptions are expected.
Sources: From *FPHC*, based on *FHC* and *FCD*

Table 1.16. The Having-Ness of Consciousness (and Other Mental States)

- **At the Micro-Physical Level**
  —Ex: The finer physics of nutrition

- **At the Chemical Level**
  —Ex: Food guide pyramid

- **At the Bio-Psychological Level**
  —Ex: The hierarchy of needs

- **At the Institutional Level**
  —Ex: The production of commodity fetishism

- **At the Organizational Level**
  —Ex: Mass standardization (and customization); bureaucratic rationality

- **At the Structural Level**
  —Ex: The divide between the haves and the have-nots

- **At the Cultural Level**
  —Ex: The bondage of greater expectations

- **At the Systemic Level**
  —Ex: The technological transformation of poverty with transvaluations

- **At the Cosmological Level**
  —Ex: Floating consciousness in deep space without the human physiological needs

Notes: These examples are solely illustrative (not exhaustive), and some of the items can be reclassified somewhere else. Nor are they always mutually exclusive. Since they are generalities, exceptions are expected.
Sources: From *FPHC*. See also *FCD* and *FHC*.

Table 1.17. The Belonging-Ness of Consciousness (and Other Mental States)

- **At the Micro-Physical Level**
  —Ex: Statistical physics and social networks

- **At the Chemical Level**
  —Ex: Olfaction and social relations

- **At the Bio-Psychological Level**
  —Ex: Dual human nature on relationships

- **At the Institutional Level**
  —Ex: Private property and social cleavage

- **At the Organizational Level**
  —Ex: Legal formalism and iron cage

- **At the Structural Level**
  —Ex: Class conflicts and social alienation

- **At the Cultural Level**
  —Ex: Ruthless society and the culture of law

- **At the Systemic Level**
  —Ex: Human extinction and post-human relations

- **At the Cosmological Level**
  —Ex: The transcendence of equality in the cosmos and beyond

Notes: These examples are solely illustrative (not exhaustive), and some of the items can be reclassified somewhere else. Nor are they always mutually exclusive. Since they are generalities, exceptions are expected.
Sources: From *FPHC*. See also *FCD* and *FHC*.

Table 1.18. The Being-Ness of Consciousness (and Other Mental States) (Part I)

- **At the Micro-Physical Level**
  —Ex: Space-time and matter-energy in classical mechanics, quantum mechanics, and the theory of relativity

- **At the Chemical Level**
  —Ex: Drugs and well being

- **At the Biological Level**
  —Ex: Neuroscience and mystical experiences

- **At the Psychological Level**
  —Ex: Reductionism and emergencism

- **At the Institutional Level**
  —Ex: The proliferation of cults and sects, and the corporate-art industry

- **At the Organizational Level**
  —Ex: The organization of ideas in the scheme of things

- **At the Structural Level**
  —Ex: Disciplinary power and control in the order of things

- **At the Systemic Level**
  —Ex: Mutualism and ecological psychology

- **At the Cultural Level**
  —Ex: The recurrence of competing spirits, and floating consciousness as a climax of evolution

(continued on next page)

Table 1.18. The Being-Ness of Consciousness (and Other Mental States) (Part II)

- **At the Cosmological Level**
  —Ex: The search for elsewherewhen, and the evolution of other consciousnesses (e.g., the hyper-spatial consciousness)

Notes: These examples are solely illustrative (not exhaustive), and some of the items can be reclassified somewhere else. Nor are they always mutually exclusive. Since they are generalities, exceptions are expected.
Sources: From *FPHC*. See also *FCD* and *FHC*.

Table 1.19. Cognitive Partiality in Different Mental States

- **The Conscious**
  - —*Biased*
    - Ex: Self-overratingness
    - Ex: Self-modesty
    - Ex: Self-fulfilling prophesy
  - —*Shallow*
    - Ex: Attribution error
    - Ex: Heuristics
    - Ex: Illusory thinking
  - —*Short-term*
    - Ex: Deindividuation
  - —*Materialistic*
    - Ex: Hierarchy of needs

- **The Subconscious**
  - Ex: mental reconstruction
  - Ex: stereotyping

- **The Unconscious**
  - Ex: automatic processing
  - Ex: cognitive intuition

Notes: These examples are solely illustrative (not exhaustive), and some of the items can be reclassified somewhere else. Nor are they always mutually exclusive. Since they are generalities, exceptions are expected. And the comparison is relative, not absolute.
Source: From *BCPC*

Table 1.20. Emotional Non-Neutrality and Behavioral Alteration in Different Mental States

- **The Emotional**
  —*Conscious*
    Ex: Feelings
  —*Subconscious*
    Ex: Moods
  —*Unconscious*
    Ex: Empathy
    Ex: Instincts

- **The Behavioral**
  —*Conscious*
    Ex: Foot-in-the-door
    Ex: Role playing
    Ex: Saying-is-believing
  —*Subconscious*
    Ex: Discrimination
  —*Unconscious*
    Ex: Non-Verbal Communication

Notes: These examples are solely illustrative (not exhaustive), and some of the items can be reclassified somewhere else. Nor are they always mutually exclusive. Since they are generalities, exceptions are expected. And the comparison is relative, not absolute.
Source: From *BCPC*

Table 1.21. The Limits of Intuition in Unconsciousness

- **Cognitive Intuition (or Intuitive Thinking)**
  —*Features*
  Ex: mental schemas
  Ex: expertise
  Ex: blindsight
  —*Problems*
  Ex: inaccurate
  Ex: non-explanatory

- **Emotional Intuition**
  —*Features*
  Ex: empathy
  Ex: instincts
  —*Problems*
  Ex: over-generalized
  Ex: non-explanatory

- **Behavioral Intuition**
  —*Features*
  Ex: prosemics
  Ex: kinesics
  —*Problems*
  Ex: unreliable
  Ex: non-explanatory

Notes: These examples are solely illustrative (not exhaustive), and some of the items can be reclassified somewhere else. Nor are they always mutually exclusive. Since they are generalities, exceptions are expected. And the comparison is relative, not absolute.
Source: From *BCPC*

Table 1.22. The Wealth/Poverty Dialectics in Different Mental States: The Case of Cognition

- **The Conscious**
  - —*Biased*
    - Ex: If richer in self-serving bias, then poorer in understanding others
  - —*Shallow*
    - Ex: If richer in heuristic thinking, then poorer in comprehending reality
  - —*Short-term*
    - Ex: If richer in short-term tendency, then poorer in long-term planning
  - —*Materialistic*
    - Ex: If richer in physiological obsession, then poorer in spiritual enlightenment

- **The Subconscious**
  - Ex: If richer in mental reconstruction, then poorer in accuracy of recall
  - Ex: If richer in stereotyping, then poorer in understanding groups

- **The Unconscious**
  - Ex: If richer in automatic processing, then poorer in depth of analysis
  - Ex: If richer in cognitive intuition, then poorer in reliability of judgments

Notes: These examples are solely illustrative (not exhaustive), and some of the items can be reclassified somewhere else. Nor are they always mutually exclusive. Since they are generalities, exceptions are expected. And the comparison is relative, not absolute.
Source: From *BCPC*

Table 1.23. The Wealth/Poverty Dialectics in Different Mental States: The Case of Emotion and Behavior

- **Emotion**
  - —*Conscious*
    - Ex: If richer in relying on familiarity and similarity for social attraction, then poorer in having diverse social relations
  - —*Subconscious*
    - Ex: If richer in moodiness, then poorer in sober thinking
  - —*Unconscious*
    - Ex: If richer in empathy, then poorer in task-oriented efficiency
    - Ex: If richer in instincts, then poorer in critical thinking

- **Behavior**
  - —*Conscious*
    - Ex: If richer in foot-in-the-door technique, then poorer in instant Gratification
    - Ex: If richer in role playing, then poorer in identity stability
  - —*Subconscious*
    - Ex: If richer in discriminating, then poorer in social harmony
  - —*Unconscious*
    - Ex: If richer in kinesics and proxemics, then poorer in verbal skills, relatively speaking.

Notes: These examples are solely illustrative (not exhaustive), and some of the items can be reclassified somewhere else. Nor are they always mutually exclusive. Since they are generalities, exceptions are expected. And the comparison is relative, not absolute.
Source: From *BCPC*

Table 1.24. The Theoretical Debate on Nature vs. Nurture (Part I)

- **The Environmental Approach**
  —*Thesis*: It focuses, relatively speaking, more on the environment (culture and society) in explaining the achievement gap among individuals and for that matter, countries or regions, when contrasted with the natural factors.
  —*Discourse*: Examples include Jose Ortega y Gasset ("Man has no nature; what he has is history"), Ashley Montagu ("Man is man because he has no instincts, because everything he is and has become he has learned from his culture, from the man-made part of the environment, from other human beings"), Stephen Jay Gould ("[The] brain [is] capable of a full range of behaviors and predisposed to none"), and Jesse Jackson (who blames white racism for the failure of blacks to close the achievement gap between whites and blacks in America). The works on dependency theory in international political economy (with a Marxian influence) and on the Protestant work ethic (by Max Weber) also point to this environmental direction.

- **The Genetic Approach**
  —*Thesis*: It focuses instead, relatively speaking again, on hereditory factors (e.g., genes and evolution) in explaining the achievement gap among individuals and for that matter, countries or regions, when contrasted with the envrionmental factors.
  —*Discourse*: Examples are Hans Eysenck and William Sheldon (in constitutional psychology), Konrad Lorenz (in his work on innate aggressive human nature), Gary Marcus (on the complexities of human thought by a tiny number of genes) and Robert Plomin, together with Michael Owen and Peter McGuffin (on the genetic basis of complex human behaviors).

(continued on next page)

Table 1.24. The Theoretical Debate on Nature vs. Nurture (Part II)

- **The Compromise Approach**
  —*Thesis*: It seeks the middle-of-the-road argument in regard to nature and nurture and regards all differences among individuals and groups as the result of the mixture of both nature and nurture, more or less equally.
  —*Discourse*: Examples include C. Murray and R. Herrnstein ("It seems highly likely to us that both genes and environment have something to do with this issue") and Dan Dennett ("Surely 'everyone knows' that the nature-nurture debate was resolved long ago, and neither side wins since everything – is – a – mixture – of – both – and - it's – all – very -complicated....")

- **The Transcendent Approach**
  —*Thesis*: It goes beyond both nature and nurture (without, however, committing the compromise fallacy) in showing their closely intertwined interactions in producing the behavioral differences as often seen in individual human endeavors on the micro scale, and for that matter, in country (or regional) endeavors on the macro one—in the context of my five theses, namely, (a) the compromise fallacy, (b) no oppression without self-oppression, (c) no success without failure, (d) the factor of randomness, and (e) the post-human vision, to be elaborated in Chapter Six.
  —*Discourse*: Peter Baofu proposed this approach on the basis of his "theory of contrastive advantages" (as an original theoretical contribution to the debate, which was first proposed and analyzed in *The Future of Capitalism and Democracy*). In the end, the human genes will not last, to be eventually superseded by post-human life forms, so the debate between genes and memes has obscured something profoundly important about the future that the world has never known. And the debate is also misleading and faulty in its dichotomy.

Source: From *BNN*. See the book for citations and details.

Table 1.25. Physical Challenges to Hyper-Spatial Consciousness

- **The Understanding of a Higher-Dimensional World of Space-Time**
  —Ex: 4 for traditional aspects of space-time (e.g., length, width, breadth and time) plus 6 more new dimensions in theory of hyper-space, with profound implications for practical applications to new forms of consciousness.

- **The Mastering of Dark Matter and Dark Energy**
  —Ex: "ordinary matter" (e.g., atoms, molecules) as a mere 4.4% of the universe, with 23% made of "cold dark matter" and the rest (about 73%) of mysterious "dark energy," with fundamental significance to questions about the limit of the speed of energy (or info), the availability of energy for use, and the nature of space-time, just to cite some examples.

- **The Exploration of Multiverses**
  —Ex: theoretical speculation of other universes (e.g., "baby universes," "gateways" in black holes, "wave function of the universe," "many worlds," "brane worlds"), with potentially seminary discoveries of different physical laws in relation to matter-energy and space-time, and vital differences to the future of post-human conquest of other universes (for the emergence of new forms of consciousness).

Notes: These examples are solely illustrative (not exhaustive), and some of the items can be reclassified somewhere else. Nor are they always mutually exclusive. Since they are generalities, exceptions are expected. The point here is to give a rough picture of the evolution of consciousness to the hyper-spatial consciousness and others totally unknown to current earthlings. As a note of clarification, it makes no difference to my argument as to whether or not the hyper-spatial consciousness may emerge before, during, and after floating consciousness.
Source: From *Table 4.5* of *FPHC*

Table 1.26. The Theory of Floating Consciousness (Part I)

- **At the Micro-Physical Level**
  —Ex: intelligent life without the human physical-chemical system

- **At the Chemical Level**
  —Ex: space radiation and toxins

- **At the Bio-Psychological Level**
  —Ex: exo-biological evolution in deep space
  —Ex: genetic engineering of new beings

- **At the Institutional Level**
  —Ex: post-capitalism
  —Ex: post-democracy

- **At the Organizational Level**
  —Ex: less legal-formalistic routines

- **At the Structural Level**
  —Ex: alien forms of violence

- **At the Cultural Level**
  —Ex: transcending freedom
  —Ex: transcending equality

(continued on next page)

Table 1.26. The Theory of Floating Consciousness (Part II)

• **At the Cosmological Level**
—Ex: parallel universes
—Ex: pocket universes

• **At the Systemic Level**
—Ex: space habitats (in zero-gravity)

Notes: Each example draws from the works of different scholars in the field. For instance, at the cosmological level, the idea of parallel universes is from the theoretical speculation in quantum cosmology by Stephen Hawking and others, while the one of pocket universes comes from the theoretical work of Allan Guth at MIT. And at the institutional level, I proposed post-capitalism and post-democracy in *FCD* (and later, from *BDPD* and *BCPC*). In addition, the examples are solely illustrative (not exhaustive), and some of the items can be reclassified somewhere else. Nor are they always mutually exclusive. Since they are generalities, exceptions are expected.
Sources: From *FPHC*—and, originally, from *FCD*

Table 1.27. The Potential of Unfolding Unconsciousness

- **Superior Senses**
  - —*Through space*
    - Ex: clair-sentience (in feeling)
    - Ex: clair-voyance (in seeing)
    - Ex: clair-audience (in hearing)
    - Ex: clair-austance (in tasting)
    - Ex: clair-alience (in smelling)
    - Ex: clair-cognizance (in knowing)
  - —*Through time*
    - Ex: pre-cognition (in knowing), pre-sentiment (in feeling)
    - Ex: retro-cognition (in knowing), retro-sentiment (in feeling)

- **Superior Contacts**
  - —*With living minds*
    - Ex: telepathy
  - —*With the dead*
    - Ex: mediumship
  - —*With other-living entities*
    - Ex: telekinesis

Notes: The categories and examples are solely illustrative (not exhaustive), and the comparison is relative (not absolute), nor are they necessarily mutually exclusive. And some can be easily re-classified elsewhere. As generalities, they allow exceptions.
Source: A summary of Ch.5 of *FPHU*

Table 1.28. The Future Exploration of Unfolding Unconsciousness

- **In the Context of Nature**
  —Ex: the higher-dimensional space-time in hyperspace theory
  —Ex: the particle-wave interactions in quantum mechanics
  —Ex: the study of brain waves in electromagnetic theory

- **In the Context of the Mind**
  —Ex: the unconscious fantasies in psychoanalysis
  —Ex: the mind-altering drugs in biochemistry
  —Ex: the manipulation of neural activity in neurobiology

- **In the Context of Culture**
  —Ex: the correlation between anomalous experience and moral/religious/aesthetic interests in culture studies
  —Ex: the role of myths and language in epistemology

- **In the Context of Society**
  —Ex: different organizational agendas in anomalous research
  —Ex: the scholarly divide on anomalous study
  —Ex: interests and propaganda in institutioanl research
  —Ex: the transformative power of technology in research on social systems
  —Ex: the natural and non-natural selection of other worlds in exo-biology
  —Ex: the role of the post-human conscious mind in qualitative demography

Notes: The examples in each category are solely illustrative (not exhaustive), and the comparison is relative (not absolute), nor are they necessarily mutually exclusive. And some can be easily re-classified elsewhere. As generalities, they allow exceptions.
Source: A summary of Chs.2-5 of *FPHU*, in conjunction with the rest of the book

Table 1.29. Creative Techniques and Traits

- **Creative Techniques**
  - —*Comprehensiveness*
    - Ex: Broadness of Scope
    - Ex: In-Depth Analysis
    - Ex: Multifaceted Taxonomy
    - Ex: Numerous Theories
  - —*Arrangement*
    - Ex: Separation (or Subtraction)
    - Ex: Combination (or Addition)
    - Ex: Replacement
    - Ex: Accommodation
    - Ex: Reversal
    - Ex: Stretching
  - —*Serendipity*
    - Ex: Play
    - Ex: Chance

- **Creative Traits**
  - —*Openness / Risk*
    - Ex: Ability to welcome new ideas or to do new things, with risk
  - —*Discipline*
    - Ex: Ability to sacrifice whatever it takes to succeed
  - —*Resilience / Confidence*
    - Ex: Ability to take failures to mind, not to heart
  - —*Trust*
    - Ex: Ability to cooperate with others for team work, if necessary

Notes: The examples in the categories are solely illustrative (not exhaustive), and the comparison is relative (not absolute), nor are they necessarily mutually exclusive. And some can be easily re-classified elsewhere. As generalities, they allow exceptions.

Source: From *Sec.4.2* of *FPHCT*. For more details, also read the rest of the book.

Table 1.30. The Desirability of Creativity, and Its Dark Sides (Part I)

- **In Relation to Instrumental Rationality**
  —*Desirability*
  - Ex: For the sake of efficiency and effectiveness in this capitalist era of our time.

  —*Undesirability*
  - Ex: What Max Weber (1930) called the "iron cage" of capitalism, where live the "sensualists without heart, specialists without spirit"—or something which the Frankfurt School has forcefully asked us to be freed from.

- **In Relation to Substantive Rationality**
  —*Desirability*
  - Ex: For the sake of God, the King, Motherland, or other comparable ideals.

  —*Undesirability*
  - Ex: Domination and oppression of various forms—something which the moderns had struggled to free themselves from in the first place, since the dawn of modernity.

- **In Relation to Autonomous Rationality**
  —*Desirability*
  - Ex: For the sake of itself, the autonomy of creative endeavor.

  —*Undesirability*
  - Ex: The decadence of its own degeneration (e.g., the donwsides of postmodernism in our postmodern times, or what Nietzsche once called the unbearable "unreality" and "falseness" of the autonomous artist's "innermost existence"—and, in other cases, the painful suffering from different mental illnesses for those highly creative individuals).

(continued on next page)

Table 1.30. The Desirability of Creativity, and Its Dark Sides (Part II)

Notes: The examples in the categories are solely illustrative (not exhaustive), and the comparison is relative (not absolute), nor are they necessarily mutually exclusive. And some can be easily re-classified elsewhere. As generalities, they allow exceptions.
Also, I already went to great lenghts to explain the desirability and its dark sides of all these three rationalities (although there may be others too) in the two volumes of *FHC*—and, for that matter, *FCD*.
Source: From *Sec.4.9* of *FPHCT*. For more details, also read the rest of the book.

Table 1.31. Posthuman-Ism, Post-Humanism, and Trans-Humanism

- **Post-Humanism**
  —The neologism "post-human" used in my books should *not* be confused with another term which looks similar but has a totally different meaning in the literature of postmodernism, namely, "post-humanism"—which constitutes a critique of "humanism" as tradionally understood (especially, though not exclusively, in relation to the idea of progress in science and reason in the Enlightenment project). (WK 2008)
  —My works reject the project of "postmodernism" and propose the future world of what I orignally called "after-postmodernity" in *FHC* and *FCD*, for instance.

- **Trans-Humanism**
  —Also, the neologism "post-human" used in my books should *not* be confused with a similar term which is used to champion the ideology of technology for the fturue co-existence between humans and cyborgs in "trans-humanism." (WK 2008a)
  —Instead, my term "post-human" in relation to "posthuman-ism" also rejects "trans-humanism" (especially, though not exclusively, in relation to the promises of technology) and refers to something else altogether, that is, the future extinction of humans and its post-human successors in deep space and beyond unto multiverses.
  —Chapter Two of my book on ethics (*BEPE*) provides a more comprehensive critique of trans-humanism.

- **Posthuman-Ism**
  —With these two clarifications in mind, the "post-humans" (as envisioned in my books) can take different forms, and I proposed, in my different books, some of them, such as "unfolding unconsciousness," "floating consciousness," "hyper-spatial consciousness," "thinking machines," "thinking robots," "genetically altered superior beings," "cyborgs," and others (as already analyzed in my previous works).
  —They will evolve in conjunction with other visions of mine in relation to nature, society, and culture.

Sources: From *Sec.1.6* of *FPHE* (and also *BEPE*). For more details, also read the rest of the books (and other books of mine).

Table 1.32. Three Great Future Transformations of the Martial Body (Part I)

- **Virtual Battle**
  - —Ex: virtual fight
  - —Ex: virtual memory

- **The Hyper-Martial Body**
  - —*In relation to the physical dimension*
    - Ex: striking (e.g., punching, kicking, trapping, acupressure-striking, and so on in an amazing way)
    - Ex: grapping (e.g., throwing, pinning, joint-locking, and so on in an amazing way)
    - Ex: running and walking (e.g., on different surfaces)
    - Ex: jumping (e.g., through walls, trees, etc.)
    - Ex: flying (e.g., through space)
    - Ex: enduring pain (e.g., absorbing punches, kicks, and so on in an amazing way)
    - Ex: using extremely advanced hi-tech weapons
    - Ex: healing (e.g., with special energy, etc.)
    - Ex: entering (e.g., into different dimensions of space)
    - Ex: sensing (e.g., in extraordinary ways through space and time, etc.)
    - Ex: wearing hi-tech suits for defensive and offensive purposes (e.g., like a hi-tech armor)
  - —*In relation to the mental dimension*
    - Ex: the ability to relax the martial body in a way that current humans cannot.
    - Ex: the ability to concentrate the martial mind in a way that current humans are not accustomed to.
    - Ex: the ability to breathe and be timely in preparing the martial body for fighting in a way that current humans are not good at.

- **The Post-Martial Floating Mind**
  - —Ex: the mind without the human body
  - —Ex: the quest for spirituality beyond martial arts

(continued on next page)

Table 1.32. Three Great Future Transformations of the Martial Body (Part II)

Notes: The examples in the categories are solely illustrative (not exhaustive), and the comparison is relative (not absolute), nor are they necessarily mutually exclusive. And some can be easily re-classified elsewhere. As generalities, they allow exceptions.
Source: From *Sec. 2.5.1* of *FPHMA*. See text for citations.

Table 1.33. Three Great Future Transformations of the Sexual Body

- **Virtual Pleasure**
  —Ex: virtual Sex
  —Ex: virtual Recall

- **The Hyper-Sexual Body**
  —*In relation to the physical dimension*
  - Ex: the ability to perform differenet sexual positions without the various physical limitations that humans currently have.
  - Ex: the ability to produce unusual amount of sexual energy to endure in sexual acts in a way that humans currently cannot.
  - Ex: the ability to engage in multiple sexual experiences comparable to (but different from) non-human sexuality (e.g., hermaphrodite, monoecious, sex-switching, and many other types) that current humans cannot.

  —*In relation to the mental dimension*
  - Ex: the ability to relax the sexual body in a way that current humans are not accustomed to.
  - Ex: the ability to concentrate the sexual mind in a way that current humans are not good at.
  - Ex: the ability to breathe and be patient in preparing the sexual body for sexuality in a way that current humans are not able to.

- **The Post-Sexual Floating Mind**
  —Ex: the mind without the human body
  —Ex: the quest for spirituality beyond sexuality

Notes: The examples in the categories are solely illustrative (not exhaustive), and the comparison is relative (not absolute), nor are they necessarily mutually exclusive. And some can be easily re-classified elsewhere. As generalities, they allow exceptions.
Source: From *Sec. 3.4.2* of *FPHS*. See text for more info.

Table 1.34. The Trinity of Pre-Modernity

- **Pre-Free-Spirited Pre-Modernity (Pre-Modernism) and Its Internal Split**
  —Competing worldviews and values both within and between linear centric (e.g., Islamic, Christian, Judaic, Imperial Roman) and cyclical-centric (e.g., Confucian, Taoist, Hindu, and Buddhist) orientations
  —Compare modernism with pre-modernism here in relation to the seven dimensions of human existence like the true and the holy (e.g., different versions of epistemic dogmas and religious superstitions), the everyday and the technological (e.g.,different versions of non-technophilism and non-consumerism), the beautiful/sublime (e.g., different versions of aesthetic non-autonomy), and the good and the just (e.g., different versions of moral particularism).

- **Pre-Capitalist Pre-Modernity (Pre-Modernization) and Its Own Discontents**
  —Competing versions of societal arrangements (e.g., feudalism, monarchism, and the holy order)

- **Hegemonic Pre-Modernity and Its Countervailing Forces**
  —Different power centers and their enemies (e.g., the Roman Empire and the "barbarian hordes," the "Holy Crusades" and the Muslims, the Middle Kingdom and the invading tribes, different social castes in India, and warring Greek city-states)

Notes: The examples in each category are solely illustrative (not exhaustive) nor necessarily mutually exclusive, and the comparison is relative (not absolute). As generalities, they allow exceptions. Also, it does not matter what the "base" era is in the analysis of any trinity. And in the present context, the "base" era is modernity (for instance, with its "free-spirited modernity" and the other two parts). So, for pre-modernity, the trinity takes the form of, say, "pre-free-spirited pre-modernity," together with the other two parts.
Sources: From Ch.2 of *BCIV* and also the 2 volumes of *FHC*

Table 1.35. The Trinity of Modernity (Part I)

- **Free-Spirited Modernity (Modernism) and Its Internal Split**
  —*On the True and the Holy*
    - The freedom from the dogmas of the past to the better understanding of, and union with, the world and self (Ch.3 of *FHC*)
    - Alternative discourses: about the true (e.g., anti-science discourses) and the holy (non-mainstream theologies) (Ch.3 of *FHC*)

  —*On the Technological and the Everyday*
    - The freedom from life harshness to the higher technophilic, consumeristic lifeform (Ch.2 of *FHC*)
    - Alternative discourses: about the everyday (e.g., transcendental mindsets) and the technological (e.g., Arcadianism) (Ch.2 of *FHC*)

  —*On the Good and the Just*
    - The freedom from the theo-aristocratic tyranny to the moral universality for a just society (Ch.5 of *FHC*)
    - Alternative discourses: about the just (e.g., Communism, Anarchism) and the good (e.g., Nazism/Fascism, and Zarathustrianism) (Chs.5-6 of *FHC*)

  —*On the Beautiful and the Sublime*
    - The freedom from the external distortion of aesthetic pleasure to the boundless infinity of totality in artistic autonomy (Ch.4 of *FHC*)
    - Alternative discourses: about the beautiful/sublime (e.g., kitsch and historical avant-gardism) (Ch.4 of *FHC*)

- **Capitalist Modernity (Modernization) and Its Own Discontents**
  —*During the Industrial Revolution*
    - Ex: Marx on the institution of inequality (Ch.1 of *FHC*)

  —*During the Modern Rational-Instrumental Epoch*
    - Ex: Weber on the politics of soft liberal institutions (Ch.5 of *FHC*)

  —*During the Great Depression*
    - Ex: Keynes on the myth of the free market (Chs.1-3 of *FHC*)

  —*During the Cold War*
    - Ex: Lasch on the narcissistic culture industry (Chs.2-3 of *FHC*)

(continued on next page)

Table 1.35. The Trinity of Modernity (Part II)

- **Hegemonic Modernity and Its Countervailing Forces**
  —*The Legacies of Colonialism and Imperialism*
    - Ex: European colonization of most of the modern world (Ch.1 of *FHC*)

  —*The Struggle for Decolonialization*
    - Ex: The countervailing forces of resentment, rechantment, and regionalism (Chs.1 & 6 of *FHC*)

Notes: The examples in each category are solely illustrative (not exhaustive) nor necessarily mutually exclusive, and the comparison is relative (not absolute). As generalities, they allow exceptions.
Sources: From the 2 volumes of *FHC*—and also from *FCD*

Table 1.36. The Trinity of Post-Modernity

- **Free-Spirited Postmodernity (Postmodernism) and Its Internal Split**

  —*On the True and the Holy*
  - Postmodern performative turn for knowing and its enemies (Ch.3 of *FHC*)
  - Postmodern comparative theology and its opponents (Ch.3 of *FHC*)

  —*On the Technological and the Everyday*
  - Postmodern corporate technological mindset and its adversaries (Ch.2 of *FHC*)
  - Postmodern postmaterialism and its critics (Ch.2 of *FHC*)

  —*On the Good and the Just*
  - Postmodern politics of difference and its foes (Ch.5 of *FHC*)

  —*On the Beautiful and the Sublime*
  - Postmodern deconstruction and its dissenters (Ch.4 of *FHC*)

- **Capitalist Postmodernity (Postmodernization) and Its Own Discontents**

  —*During the Post-Cold War and Beyond*
  - Ex: post-Fordism and its shortcomings (Ch.6 of *FHC*; Chs.6-7 of *FCD*)

- **Hegemonic Postmodernity and Its Countervailing Forces**

  —*The Debate on the Global Village*
  - Ex: uni-civilizationalism vs. multi-civilizationalism (Ch.6 of *FHC*)

  —*The Resistance Movement*
  - Ex: rechantment and the politics of civilizational claims (e.g., Islamic, Confucian and other ethos in relation to the Same) (Ch.6 of *FHC*; Ch.10 of *FCD*)
  - Ex: resentment and the politics of resurgence (e.g., the rising Chinese superpower, the growing EU, and other players in relation to the U.S. and her allies) (Ch.6 of *FHC*; Ch.8 of *FCD*)
  - Ex: regionalism and the politics of inequality (e.g., trans- or international blocs, the North-South divide, NGO's) (Ch.6 of *FHC*; Ch.5 of *FCD*)

Notes: The examples in each category are solely illustrative (not exhaustive) nor necessarily mutually exclusive, and the comparison is relative (not absolute). As generalities, they allow exceptions.
Sources: From *FCD* and the 2 volumes of *FHC*

Table 1.37. The Trinity of After-Postmodernity

- **Free-Spirited After-Postmodernity (After-Postmodernism) and Its Internal Split**
  —The discourse of naked contingency (Ch.10 of *FCD*; Ch.4 of *FPHC*)

- **Post-Capitalist After-Postmodernity (After-Postmodernization) and Its Own Discontents**
  —Different versions of post-capitalism and post-democracy, and their enemies (Ch.10 of *FCD*; Chs.3-4 of *FPHC*)

- **Hegemonic After-Postmodernity and Its Countervailing Forces**
  —The Cyclical Progression of Hegemony in Multiverses (Chs.9-10 of *FCD*; Ch.4 of *FPHC*)

Notes: The examples in each category are solely illustrative (not exhaustive) nor necessarily mutually exclusive, and the comparison is relative (not absolute). As generalities, they allow exceptions.
Sources: From *FCD* and also *FHC*

• Part Two •

Quantitative Research

CHAPTER 2
QUANTITATIVE RESEARCH AND ITS INADEQUACY

> Statistics are like a bikini. What they reveal is suggestive, but what they conceal is vital.
> —Aaron Levenstein (TEC 2011)

The Preciseness of Quantitative Research

Data analysis involves both quantitative and qualitative research—as already analyzed in *Sec. 1.2*.

This distinction between quantitative and qualitative research is not absolute of course, because both can be combined for research. For instance, "[q]uantitative methods might be used with a global qualitative frame. Qualitative methods might be used to understand the meaning of the numbers produced by quantitative methods. Using quantitative methods, it is possible to give precise and testable expression to qualitative ideas. This combination of quantitative and qualitative data gathering is often referred to as mixed-methods research." (WK 2011d)

This then suggests that any study of quantitative research can be relevant to qualitative reserch too, as one cannot be completely separated from the other—and vice versa.

Because of the dialectic treatment of quantitative and qualitative research, the issues concerning quantitative research in this chapter (Chapter Two) and qualitative research in the next chapter (Chapter Three) often overlap, as each issue on one can be relevant to the other too.

With this dialectic treatment in mind—a good way to examine quantitative research and its inadequacy is by way of the evaluation of the extent to which it is in fact both possible and desirabe. And this can be done by way of a comprehensive analysis of quantitative research in the context of data analysis from the four perspectives of the mind, nature, society, and culture (in accordance to my sophisticated methodological holism, as explained in *Sec. 1.8*).

In other words, this chapter is organized in four main sections, in relation to (2.2) quantitative research and the mind, (2.3) quantitative research and nature, (2.4) quantitative research and society, and (2.5) quantitative research and culture—to be examined hereafter, in that order (and summarized in *Table 2.1*).

Quantitative Research and the Mind

Quantitative research, when examined from the perspective of the mind, can reveal its possibility and desirability in an intriguing way, and this can be shown by way of two case studies, namely, (2.2.1) quantitative research, and the challenge of action science, and (2.2.2) quantitative research, and the controversy of cognitive biases—to be addressed in what follows, respectively.

Quantitative Research, and the Challenge of Action Science

One good way to evaluate the possibility and desirability of quantitative research from the perspective of the mind is to explore quantitative research and the challenge of action science.

The Challenge of Action Science to Go beyond Data Collection

The term "action science" (aka "action research") refers to "a reflective process of progressive problem solving led by individuals working with others in teams or as part of a 'community of practice' to improve the way they address issues and solve problems. Action research is done simply by action, hence the name." (WK 2011ee)

It was "Kurt Lewin, then a professor at MIT," who "first coined the term 'action research' in about 1944. In his 1946 paper 'Action Research and Minority Problems' he described action research as 'a comparative research on the conditions and effects of various forms of social action and research leading to social action' that uses 'a spiral of steps, each of which is composed of a circle of planning, action, and fact-finding about the result of the action.'" (WK 2011ee)

In this way, "[a]ction research challenges traditional social science, by moving beyond reflective knowledge created by outside experts sampling variables to an active moment-to-moment theorizing, data collecting, and inquiry occurring in the midst of emergent structure," ultimately for social change. (WK 2011ee)

Action research can also be exploited "by larger organizations or institutions, assisted or guided by professional researchers, with the aim of improving their strategies, practices, and knowledge of the environments within which they practice. As designers and stakeholders, researchers work with others to propose a new course of action to help their community improve its work practices." (WK 2011ee)

Thus, the challenge of action science is its innovative combination of "data-driven collaborative analysis or research" and "problem solving actions," ultimately for social change.

The General Framework of Action Science

Kurt Lewin (1958) was concerned "with social change and, more particularly, with effective, permanent social change" and "believed that the motivation to change was strongly related to action: If people are active in decisions affecting them, they are more likely to adopt new ways. 'Rational social management,' he said, 'proceeds in a spiral of steps, each of which is composed of a circle of planning, action, and fact-finding about the result of action.'" (WK 2011ee)

More specifically, Lewin proposed three major steps to understand the action-research process, as shown below (and summarized in *Table 2.2*): (WK 2011ee)

- "Input" (for "Unfreezing")—is the first stage of the action-research process where, "[f]aced with a dilemma or disconfirmation, the individual or group becomes aware of a need to change," so the individual or group engages in "preliminary diagnosis, data gathering, feedback of results, action planning," etc.
- "Transformation" (for "Changing")—is the second stage of the action-research process where "the situation is diagnosed and new models of behavior are explored and tested," so the individual or group implements "action planning and action steps."
- "Output" (for "Refreezing")— is the third (and last) stage of the action-research process where "application of new behavior is evaluated, and, if reinforcing, adopted."

It should be noted that three feedbacks occur throughout the process, they are, namely, (a) "feedback loop A" (from transformation to input), (b) "feedback loop B (from output to transformation), and (c) "feedback loop C" (from output to input). (WK 2011ee)

So, the cycle of the process continues, as "action research…sets in motion a long-range, cyclical, self-correcting mechanism for maintaining and enhancing the effectiveness of the client's system by leaving the system with practical and useful tools for self-analysis and self-renewal." (WK 2011ee; R. Johnson 1976)

Different Versions of Action Science

Since 1944 when Kurt Lewin at MIT first proposed action research to challenge traditional social science, there has been different versions of action science.

For illustration, consider the following versions.

(a) Chris Argyris' Action Science

The first version of action science to be introduced here concerns Chris Argyris' action science.

Chris Argyris proposed his "Action Science" as "the study of how human beings design their actions in difficult situations. Human actions are designed to achieve intended consequences and governed by a set of environment variables. How those governing variables are treated in designing actions are the key differences between single loop learning and double loop learning." (WK 2011ee)

In regard to single loop learning, "when actions are designed to achieve the intended consequences and to suppress conflict about the governing variables, a single loop learning cycle usually ensues." (WK 2011ee)

And in regard to double loop learning, "when actions are taken, not only to achieve the intended consequences, but also to openly inquire about conflict and to possibly transform the governing variables, both single loop and double loop learning cycles usually ensue." (WK 2011ee)

Argyris used both "single loop and double loop learning concepts not only to personal behaviors but also to organizational behaviors in his models....This is different from experimental research in which environmental variables are controlled and researchers try to find out cause and effected in isolated environment." (WK 2011ee)

(b) John Heron and Peter Reason's Cooperative Inquiry

The second version of action science to be introduced here concerns John Heron and Peter Reason's cooperative inquiry.

This version, "cooperative inquiry, also known as collaborative inquiry, was first proposed by John Heron in 1971 and later expanded with Peter Reason." (WK 2011ee)

Their main idea is to focus on "research 'with' rather than 'on' people," in that "[i]t emphasizes that all active participants are fully involved in research decisions as co-researchers. Cooperative inquiry creates a research cycle among four different types of knowledge," as shown below: (WK 2011ee)

- "Propositional Knowing"—"as in contemporary science"
- "Practical Knowing"—"the knowledge that comes with actually doing what you propose"
- "Experiential Knowing"—"the feedback we get in real time about our interaction with the larger world"
- "Presentational Knowing"—"the artistic rehearsal process through which we craft new practices"

The research process of cooperative inquiry is to "include…these four stages at each cycle with deepening experience and knowledge of the initial proposition, or of new propositions, at every cycle." (WK 2011ee)

(c) Paulo Freire's Participatory Action Research (PAR)

The third version of action science to be introduced here concerns Paulo Freire's Participatory Action Research (PAR).

More recently, "participatory action research has emerged…as a significant methodology for intervention, development and change within communities and groups. It is now promoted and implemented by many international development agencies and university programs, as well as countless local community organizations around the world. PAR builds on the critical pedagogy put forward by Paulo Freire as a response to the traditional formal models of education where the 'teacher' stands at the front and 'imparts' information to the 'students' who are passive recipients." (WK 2011ee)

In fact, this version "was further developed in 'adult education' models throughout Latin America. Orlando Fals-Borda (1925–2008), Colombian sociologist and political activist, was one of principal promoters of 'participatory action research' (IAP in Spanish) in Latin America." (WK 2011ee)

(d) William Torbert's Developmental Action Inquiry

The fourth version of action science to be introduced here concerns William Torbert's developmental action inquiry.

This version of action science is a "way of simultaneously conducting action and inquiry as a disciplined leadership practice that increases the wider effectiveness of our actions. Such action helps individuals, teams, organizations become more capable of self-transformation and thus more creative, more aware, more just and more sustainable." (WK 2011ee; W. Torbert 2004)

More specifically, "action Inquiry challenges our attention to span…different territories of experience (at the personal, group, or organizational scales) in the midst of actions. This practice promotes timeliness—learning with moment to moment intentional awareness—among individuals and with regard to the outside world of nature and human institutions. It studies the 'pre-constituted internalized and externalized universe in the present, both as it resonates with and departs from the past, and as it resonates with and potentiates the future.'" (WK 2011ee)

(e) J. Whitehead's Living Theory and J. McNiff's Action Research

And the fifth version of action science to be introduced here concerns Jack Whitehead's Living Theory and Jean McNiff's Action Research approaches.

In this version, "individuals generate explanations of their educational influences in their own learning, in the learning of others and in the learning of social formations. They generate the explanations from experiencing themselves as living contradictions in enquiries of the kind, 'How do I improve what I am doing?' They use action reflection cycles of expressing concerns (saying why you are concerned in relation to values), imagining possibilities in developing action plans, acting and gathering data, evaluating the influences of action, [and] modifying concerns, ideas and action in the light of the evaluations. The explanations include life-affirming, energy-flowing values as explanatory principles. (WK 2011ee; J. Whitehead 1989 & 2006)

Problems with Action Science

Yet, action science has its own problems. For illustration, consider a few major criticisms below (as summarized in *Table 2.2*).

Firstly, one major criticism of action science is that the tension among its conflicting goals remains unresolved.

For instance, "[t]his tension exists between" the conflicts in the following dilemmas: (WK 2011ee)

- Ex: "those who are more driven by the researcher's agenda and those more driven by participants"
- Ex: "those who are motivated primarily by instrumental goal attainment and those motivated primarily by the aim of personal, organizational, or societal transformation"
- Ex: "1st-, to 2nd-, to 3rd-person research, that is, my research on my own action, aimed primarily at personal change; our research on our

> group (family/team), aimed primarily at improving the group; and 'scholarly' research aimed primarily at theoretical generalization and/or large scale change"

Of course, there can be other dilemmas, so the examples above are illustrative (not exhaustive).

Secondly, another major criticism of action science is that its innovative combination of "problem solving actions" and "data-driven collaborative analysis or research" remains to be reconciled.

In other words, the problem here is that "problem solving actions" are more praxis-oriented but "data-driven collaborative analysis or research" is more reflection-oriented. Although the two types of research do not have to be mutually exclusive, it is not clear how or to what extent their integration is achieved in the context of action science.

And thirdly, still another major criticism of action science is that it is not clear why the focus on action theorizing is necessarily more desirable than the conventional attention to reflective knowledge instead.

The point here does not mean to favor reflective knowledge but to question the convincingness of the main claim in action science.

Beyond Action Research

These criticisms of action research, however, do not mean to reject its usefulness to challenge traditional "data-driven" analysis but are to show us the two opposing sides of the debate, such that the possibility and desirability of quantitative research (from the perspective of the mind with quantitative research and the challenge of action science as a case study here) are not to the extent that the respective defenders would like us to believe.

More importantly, the analysis of quantitative research and the challenge of action science can teach us something valuable about the ontological principles in existential dialectics, and good examples include the formalness-informalness principle, the absoluteness-relativeness principle, the predictability-unpredictability principle, the explicability-inexplicability principle, the fiction-reality principle, the preciseness-vagueness principle, the simpleness-complicatedness principle, the openness-hiddenness principle, the denseness-emptiness principle, the change-constancy principle, the slowness-quickness principle, the expansion-contraction principle, the theory-praxis principle, the convention-novelty principle, the evolution-transformation principle, the symmetry-asymmetry principle, the softness-hardness principle, the seriousness-playfulness principle, the regression-progression principle, the same-difference principle, and the stability-reaction principle.

For instance, in relation to the formalness-informalness principle, if there is formalness (e.g., the formal logical requirement of completeness and soundness for a system of ideas, as in action science), there is informalness (e.g., the nonformal existence of incompleteness and unsoudness in action science, as shown in the tension among its conflicting goals pointed out by the critics). And the reverse direction also holds true.

In relation to the absoluteness-relativeness principle, if there is absoluteness (e.g., the absolute view about propositional knowing as the legitimate form of knowing in contemporary mainstream science), there is relativeness (e.g., what is legitimate for the advocates of contemporary mainstream science in regard to propositional knowing is not necessarily sufficient for the advocates of practical knowing, as shown in John Heron and Peter Reason's cooperative inquiry). And the reverse direction also holds true.

In relation to the predictability-unpredictability principle, if there is predictability (e.g., the predictable tendency of those in action research to stress "problem solving actions" in their research), there is unpredictability (e.g, the more difficult task to predict exactly when, or the extent to which, a particular application of action research will be successful in resolving the inherent tension among its conflicting goals, as pointed out by the critics). And the reverse direction also holds true.

In relation to the explicability-inexplicability principle, if there is explicability (e.g., the explanation by the advocates of action research that it is desirable to move "beyond reflective knowledge" for "an active moment-to-moment theorizing, data collecting, and inquiry occurring in the midst of emergent structure," which contributes to social change), there is inexplicability (e.g., the lack of sufficient explanation by the advocates of action research of why it is necessarily desirable to focus more on action theorizing than on reflective knowledge, since not every researcher is driven by the desire for social change). And the reverse direction also holds true.

In relation to the fiction-reality principle, if there is fiction (e.g, the subjective experience of "living contradictions" in 1st-person action research, as in living theory), there is reality (e.g., the better understanding of reality after going through "action reflection cycles" and learning from the evaluations). And the reverse direction also holds true.

In relation to the preciseness-vagueness principle, if there is preciseness (e.g., the precise identification of "four different types of knowledge" by Heron and Reason in the context of cooperative inquiry), there is vagueness (e.g., the vagueness in the identificaiton, since it is not clear why there must be only "four different types of knowledge," not five, six, seven, and so on). And the reverse direction also holds true.

In relation to the simpleness-complicatedness principle, if there is simpleness (e.g., the relatively simple portrayal of action research by its advocates), there is complicatedness (e.g., the relatively more complicated analysis of action resarch by questionings its claims and assumptions, as shown in the criticisms pointed out by the critics). And the reverse direction also holds true.

In relation to the openness-hiddenness principle, if there is openness (e.g., the open exploration of the need to combine "problem solving actions" with "data-driven collaborative analysis or research" in action research), there is hiddenness (e.g., the hidden bias in action research, because of its favoring action theorizing over reflective knowledge). And the reverse direction also holds true.

In relation to the denseness-emptiness principle, if there is denseness (e.g., the relatively denser concern with action in "action research"), there is emptiness (e.g., the relatively less dense, or more empty, concern with action in "contemporary science," as pointed out by the advocates of action research). And the reverse direction also holds true.

In relation to the change-constancy principle, if there is change (e.g., the ever new ways to do research, as shown in the emergence of action research), there is constancy (e.g., the ever constant existence of different problems with different ways of doing research, as shown in the problems with action research pointed out by the critics). And the reverse direction also holds true.

In relation to the slowness-quickness principle, if there is slowness (e.g., the relatively slower speed of learning, if a double loop learning process is used, since what is to learn is more complex, by trying " not only to achieve the intended consequences, but also to openly inquire about conflict and to possibly transform the governing variables"), there is quickness (e.g., the relatively quicker speed of learning, if a single loop learning process is used, since what is to learn is less complex, by trying to achieve only "the intended consequences and to suppress conflict about the governing variables"). And the reverse direction also holds true.

In relation to the expansion-contraction principle, if there is expansion (e.g., the relatively more developed ability of those in action research to cope with the problems relating to "action" in practice), there is contraction (e.g., the relatively less developed ability of those in action research to cope with the problems relating to "reflective knowledge"). And the reverse direction also holds true.

In relation to the theory-praxis principle, if there is theory (e.g., the theoretical construction of "participatory action research" by Paulo Freire), there is praxis (e.g., the practical application of "participatory action research" by Orlando Fals-Borda to the field of adult education in Latin America). And the reverse direction also holds true.

In relation to the convention-novelty principle, if there is convention (e.g., the conventional wisdom of doing research by focusing on "reflective knowledge"), there is novelty (e.g., the alternative novel challenge to the conventional wisdom about "reflective knowledge" by the new idea of "moving beyond reflective knowledge…to an active moment-to-moment theorizing"). And the reverse direction also holds true.

In relation to the evolution-transformation principle, if there is evolution (e.g., the natural evolution of humans to learn from each other in small groups), there is transformation (e.g., the technical transformation of humans to learn by the invention of "international development agencies and university programs, as well as countless local community organizations around the world" for Participatory Action Research). And the reverse direction also holds true.

In relation to the symmetry-asymmetry principle, if there is symmetry (e.g., the co-existence of different kinds of knowing), there is asymmetry (e.g., the popular pursuit of propositional knowing in contemporary science—but the popular pursuit of action knowing in action science). And the reverse direction also holds true.

In relation to the softness-hardness principle, if there is softness (e.g., the praise of action research by its advocates as a new form of doing research to challenge "traditional social science"), there is hardness (e.g., the critique of action research by its critics, as shown in the criticisms aforediscussed). And the reverse direction also holds true.

In relation to the seriousness-playfulness principle, if there is seriousness (e.g., the serious business of action research for social change, as shown in the version by John Heron and Peter Reason), there is playfulness (e.g., the playful dimension of action research, as shown in the version by John Heron and Peter Reason for "presentational knowing," where the researcher can rehearse or "craft" in a creative or "artistic" way different possible "new practices").

In relation to the regression-progression principle, if there is regression (e.g., the regression made by action research, as shown in the problems pointed out by the critics), there is progression (e.g., the progress made by action research, as shown in the successes of different versions in the last six decades). And the reverse direction also holds true.

In relation to the same-difference principle, if there is similarity in outcome (e.g., the contribution to the molding and control of beliefs, values, and behaviors, regardless of whether this be done by way of action teaching, as in PAR, or by way of traditional teaching), there is difference in outcome (e.g., the contribution to the molding and control of beliefs, values, and behaviors by way of action teaching for a relatively more egalitarian lifeworld—but the contribution to the molding and control of beliefs, values, and behaviors by way of traditional teaching for a relatively more hierarchical lifeworld). And the reverse direction also holds true.

And in relation to the stability-reaction principle, if there is stability (e.g., the relatively more stable condition of an "output" when it is ready for "refreezing," in the general framework of action research by Lewin), there is reaction (e.g., output can be subject to further changes, once it is "unfreezed" and "transformed" in the cycle of the action research process, so that "action research…sets in motion a long-range, cyclical, self-correcting mechanism). And the reverse direction also holds true.

Quantitative Research, and the Controversy of Cognitive Biases

Another way to evaluate the possibility and desirability of quantitative research from the perspective of the mind is to challenge quantitative research in the context of the controversy of cognitive biases.

In other words, one assumption in quantitative research is that a researcher in question can handle data with minimal (or no) cognitive biases, with implications for the deeper question of rationality. But is this really true?

The Meaning of Cognitive Bias

The term "cognitive bias" refers to "a pattern of deviation in judgment that occurs in particular situations. Implicit in the concept of a 'pattern of deviation' is a standard of comparison; this may be the judgment of people outside those particular situations, or may be a set of independently verifiable facts." (WK 2011ii)

Although "cognitive biases" can be "presumably adaptive" in some cases, for example, when "they lead to more effective actions in given contexts or enable faster decisions when faster decisions are of greater value"—in many other cases, they "can lead to perceptual distortion, inaccurate judgment, or illogical interpretation. It is a phenomenon studied in cognitive science and social psychology." (WK 2011ii; D. Kahneman 1972)

Categories of Cognitive Biases

In general, cognitive biases can be classified in different categories, as shown below: (WK 2011m)

- Biases in Probability and Belief
  - –"Ambiguity Effect"—"the tendency to avoid options for which missing information makes the probability seem 'unknown.'" (J. Baron 1994)
  - –"Anchoring Effect"—"the tendency to rely too heavily, or 'anchor,' on a past reference or on one trait or piece of information when making decisions (also called 'insufficient adjustment')"
  - –"Attentional Bias"—"the tendency to neglect relevant data when making judgments of a correlation or association"
  - –"Authority Bias"—"the tendency to value an ambiguous stimulus…according to the opinion of someone who is seen as an authority on the topic"
  - –"Availability Heuristic"— "estimating what is more likely by what is more available in memory, which is biased toward vivid, unusual, or emotionally charged examples"
  - –"Availability Cascade"—"a self-reinforcing process in which a collective belief gains more and more plausibility through its increasing repetition in public discourse (or 'repeat something long enough and it will become true')"
  - –"Base Rate Neglect or Base Rate Fallacy"—"the tendency to base judgments on specifics, ignoring general statistical information" (J. Baron 1994)
  - –"Belief Bias"—"an effect where someone's evaluation of the logical strength of an argument is biased by the believability of the conclusion" (K. Klauer 2000)
  - –"Clustering Illusion"—"the tendency to see patterns where actually none exist"
  - –"Conjunction Fallacy"—"the tendency to assume that specific conditions are more probable than general ones" (J. Fisk 2004)
  - –"Forward Bias"—"the tendency to create models based on past data which are validated only against that past data"
  - –"Gambler's Fallacy"—"the tendency to think that future probabilities are altered by past events, when in reality they are unchanged. [This] results from an erroneous conceptualization

of the Law of large numbers. For example, 'I've flipped heads with this coin five times consecutively, so the chance of tails coming out on the sixth flip is much greater than heads.'"

–"Hindsight Bias"—"sometimes called the 'I-knew-it-all-along' effect, the tendency to see past events as being predictable at the time those events happened" (R. Pohl 2004)

–"Illusory Correlation"—"inaccurately perceiving a relationship between two events, either because of prejudice or selective processing of information" (A. Tversky 1974)

–"Neglect of Probability"—"the tendency to completely disregard probability when making a decision under uncertainty" (J. Baron 1994)

–"Observer-Expectancy Effect"—"when a researcher expects a given result and therefore unconsciously manipulates an experiment or misinterprets data in order to find it"

–"Optimism Bias"—"the tendency to be over-optimistic about the outcome of planned actions" (D. Hardman 2009)

–"Ostrich Effect"—"ignoring an obvious (negative) situation"

–"Overconfidence Effect"—"excessive confidence in one's own answers to questions"

–"Positive Outcome Bias"— "the tendency of one to overestimate the probability of a favorable outcome coming to pass in a given situation"

–"Pareidolia"—"a vague and random stimulus (often an image or sound) is perceived as significant, e.g., seeing images of animals or faces in clouds, the man in the moon, and hearing hidden messages on records played in reverse"

–"Pessimism Bias"—"the tendency for some people, especially those suffering from depression, to overestimate the likelihood of negative things happening to them"

–"Primacy Effect"—"the tendency to weigh initial events more than subsequent events" (J. Baron 1994)

–"Recency Effect"—"the tendency to weigh recent events more than earlier events"

–"Disregard of Regression toward the Mean"—"the tendency to expect extreme performance to continue"

–"Stereotyping"—"expecting a member of a group to have certain characteristics without having actual information about that individual"

–"Subadditivity Effect"—"the tendency to judge probability of the whole to be less than the probabilities of the parts"

- –"Subjective Validation"—"perception that something is true if a subject's belief demands it to be true. Also assigns perceived connections between coincidences."

- Decision-Making and Behavioral Biases
  - –"Anchoring"—"the common human tendency to rely too heavily, or 'anchor,' on one trait or piece of information when making decisions"
  - –"Attentional Bias"—"implicit cognitive bias defined as the tendency of emotionally salient stimuli in one's environment to preferentially draw and hold attention"
  - –"Confirmation Bias"—"the tendency to search for or interpret information in a way that confirms one's preconceptions." (M. Oswald 2004)
  - –"Congruence Bias"—"the tendency to test hypotheses exclusively through direct testing, in contrast to tests of possible alternative hypotheses"
  - –"Contrast Effect"—"the enhancement or diminishing of a weight or other measurement when compared with a recently observed contrasting object" (S. Plous 1993)
  - –"Distinction Bias"—"the tendency to view two options as more dissimilar when evaluating them simultaneously than when evaluating them separately" (C. Hsee 2004)
  - –"Experimenter's or Expectation Bias"—"the tendency for experimenters to believe, certify, and publish data that agree with their expectations for the outcome of an experiment, and to disbelieve, discard, or downgrade the corresponding weightings for data that appear to conflict with those expectations" (M. Jeng 2006)
  - –"Focusing Effect"—"the tendency to place too much importance on one aspect of an event; causes error in accurately predicting the utility of a future outcome" (D. Kahneman 2006)
  - –"Hostile Media Effect"—"the tendency to see a media report as being biased due to one's own strong partisan views"
  - –"Information Bias"—"the tendency to seek information even when it cannot affect action" (J. Baron 1994)
  - –"Interloper effect"—"the tendency to value third party consultation as objective, confirming, and without motive"
  - –"Mere Exposure Effect"—"the tendency to express undue liking for things merely because of familiarity with them" (R. Bornstein 2004)

–"Money Illusion"—"the tendency to concentrate on the nominal (face value) of money rather than its value in terms of purchasing power" (E. Shafir 2000)
–"Negativity Bias"——"the tendency to pay more attention and give more weight to negative than positive experiences or other kinds of information"
–"Omission Bias"—"the tendency to judge harmful actions as worse, or less moral, than equally harmful omissions (inactions)" (J. Baron 1994)
–"Planning Fallacy"—"the tendency to underestimate task-completion times" (L. Sanna 2004)
–"Reactance"— "the urge to do the opposite of what someone wants you to do out of a need to resist a perceived attempt to constrain your freedom of choice"
–"Restraint Bias"—"the tendency to overestimate one's ability to show restraint in the face of temptation"
–"Selective Perception"—"the tendency for expectations to affect perception"
–"Semmelweis Reflex"— "the tendency to reject new evidence that contradicts an established paradigm" (W. Edwards 1968)
–"Status Quo Bias"—"the tendency to like things to stay relatively the same" (J. Baron 1994; D. Kahneman 1991)
–"Wishful Thinking"—"the formation of beliefs and the making of decisions according to what is pleasing to imagine instead of by appeal to evidence or rationality" (J. Baron 1994)
–"Zero-Risk Bias"—"preference for reducing a small risk to zero over a greater reduction in a larger risk"

• Memory Errors
–"Choice-Supportive Bias"—"the tendency to remember one's choices as better than they actually were"
–"Consistency Bias"—"incorrectly remembering one's past attitudes and behavior as resembling present attitudes and behavior"
–"Cryptomnesia"—"a form of misattribution where a memory is mistaken for imagination"
–"Egocentric Bias"—"recalling the past in a self-serving manner, e.g. remembering one's exam grades as being better than they were, or remembering a caught fish as being bigger than it was"
–"False Memory"—"confusion of imagination with memory, or the confusion of true memories with false memories"
–"Hindsight Bias"—"filtering memory of past events through present knowledge, so that those events look more predictable than they

actually were; also known as the 'I-knew-it-all-along effect.'" (R. Pohl 2004)

- –"Reminiscence Bump"—"the effect that people tend to recall more personal events from adolescence and early adulthood than from other lifetime periods"
- –"Rosy Retrospection"—"the tendency to rate past events more positively than they had actually rated them when the event occurred"
- –"Self-Serving Bias"—"perceiving oneself responsible for desirable outcomes but not responsible for undesirable ones"
- –"Suggestibility"— "a form of misattribution where ideas suggested by a questioner are mistaken for memory"
- –"Telescoping Effect"—"the effect that recent events appear to have occurred more remotely and remote events appear to have occurred more recently"
- –"Von Restorff Effect"—"the tendency for an item that 'stands out like a sore thumb' to be more likely to be remembered than other items"

• Social Biases

- –"Actor–Observer Bias"—"the tendency for explanations of other individuals' behaviors to overemphasize the influence of their personality and underemphasize the influence of their situation….However, this is coupled with the opposite tendency for the self in that explanations for our own behaviors overemphasize the influence of our situation and underemphasize the influence of our own personality."
- –"Bandwagon Effect"—"the tendency to do (or believe) things because many other people do (or believe) the same. Related to groupthink and herd behavior."
- –"Bias Blind Spot"—"the tendency to see oneself as less biased than other people" (E. Pronin 2007)
- –"Egocentric Bias"—this "occurs when people claim more responsibility for themselves for the results of a joint action than an outside observer would."
- –"Forer Effect (aka Barnum Effect)"—"the tendency to give high accuracy ratings to descriptions of their personality that supposedly are tailored specifically for them, but are in fact vague and general enough to apply to a wide range of people. For example, horoscopes"

–"False Consensus Effect"—"the tendency for people to overestimate the degree to which others agree with them" (G. Marks 1987)

–"Fundamental Attribution Error"—"the tendency for people to over-emphasize personality-based explanations for behaviors observed in others while under-emphasizing the role and power of situational influences on the same behavior" (S. Sutherland 2007)

–"Halo Effect"—"the tendency for a person's positive or negative traits to 'spill over' from one area of their personality to another in others' perceptions of them" (J. Baron 1994)

–"Herd Instinct"—"common tendency to adopt the opinions and follow the behaviors of the majority to feel safer and to avoid conflict"

–"Illusion of Asymmetric Insight"—"people perceive their knowledge of their peers to surpass their peers' knowledge of them." (E. Pronin 2001)

–"Illusion of Control"—"the tendency to overestimate one's degree of influence over other external events" (S. Thompson 1999)

–"Illusory Superiority"—"overestimating one's desirable qualities, and underestimating undesirable qualities, relative to other people....Also known as 'Lake Wobegon effect,' 'better-than-average effect,' or 'superiority bias.'" (V. Hoorens 1993)

–"Ingroup Bias"—"the tendency for people to give preferential treatment to others they perceive to be members of their own groups"

–"Outgroup Homogeneity Bias"—"individuals see members of their own group as being relatively more varied than members of other groups." (S. Plous 1993)

–"Projection Bias"—"the tendency to unconsciously assume that others (or one's future selves) share one's current emotional states, thoughts and values" (C. Hsee 2006)

–"Self-Serving Bias—"the tendency to claim more responsibility for successes than failures"

–"System Justification"—"the tendency to defend and bolster the status quo"

–"Trait Ascription Bias"—"the tendency for people to view themselves as relatively variable in terms of personality, behavior and mood while viewing others as much more predictable"

Surely, there can be other examples, so the list above is illustrative (not exhaustive).

Problems with the Study of Cognitive Biases

However, the study of cognitive biases has its own problems. For illustration, consider a few major criticisms below.

Firstly, one major criticism against the study of cognitive biases is that there is no consensus on the explanations of their causes.

For instance, over the decades, there have been different theories to explain them, and good examples are "bounded rationality" (by H. Simon and others), "attribution theory" (by Fritz Heider, Harold Kelley, Edward E. Jones, Lee Ross, Bernard Weiner, and others), "cognitive dissonance" (by L. Festinger and others), "heuristics" (by A. Tversky, Paul Slovic, Melissa Finucane, Ellen Peters, Donald G. MacGregor, and others), "adaptive bias" (by Martie G. Haselton, David M. Buss, and others), "misinterpretations or misuse of statistics" (by many different scholars), and so on. (WK 2011o, 2011p, 2011q, 2011r, 2011s, 2011t & 2011u)

Secondly, another major criticism against the study of cognitive biases is that it is not clear to what extent cognitive biases can be cured.

For instance, Daniel Kahneman, Amos Tversky, and Bent Flyvbjerg (2008) proposed "reference class forecasting…to eliminate or reduce the impact of cognitive biases on decision making." (WK 2011m) But for the critics, cognitive biases are too numerous and complicated to be eliminated by the application of a single method.

And thirdly, still another major criticism against the study of cognitive biases is that it is not clear to what extent cognitive biases are undesirable, since "cognitive biases" can be "presumably adaptive," for example, when "they lead to more effective actions in given contexts or enable faster decisions when faster decisions are of greater value."

Beyond Cognitive Biases

These criticisms of the study of cognitive biases do not mean to reject its usefulness to question the assumption of the freedom from cognitive biases in conventional quantitative research but are to show us the two opposing sides of the debate, such that the possibility and desirability of quantitative research (from the perspective of the mind with quantitative research and the controversy of cognitive biases as a case study here) are not to the extent that the respective defenders would like us to believe.

Furthermore, the analysis of quantitative and the controversy of cognitive biases can teach us something revealing about the ontological principles in existential dialectics, and good examples include the formalness-informalness principle, the absoluteness-relativeness principle, the predictability-unpredictability principle, the explicability-inexplicability principle, the fiction-reality principle, the preciseness-vagueness principle, the simpleness-complicatedness principle, the openness-hiddenness principle, the denseness-emptiness principle, the change-constancy principle, the slowness-quickness principle, the expansion-contraction principle, the theory-praxis principle, the convention-novelty principle, the evolution-transformation principle, the symmetry-asymmetry principle, the softness-hardness principle, the seriousness-playfulness principle, the regression-progression principle, the same-difference principle, and the stability-reaction principle.

For instance, in relation to the formalness-informalness principle, if there is formalness (e.g., the formal logical requirement of soundness for a system of ideas, as in the assumption of the freedom from cognitive biases in quantitative research), there is informalness (e.g., the nonformal existence of unsoudness in quantitative research, as shown in the existence of numerous cognitive biases pointed out by the critics). And the reverse direction also holds true.

In relation to the absoluteness-relativeness principle, if there is absoluteness (e.g., the absolute view of those with "egocentric bias" to "claim more responsibility for themselves for the results of a joint action than an outside observer would"), there is relativeness (e.g., what is right for those with "egocentric bias" in regard to claiming more responsibility for themselves is not necessarily so for an outside observer). And the reverse direction also holds true.

In relation to the predictability-unpredictability principle, if there is predictability (e.g., the predictable tendency of someone with "optimism bias" to be "over-optimistic about the outcome of planned actions"), there is unpredictability (e.g., the more difficult task to predict exactly the extent to which someone with "optimism bias" will be "over-optimistic about the outcome of planned actions" on a particular occassion). And the reverse direction also holds true.

In relation to the explicability-inexplicability principle, if there is explicability (e.g., the explanation by those in the study of cognitive biases to propose "reference class forecasting" as a way "to eliminate or reduce the impact of cognitive biases on decision making"), there is inexplicability (e.g., the lack of sufficient explanation by those in the study of cognitive biases of why "reference class forecasting" is necessarily desirable, since "cognitive biases" can be "presumably adaptive," for example, when "they lead to more effective actions in given contexts or enable faster decisions when faster decisions are of greater value"). And the reverse direction also holds true.

In relation to the fiction-reality principle, if there is fiction (e.g., the fictional aspects of cognitive biases when they "can lead to perceptual distortion, inaccurate judgment, or illogical interpretation"), there is reality (e.g., the realistic aspects of cognitive biases whey "they lead to more effective actions in given contexts or enable faster decisions when faster decisions are of greater value"). And the reverse direction also holds true.

In relation to the preciseness-vagueness principle, if there is preciseness (e.g., the precise identification of the examples of different categories of cognitive biases aforecited), there is vagueness (e.g., the vagueness in the precise identification of the examples of different categories of cognitive biases aforecited, since it is not clear why there must be these specific examples, but not others). And the reverse direction also holds true.

In relation to the simpleness-complicatedness principle, if there is simpleness (e.g., the relatively simple analysis of cognitive biases by the scholars who study them), there is complicatedness (e.g., the relatively more complicated analysis of cognitive biases by questioning the claims and assumptions in the study, as shown in the problems pointed out by the critics as cited). And the reverse direction also holds true.

In relation to the openness-hiddenness principle, if there is openness (e.g., the open exploration of the relationship between personality and behaviors), there is hiddenness (e.g., the hidden bias of those with "fundamental attribution error" to "over-emphasize personality-based explanations for behaviors observed in others while under-emphasizing the role and power of situational influences on the same behavior"). And the reverse direction also holds true.

In relation to the denseness-emptiness principle, if there is denseness (e.g., the relatively denser concentration of concerns, by those with "system justification bias," with defending and bolstering "the status quo"), there is emptiness (e.g., the relatively less dense, or more empty, concentration of concerns, by those with "negativity bias," with defending and bolstering the status quo, since these individuals tends "to pay more attention and give more weight to negative than positive experiences or other kinds of information"). And the reverse direction also holds true.

In relation to the change-constancy principle, if there is change (e.g., the ever new ways that a person can be cognitively biased, as shown in the numerous examples as cited), there is constancy (e.g., the ever constant existence of different problems with the examples, as pointed out by the critics as cited). And the reverse direction also holds true.

In relation to the slowness-quickness principle, if there is slowness (e.g., the relatively slower readiness of someone with "confirmation bias" to "search for or interpret information in a way" that contradicts his "preconceptions"), there is quickness (e.g., the relatively quicker readiness of someone with "confirmation bias" to "search for or interpret information in a way" that confirms his "preconceptions"). And the reverse direction also holds true.

In relation to the expansion-contraction principle, if there is expansion (e.g., the relatively more developed ability of someone with "overconfidence effect" to answer questions), there is contraction (e.g., the relatively less developed ability of someone with "overconfidence effect" to realize that his answers to questions are not really that great). And the reverse direction also holds true.

In relation to the theory-praxis principle, if there is theory (e.g., the theoreitcal construction of cognitive biases in psychology), there is praxis (e.g., the practical application of the theoreitcal construction of cognitive biases in psychology to the field of jurisprudence, when "a fair jury trial" now requires that "the jury ignore irrelevant features of the case, weigh the relevant features appropriately, consider different possibilities open-mindedly and resist fallacies"). (WK 2011ii) And the reverse direction also holds true.

In relation to the convention-novelty principle, if there is convention (e.g., the conventional view that quantitative research can be conducted in a way free of cognitive biases), there is novelty (e.g., the alternative novel challenge to the convnetional wisdom about quantitative research, by the new view in modern psychology that people have all kinds of cognitive biases). And the reverse direction also holds true.

In relation to the evolution-transformation principle, if there is evolution (e.g., the natural evolution of humans to engage in all kinds of cognitive biases in the state of nature), there is transformation (e.g, the technical transformation of humans to undestand cognitive biases with the invention of modern psychology to provide a systematic study of them, together with some scientific methods like "reference class forecasting" to deal with them). And the reverse direction also holds true.

In relation to symmetry-asymmetry principle, if there is symmetry (e.g., the co-existence of different cogntive biases), there is asymmetry (e.g., some cognitive biases "are presumably adaptive"—but some others are not, as they are "presumably result from a lack of appropriate mental mechanisms, or from the misapplication of a mechanism that is adaptive under different circumstances"). And the reverse direction also holds true.

In relation to the softness-hardness principle, if there is softness (e.g., the defense of some cognitive biases as "presumably adaptive"), there is hardness (e.g., the critique of some other cognitive biases as "presumably…a lack of appropriate mental mechanisms"). And the reverse direction also holds true.

In relation to the seriousness-playfulness principle, if there is seriousness (e.g., the seriousness of someone with "hindsight bias" to understand past events), there is playfulness (e.g., the playfulness of someone with "hindsight bias" to "filter…memory of past events through present knowledge, so that those events look more predictable than they actually were"). And the reverse direction also holds true.

In relation to the regression-progression principle, if there is regression (e.g., the regression made by the study of cognitive biases, as shown in the problems pointed out by the critics as cited), there is progression (e.g., the progress made by the study of cognitive biases, as shown in the enhancement of our understanding of different categories of cognitive biases). And the reverse direction also holds true.

In relation to the same-difference principle, if there is similarity in outcome (e.g., the contribution to the molding and control of beliefs, values, and behaviors, regardless of whether this be done by the use of reference class forecasting or by the appeal to bounded rationality), there is difference in outcome (e.g., the contribution to the molding and control of beliefs, values, and behaviors by the use of reference class forecasting for a relatively more active treatment of human rationality—but the contribution to the molding and control of beliefs, values, and behaviors by the appeal to bounded rationality for a relatively more acceptance of the limits on human rationality). And the reverse direction also holds true.

And in relation to the stability-reaction principle, if there is stability (e.g., the relatively more stable condition of cognitive biases when they are formed and internalized in an individiual's psychology), there is reaction (e.g., cognitive biases can be subject to further changes, once the individuals interact with others in the larger environment, such that some biases can be modified, some new ones are formed, or some others ones are eliminated with the application of "reference class forecasting", and so on). And the reverse direction also holds true.

Quantitative Research and Nature

Quantitative research, when examined from the perspective of nature, can show us in a wonderful way its possibility and desirability, and this can be shown by way of two case studies, namely, (2.3.1) quantitative research, confirmation bias, and paradigm shifts, and (2.3.2) quantitative research, and the limits of scientific modeling—to be addressed in what follows, respectively.

Quantitative Research, Confirmation Bias, and Paradigm Shifts

The first case study here to evaluate the possibility and desirability of quantitative research from the perspective of nature is to explore quantitative research, confirmation bias, and paradigm shifts.

Definitional Issue of "Paradigm"

The word "paradigm" needs to be clarified at the outset, as it "has been used in science to describe distinct concepts. It comes from Greek 'παράδειγμα' (*paradeigma*), 'pattern, example, sample,' from the verb 'παραδείκνυμι' (*paradeiknumi*), 'exhibit, represent, expose,' and…from 'παρά' (*para*), 'beside, by' + 'δείκνυμι' (*deiknumi*), 'to show, to point out.'" (WK 2011gg)

The word had been used in different contexts; for example, "[t]he original Greek term παραδείγματι (*paradeigma*) was used in Greek texts such as Plato's *Timaeus* (28A) as the model or the pattern that the Demiurge (god) used to create the cosmos. The term had a technical meaning in the field of grammar: the 1900 Merriam-Webster dictionary defines its technical use only in the context of grammar or, in rhetoric, as a term for an illustrative parable or fable. In linguistics, Ferdinand de Saussure used paradigm to refer to a class of elements with similarities." (WK 2011gg)

Nowadays, "the word has come to refer very often now to a thought pattern in any scientific discipline or other epistemological context. The Merriam-Webster Online dictionary [2011] defines this usage as 'a philosophical and theoretical framework of a scientific school or discipline within which theories, laws, and generalizations and the experiments performed in support of them are formulated; broadly: a philosophical or theoretical framework of any kind.'" (WK 2011gg)

Comparable Terms for "Paradigm"

The word "paradigm" has been compared with some related terms like "groupthink," "mindset," "episteme," "discourse," "mathesis," "taxinomia," "thinking inside the box," "archetype," and the like.

For instance, the word has been used in relation to terms like "*groupthink*…and…*mindset*," which "have somewhat similar meanings that apply to smaller and larger scale examples of disciplined thought. Michel Foucault used the terms *episteme* and *discourse*, *mathesis* and *taxinomia*, for aspects of a 'paradigm' in [Thomas] Kuhn's original sense." (WK 2011gg)

Another comparable term is "'the box' in the commonly used phrase 'thinking outside the box.' Thinking inside the box is analogous with normal science. The box encompasses the thinking of normal science and thus the box is analogous with paradigm. 'Thinking outside the box' would be what Kuhn calls revolutionary science. Revolutionary science is usually unsuccessful, and very rarely leads to new paradigms. However, when they are successful they lead to large scale changes in the scientific worldview. When these large scale shifts in the scientific view are implemented and accepted by the majority, it will then become 'the box' and science will progress within it." (WK 2011gg)

Still another comparable term is "archetype," which refers to "a pattern or model or an outstandingly clear or typical example"; for instance, "archetype…is frequently used in this sense in the design professions. Design Paradigms or archetypes comprise functional precedents for design solutions. The best known references on design paradigms are *Design Paradigms: A Sourcebook for Creative Visualization*, by [Warren] Wake, and *Design Paradigms* by [Henry] Petroski. In philosophy, C. S. Herrman applies the archetypal indication to methodology." (WK 2011gg)

Scientific Paradigm

A good example of this contemporary usage is the one offered by Thomas Kuhn, "when he adopted the word to refer to the set of practices that define a scientific discipline at any particular period of time. Kuhn himself came to prefer the terms exemplar and normal science, which have more precise philosophical meanings. However in his book *The Structure of Scientific Revolutions* Kuhn defines a scientific paradigm" in the following way: (WK 2011gg)

- Ex: "*what* is to be observed and scrutinized"
- Ex: "the kind of *questions* that are supposed to be asked and probed for answers in relation to this subject"
- Ex: "*how* these questions are to be structured"
- Ex: "*how* the results of scientific investigations should be interpreted"
- Ex: "*how* is an experiment to be conducted, and what equipment is available to conduct the experiment"

In this way, "within normal science, the paradigm is the set of exemplary experiments that are likely to be copied or emulated. In this scientific context, the prevailing paradigm often represents a more specific way of viewing reality, or limitations on acceptable programs for future research, than the more general scientific method." (WK 2011gg)

For example, "the standard model of physics" illustrates this paradigm in action, since, although "[t]he scientific method would allow for orthodox scientific investigations into phenomena which might contradict or disprove the standard model,...grant funding would be proportionately more difficult to obtain for such experiments, depending on the degree of deviation from the accepted standard model theory which the experiment would be expected to test for. To illustrate the point, an experiment to test for the mass of neutrinos or the decay of protons (small departures from the model) would be more likely to receive money than experiments to look for the violation of the conservation of momentum, or ways to engineer reverse time travel." (WK 2011gg)

For Kuhn, there are different paradigms in different historical eras, and "the paradigms are incommensurable, meaning two paradigms cannot be reconciled with each other because they cannot be subjected to the same common standard of comparison. That is, no meaningful comparison between them is possible without fundamental modification of the concepts that are an intrinsic part of the paradigms being compared." (WK 2011gg)

However, there is a problem in "this way of looking at the concept of 'paradigm,'" because it "creates a paradox of sorts, since competing paradigms are in fact constantly being measured against each other....Nonetheless, competing paradigms are not fully intelligible solely within the context of their own conceptual frameworks....For this reason, paradigm as a concept in the philosophy of science might more meaningfully be defined as a self-reliant explanatory model or conceptual framework. This definition makes it clear that the real barrier to comparison is not necessarily the absence of common units of measurement, but an absence of mutually compatible or mutually intelligible concepts. Under this system, a new paradigm which replaces an old paradigm is not necessarily better, because the criteria of judgment are controlled by the paradigm itself, and by the conceptual framework which defines the paradigm and gives it its explanatory value." (WK 2011gg; M. Slattery 2003; T. Nickles 2002)

Paradigm Shifts

In a dynamic sense, paradigm can be understood in terms of "paradigm shifts." For instance, in *The Structure of Scientific Revolutions*, Kuhn argued that "Successive transition from one paradigm to another via revolution is the usual developmental pattern of mature science." (WK 2011gg)

Thus, "paradigm shifts tend to be most dramatic in sciences that appear to be stable and mature, as in physics at the end of the 19th century. At that time, a statement generally attributed to physicist Lord Kelvin famously claimed, 'There is nothing new to be discovered in physics now. All that remains is more and more precise measurement.' Five years later, Albert Einstein published his paper on special relativity, which challenged the very simple set of rules laid down by Newtonian mechanics, which had been used to describe force and motion for over two hundred years. In this case, the new paradigm reduces the old to a special case in the sense that Newtonian mechanics is still a good model for approximation for speeds that are slow compared to the speed of light. Philosophers and historians of science, including Kuhn himself, ultimately accepted a modified version of Kuhn's model, which synthesizes his original view with the gradualist model that preceded it. Kuhn's original model is now generally seen as too limited." (WK 2011gg)

However, "Kuhn's idea was itself revolutionary in its time, as it caused a major change in the way that academics talk about science. Thus, it could be argued that it caused or was itself part of a 'paradigm shift' in the history and sociology of science. However, Kuhn would not recognize such a paradigm shift. Being in the social sciences, people can still use earlier ideas to discuss the history of science." (WK 2011gg)

Paradigm Paralysis and Confirmation Bias in Data Analysis

This discussion of "paradigm" is important for the field of data analysis, especially in relation to "confirmation bias" in quantitative research (as already introduced in *Sec. 1.3* and *Sec. 1.4*).

A good example is the idea of "paradigm paralysis," which refers to "the inability or refusal to see beyond the current models of thinking. This is similar to what psychologists term *confirmation bias*." (WK 2011gg; J. Harrison 1994)

Paradigm paralysis (or confirmation bias in the current context of data analysis) thus constitutes "the greatest barrier to a paradigm shift," and good examples include "the rejection of Galileo's theory of a heliocentric universe, the discovery of electrostatic photography, xerography and the quartz clock." (WK 2011gg)

Problems with the Study of Paradigm

However, the study of paradigm so understood (to challenge confirmation bias in data analysis) is not without its own problems. Consider, for illustration, some major criticisms below.

Firstly, one major criticism of paradigm study is that it can degeneratee into a form of relativism, with Kuhn's work as a good example.

In reply, Kuhn responded to this critique in an essay titled "Objectivity, Value Judgment, and Theory Choice" (1977), in which "he reiterates five criteria…that…help determine…theory choice," as shown below: (WK 2011hh)

- "Accurate"—empirically adequate with experimentation and observation"
- "Consistent"—"internally consistent, but also externally consistent with other theories"
- "Broad Scope"—"a theory's consequences should extend beyond that which it was initially designed to explain"
- "Simple"—"the simplest explanation, principally similar to Occam's razor"
- "Fruitful"—"a theory should disclose new phenomena or new relationships among phenomena"

But Kuhn's response did not satisfy his critics, since he also acknowledged that "these criteria…are imprecise in practice and relative to individual scientists. According to Kuhn, 'When scientists must choose between competing theories, two men fully committed to the same list of criteria for choice may nevertheless reach different conclusions.'" (WK 2011hh)

Thus, for the critics, "the criteria still are not 'objective' in the usual sense of the word because individual scientists reach different conclusions with the same criteria due to valuing one criterion over another or even adding additional criteria for selfish or other subjective reasons." (WK 2011hh)

And secondly, another major criticism of paradigm study is that its validity is limited, as it does not quite fit well the situations in the socal sciences.

In fact, even "Kuhn himself did not consider the concept of paradigm as appropriate for the social sciences….He wrote this book at the Palo Alto Center for Scholars, surrounded by social scientists, when he observed that they were never in agreement on theories or concepts. He explains that he wrote this book precisely to show that there are no, nor can there be any, paradigms in the social sciences." (WK 2011hh)

Beyond Paradigm Shifts

These criticisms of paradigm study, however, do not suggest its total invalidity to challenge the confirmation bias in data analysis but are to show us the two opposing sides of the debate, such that the possibility and desirability of quantitative research (from the perspective of nature with quantitative research, confirmation bias, and paradign shifts as a case study here) are not to the extent that the respective defenders would like us to believe.

In addition, the analysis of quantitative research, confirmation bias, and paradign shifts can reveal valuable info about the ontological principles in existential dialectics, and good examples include the formalness-informalness principle, the absoluteness-relativeness principle, the partiality-totality principle, the predictability-unpredictability principle, the explicability-inexplicability principle, the fiction-reality principle, the preciseness-vagueness principle, the simpleness-complicatedness principle, the openness-hiddenness principle, the denseness-emptiness principle, the change-constancy principle, the order-chaos principle, the slowness-quickness principle, the expansion-contraction principle, the theory-praxis principle, the convention-novelty principle, the evolution-transformation principle, the symmetry-asymmetry principle, the softness-hardness principle, the regression-progression principle, the same-difference principle, and the stability-reaction principle.

For instance, in relation to the formalness-informalness principle, if there is formalness (e.g., the formal logical requirement of completeness in a system of ideas, as a "paradigm"), there is informalness (e.g., the non-formal existence of incompleteness in a "paradigm," since a paradigm cannot justify itself to be "better" than an old one, because "the paradigms are incommensurable" and "no meaningful comparison between them is possible without fundamental modification of the concepts that are an intrinsic part of the paradigms being compared"). And the reverse direction also holds true.

In relation to the absoluteness-relativeness principle, if there is absoluteness (e.g., the absolute view within each paradigm in relation to its own worldview), there is relativeness (e.g., what is the right worldview within an older paradigm is not necessarily so for a newer paradigm which is to replace it later). And the reverse direction also holds true.

In relation to the partiality-totality principle, if there is partiality (e.g., the partial view within each paradigm in regard to its own worldview), there is totality (e.g., the more holistic view of different paradigms over time, in the history of "paradigm shifts," such that the whole is not the sum of its parts, that is, the implications of the history of "paradigm shifts," which are still subject to a heated debate among historians of science, are not reducible to the sum of the individual paradigms). And the reverse direction also holds true.

In relation to the predictability-unpredictability principle, if there is predictability (e.g., the predictable tendency of a paradigm to resolve some unresolved problems that an old paradigm cannot), there is unpredictability (e.g., the more difficult task to predict exactly how long a new paradigm will last, before it is to be replaced by another one later). And the reverse direction also holds true.

In relation to the explicability-inexplicability principle, if there is explicability (e.g., the explanation by Kuhn that its work on paradigm shifts does not suggest relativism, since "he reiterates five criteria…that…help determine…theory choice"), there is inexplicability (e.g., the lack of sufficient explanation by Kuhn, since, for his critics, "the criteria still are not 'objective' in the usual sense of the word because individual scientists reach different conclusions with the same criteria due to valuing one criterion over another or even adding additional criteria for selfish or other subjective reasons"). And the reverse direction also holds true.

In relation to the fiction-reality principle, if there is fiction (e.g., the fictional aspect of a paradigm, as shown in "paradigm paralysis," which refers to "the inability or refusal to see beyond the current models of thinking," that is, the inability to see what is wrong with one's worldview), there is reality (e.g., the realistic aspect of a paradigm, in that it can resolve some unresolved problems that an old paradigm cannot). And the reverse direction also holds true.

In relation to the preciseness-vagueness principle, if there is preciseness (e.g., the precise identification of the "five criteria…that…help determine…theory choice" by Kuhn), there is vagueness (e.g., the vagueness in the identification of the five criteria, since it is not clear why there must be only five criteria or whether or not they should be valued equally, as, for the critics, "individual scientists reach different conclusions with the same criteria due to valuing one criterion over another or even adding additional criteria for selfish or other subjective reasons"). And the reverse direction also holds true.

In relation to the simpleness-complicatedness principle, if there is simpleness (e.g., the relatively simple analysis of "paradigm" by Kuhn), there is complicatedness (e.g., the relatively more complicated analysis of "paradigm" by challenging the claims and assumptions, as shown in the criticisms pointed out by the critics as cited). And the reverse direction also holds true.

In relation to the openness-hiddenness principle, if there is openness (e.g., the open exploration of "new phenomena or new relationships among phenomena" by a new paradigm), there is hiddenness (e.g., the hidden bias in a new paradigm, because of its tendency to be eventually consolidated inside "the box"). And the reverse direction also holds true.

In relation to the denseness-emptiness principle, if there is denseness (e.g., the relatively denser concentration of "grant funding" in "experiments" which do not deviate too much from "the standard model of physics"), there is emptiness (e.g., the relatively less dense, or more empty, concentration of "grant funding" in "experiments" which deviate greatly from "the standard model of physics"). And the reverse direction also holds true.

In relation to the change-constancy principle, if there is change (e.g., the ever new paradigms to emerge in science over time), there is constancy (e.g., the ever constant existence of "paradigm shifts" in science over time). And the reverse direction also holds true.

In relation to the order-chaos principle, if there is order (e.g., in cybernetics, the use of "protoprogramme for reducing the chaotic mass to some form of order," as "[a] paradigm there would be a sort of prohibition to proceed with any action that would increase the total entropy of the system"), there is chaos (e.g., the result in chaotic mass, when there is no "paradigm," that is, when there is no "prohibition to proceed with any action that would increase the total entropy of the system"). (WK 2011gg) And the reverse direction also holds true.

In relation to the slowness-quickness principle, if there is slowness (e.g., the relatively slower readiness of a scientist to question the existing paradigm in the stage of "normal science," where the paradigm dominates science), there is quickness (e.g., the relatively quicker readiness of a scientist to question the existing paradigm and to look for a new paradigm in the stage of "revolutionary science," where the existing paradigm is in crisis). (WK 2011hh) And the reverse direction also holds true.

In relation to the expansion-contraction principle, if there is expansion (e.g., the relatively more developed ability of a scientist who accepts a new paradigm in normal science to solve unresolved problems that an older paradigm could not), there is contraction (e.g., the relatively less developed ability of a scientist to solve new anomalous or unresolved problems that the existing paradigm cannot, until another new paradigm in revolutionary science emerges later). And the reverse direction also holds true.

In relation to the theory-praxis principle, if there is theory (e.g., the theoretical construction of "paradigm" study by Kuhn and others), there is praxis (e.g.,the practical application of Kuhn's idea to the field of the social sciences, as Kuhn was honored "as a foundational force behind the post-Mertonian Sociology of Scientific Knowledge"). (WK 2011hh) And the reverse direction also holds true.

In relation to the convention-novelty principle, if there is convention (e.g., the conventional wisdom, before Kuhn's work, about the quantitiative modeling of science as helping us understand reality "better" over time), there is novelty (e.g., the alternative novel challenge to the conventional wisdom about the progress in the quantitative modeling of science, by Kuhn's new idea that "a new paradigm which replaces an old paradigm is not necessarily better, because the criteria of judgment are controlled by the paradigm itself, and by the conceptual framework which defines the paradigm and gives it its explanatory value"). And the reverse direction also holds true.

In relation to the evolution-transformation principle, if there is evolution (e.g., the natural evolution of humans to study nature in the state of nature, albeit in a primitive way), there is transformation (e.g., the technical transformation of human ability to study nature by the invention of systemtic fields of knowledge, like quantitative research, philosophy of science, and sociology of knowledge). And the reverse direction also holds true.

In relation to the symmetry-asymmetry principle, if there is symmetry (e.g., the existence of different views about the scientifiic way of thinking), there is asymmetry (e.g., the popularity of the term "Scientific Revolution" in the singular "in the late Renaissance"—but the popularity of "the term 'scientific revolutions' in the plural, taking place at widely different periods of time and in different disciplines" nowadays, because of the work of Kuhn and others). And the reverse direction also holds true.

In relation to the softness-hardness principle, if there is softness (e.g., the praise of Kuhn's work, as shown in honoring him "as a foundational force behind the post-Mertonian Sociology of Scientific Knowledge"), there is hardness (e.g., the critique of Kuhn's work, as shown by the criticism of "relativism" pointed out by the critics as cited). And the reverse direction also holds true.

In relation to the regression-progression principle, if there is regression (e.g., the regression made by Kuhn's work, as shown in the problems pointed out by the critics), there is progression (e.g., the progress made by Kuhn's work, as shown in "the enormous impact of Kuhn's work" in "the philosophy of science"). (WK 2011hh) And the reverse direction also holds true.

In relation to the same-difference principle, if there is similarity in outcome (e.g., the contribution to the molding and control of beliefs, values, and behaviors, regardless of whether this be done by the belief in a logic of discovery, as in modern science before Kuhn's work, or by the belief in a psychology of discovery, as in Kuhn's work), there is difference in outcome (e.g., the contribution to the molding and control of beliefs, values, and behaviors by the belief in a logic of discovery for a relatively more progressive worldview—but the contribution to the molding and control of beliefs, values, and behaviors by the belief in a psychology of discovery for a relatively more relativist worldview). And the reverse direction also holds true.

And in relation to the stability-reaction principle, if there is stability (e.g., the relatively more stable condition of "normal science" when its problem-solving ability is enhanced by the acceptance of a central, dominant paradigm), there is reaction (e.g., normal science can be subject to further changes, as it increasingly shifts to the stage of "revolutionary science," with the accumulation of "anomalous problems"). (WK 2011hh) And the reverse direction also holds true.

Quantitative Research, and the Limits of Scientific Modeling

Another case study here to evaluate the possibility and desirability of quantitative research from the perspective of nature is to explore quantitative research and the limits of scientific modeling.

The term "scientific modeling" refers to "the process of generating abstract, conceptual, graphical and/or mathematical models. Science offers a growing collection of methods, techniques and theory about all kinds of specialized scientific modeling. A scientific model can provide a way to read elements easily which have been broken down to a simpler form." (WK 2011jj)

Basics of Scientific Modeling

Although "there is little general theory about scientific modeling," it can be summarized in terms of some basic aspects, as addressed below (and outlined in *Table 2.3*).

(a) Language

The first basic aspect of scientific modeling to be introduced here concerns "language" (or more precisely, "modeling language").

By definition, "a modeling language is any artificial language that can be used to express information or knowledge or systems in a structure that is defined by a consistent set of rules. The rules are used for interpretation of the meaning of components in the structure." (WK 2011jj)

Good examples of modeling languages include "the Architecture Description Language (ADL) for description and representation of system architectures, Unified Modeling Language (UML) for software systems, IDEF for processes, and the VRML for 3-D computer graphics models designed particularly with the World Wide Web in mind." (WK 2011jj)

(b) Simulation

The second basic aspect of scientific modeling to be introduced here concerns "simulation."

By definitioin, "a simulation is the implementation of a model." (WK 2011jj)

Simulation can be static or dynamic: "A steady state simulation provides information about the system at a specific instant in time (usually at equilibrium, if it exists). A dynamic simulation provides information over time. A simulation brings a model to life and shows how a particular object or phenomenon will behave. It is useful for testing, analysis or training where real-world systems or concepts can be represented by a model." (WK 2011jj)

(c) Structure

The third basic aspect of scientific modeling to be introduced here concerns "structure."

By definition, "structure is a fundamental and sometimes intangible notion covering the recognition, observation, nature, and stability of patterns and relationships of entities. From a child's verbal description of a snowflake, to the detailed scientific analysis of the properties of magnetic fields, the concept of structure is an essential foundation of nearly every mode of inquiry and discovery in science" and, for that matter, non-science. (WK 2011jj; W. Pullan 2000)

(d) Systems

The fourth basic aspect of scientific modeling to be introduced here concerns "systems."

By definition, "a system is a set of interacting or interdependent entities, real or abstract, forming an integrated whole. In general, a system is a construct or collection of different elements that together can produce results not obtainable by the elements alone." (WK 2011jj; P. Fishwick 1995)

In fact, "the concept of an 'integrated whole' can also be stated in terms of a system embodying a set of relationships which are differentiated from relationships of the set to other elements, and from relationships between an element of the set and elements not a part of the relational regime. There are two types of systems: [a] discrete in which the variables change instantaneously at separate points in time and [b] continuous where the state variables change continuously with respect to time." (WK 2011jj; J. Sokolowski 2009)

(e) The Process of Generating a Model

The fifth basic aspect of scientific modeling to be introduced here concerns "the process of generating a model."

By definition, the process of generating a model refers to "a conceptual representation of some phenomenon. Typically a model will refer only to some aspects of the phenomenon in question, and two models of the same phenomenon may be essentially different, that is in which the difference is more than just a simple renaming."(WK 2011jj)

This difference "may be due to differing requirements of the model's end users or to conceptual or aesthetic differences by the modellers and decisions made during the modelling process. Aesthetic considerations that may influence the structure of a model might be the modeler's preference for a reduced ontology, preferences regarding probabilistic models vis-a-vis deterministic ones, discrete vs continuous time etc. For this reason users of a model need to understand the model's original purpose and the assumptions of its validity." (WK 2011jj)

In mathematical modeling, for example, a modeler has to decide the following choices: (WK 2011kk)

- "Linear vs. Nonlinear"—"Nonlinearity, even in fairly simple systems, is often associated with phenomena such as chaos and irreversibility. Although there are exceptions, nonlinear systems and models tend to be more difficult to study than linear ones. A common approach to nonlinear problems is linearization, but this can be problematic if one is trying to study aspects such as irreversibility, which are strongly tied to nonlinearity."

- "Deterministic vs. Probabilistic (Stochastic)"—"A deterministic model is one in which every set of variable states is uniquely determined by parameters in the model and by sets of previous states of these variables....Conversely, in a stochastic model, randomness is present, and variable states are not described by unique values, but rather by probability distributions."
- "Static vs. Dynamic"—"A static model does not account for the element of time, while a dynamic model does. Dynamic models typically are represented with difference equations or differential equations."
- "Discrete vs. Continuous"—"A discrete model does not take into account the function of time and usually uses time-advance methods, while a Continuous model does. Continuous models typically are represented with f(t) and the changes are reflected over continuous time intervals."

Of course, there can be other choices, so the list above is illustrative (not exhaustive).

(f) The Process of Evaluating a Model

The sixth basic aspect of scientific modeling to be introduced here concerns "the process of evaluating a model."

The process of evaluating a model can have different criteria. For instance, "a model is evaluated first and foremost by its consistency to empirical data; any model inconsistent with reproducible observations must be modified or rejected. However, a fit to empirical data alone is not sufficient for a model to be accepted as valid." (WK 2011jj)

In general, the following factors are important in evaluating a model: (WK 2011jj)

- "Ability to explain past observations"
- "Consistency to empirical data"
- "Ability to predict future observations"
- "Cost of use, especially in combination with other models"
- "Refutability, enabling estimation of the degree of confidence in the model"
- "Simplicity, or even aesthetic appeal"

Of course, there can be other factors, so the list above is illustrative (not exhaustive).

(g) Visualization

And the seventh basic aspect of scientific modeling to be introduced here concerns "visualization."

By definition, "visualization is any technique for creating images, diagrams, or animations to communicate a message. Visualization through visual imagery has been an effective way to communicate both abstract and concrete ideas since the dawn of man." (WK 2011jj)

Good examples of visualization include "cave paintings, Egyptian hieroglyphs, Greek geometry, and Leonardo da Vinci's revolutionary methods of technical drawing for engineering and scientific purposes." (WK 2011jj)

Limits of Scientific Modeling

However, scientific modeling has its inherent limits. Consider, for illustration, some main limits below (and summarized in *Table 2.3*).

(a) Simplification

The first limit of scientific modeling to be introduced here is simplification.

After all, "[m]odels are typically used when it is either impossible or impractical to create experimental conditions in which scientists can directly measure outcomes. Direct measurement of outcomes under controlled conditions will always be more accurate than modeled estimates of outcomes. When predicting outcomes, models use assumptions, while measurements do not....As the number of assumptions in a model increases, the accuracy and relevance of the model will likely diminish." (WK 2011jj)

This use of assumptions then means that "[a]ll models are in simulacra, that is, simplified reflections of reality, but despite their inherent falsity, they are nevertheless extremely useful. Building and disputing models is fundamental to the scientific enterprise. Complete and true representation may be impossible, but scientific debate often concerns which is the better model for a given task, such as the most accurate climate model for seasonal forecasting." (WK 2011jj; G. Box 1987)

This does not mean that complexity should be ignored in scientific modeling, since the point here is that "model complexity involves a trade-off between simplicity and accuracy of the model. Occam's razor is a principle particularly relevant to modelling; the essential idea being that among models with roughly equal predictive power, the simplest one is the most desirable. While added complexity usually improves the realism of a model, it can make the model difficult to understand and analyze, and can also pose computational problems, including numerical instability. Thomas Kuhn argues that as science progresses, explanations tend to become more complex before a paradigm shift offers radical simplification." (WK 2011kk)

In the end, the aim of scientific modeling is "to construct a formal system for which reality is the only interpretation. The world is an interpretation (or model) of these sciences, only insofar as these sciences are true. For the scientist, a model is also a way in which the human thought processes can be amplified. For instance, models that are rendered in software allow scientists to leverage computational power to simulate, visualize, manipulate and gain intuition about the entity, phenomenon or process being represented. Such computer models are *in silico*. Other types of scientific model are *in vivo* (living models, such as laboratory rats) and *in vitro* (in glassware, such as tissue culture)." (WK 2011jj; H. Freudenthal 1951; C. Churchman 1968; E. Griffiths 2009)

(b) Subjectivity

The second limit of scientific modeling to be introduced here is subjectivity.

Scientific modeling can also use subjective information, especially when necessarily information for modeling is not available. For instance, "[s]ometimes it is useful to incorporate subjective information into a mathematical model. This can be done based on intuition, experience, or expert opinion, or based on convenience of mathematical form." (WK 2011kk)

A good example is the use of "Bayesian statistics" which "provides a theoretical framework for incorporating such subjectivity into a rigorous analysis: one specifies a prior probability distribution (which can be subjective) and then updates this distribution based on empirical data. An example of when such approach would be necessary is a situation in which an experimenter bends a coin slightly and tosses it once, recording whether it comes up heads, and is then given the task of predicting the probability that the next flip comes up heads. After bending the coin, the true probability that the coin will come up heads is unknown, so the experimenter would need to make an arbitrary decision (perhaps by looking at the shape of the coin) about what prior distribution to use. Incorporation of the subjective information is necessary in this case to get an accurate prediction of the probability, since otherwise one would guess 1 or 0 as the probability of the next flip being heads, which would be almost certainly wrong." (WK 2011kk; D. MacKay 2003)

(c) Extrapolation

The third limit of scientific modeling to be introduced here is extrapolation.

The problem of extrapolation is an issue about "the scope of a model, that is, determining what situations the model is applicable to….If the model was constructed based on a set of data, one must determine for which systems or situations the known data is a "typical" set of data." (WK 2011kk)

The problem of "extrapolation" should not be confused with that of "interpolation," which refers to "the question of whether the model describes well the properties of the system between data points"—whereas "extrapolation" refers to "the same question for events or data points outside the observed data." (WK 2011kk)

For instance, "in evaluating Newtonian classical mechanics, we can note that Newton made his measurements without advanced equipment, so he could not measure properties of particles travelling at speeds close to the speed of light. Likewise, he did not measure the movements of molecules and other small particles, but macro particles only. It is then not surprising that his model does not extrapolate well into these domains, even though his model is quite sufficient for ordinary life physics." (WK 2011kk)

(d) Causality

And the fourth limit of scientific modeling to be introduced here is causality or cause-effect relationship (not just correlation).

As the old saying goes, "correlation does not imply causation." (WK 2011mm) The problem of causality here is also related to the issue of extrapolation (although the two are not the same), because true causality between two variables makes the task of extrapolation (of past data for future events outside them) easier.

In any event, "many types of modelling implicitly involve claims about causality….As the purpose of modelling is to increase our understanding of the world, the validity of a model rests not only on its fit to empirical observations, but also on its ability to extrapolate to situations or data beyond those originally described in the model. One can argue that a model is worthless unless it provides some insight which goes beyond what is already known from direct investigation of the phenomenon being studied." (WK 2011kk)

A good example of this limit concerns "the argument that the mathematical models of Optimal foraging theory do not offer insight that goes beyond the common-sense conclusions of evolution and other basic principles of ecology." (WK 2011kk; G. Pyke 1984)

Beyond Scientific Modeling

These criticisms of scientific modeling should not be misunderstood as a total rejection of its usefulness but serve to show us the two opposing sides of the debate, such that the possibility and desirability of quantitative research (from the perspective of nature with quantitative research and the limits of scientific modeling as a case study here) are not to the extent that the respective defenders would like us to believe.

Furthermore, the analysis of quantitative research and the limits of scientific modeling can teach us something important about the ontological principles in existential dialectics, and good examples include the formalness-informalness principle, the absoluteness-relativeness principle, the partiality-totality principle, the predictability-unpredictability principle, the explicability-inexplicability principle, the fiction-reality principle, the finiteness-transfiniteness principle, the preciseness-vagueness principle, the simpleness-complicatedness principle, the openness-hiddenness principle, the denseness-emptiness principle, the change-constancy principle, the order-chaos principle, the slowness-quickness principle, the expansion-contraction principle, the theory-praxis principle, the convention-novelty principle, the evolution-transformation principle, the symmetry-asymmetry principle, the softness-hardness principle, the seriousness-playfulness principle, the regression-progression principle, the same-difference principle, and the stability-reaction principle.

For instance, in relation to the formalness-informalness principle, if there is formalness (e.g., the formal logical requirement of soundness in a system of ideas, as scientific modeling), there is informalness (e.g., the non-formal existence of unsoundness in scientific modeling, as shown in the problem of subjectivity in scientifiic modeling especially when necessary information is not available). And the reverse direction also holds true.

In relation to the absoluteness-relativeness principle, if there is absoluteness (e.g., the absolute view of a scientific modeling which is "deterministic"), there is relativeness (e.g., what is right for a scientific modeling which is "deterministic" is not necessarily so for another which is "probabilistic" instead). And the reverse direction also holds true.

In relation to the partiality-totality principle, if there is partiality (e.g., the partial view of a specific element in "a system"), there is totality (e.g., the more holistic view of the entire "collection of different elements" in a system, such that the whole is not the sum of its parts, that is, the whole collection with their multifaceted relationships "can produce results not obtainable by the elements alone" or the mere summation of them). And the reverse direction also holds true.

In relation to the predictability-unpredictability principle, if there is predictability (e.g., it is easier to do "interpolation"), there is unpredictability (e.g., it is more difficult to do "extrapolation"). And the reverse direction also holds true.

In relation to the explicability-inexplicability principle, if there is explicability (e.g., the explanation of the relationship between two different variables in a scientific modeling on the basis of "correlation"), there is inexplicability (e.g., the more difficult task to explain "causality" between two different variables in a scientific modeling, as shown in the problem of inability to extrapolate in scientific modeling, insofar as the causal relation is not well understood). And the reverse direction also holds true.

In relation to the fiction-reality principle, if there is fiction (e.g., the fictional aspect of a scientific modeling when it relies on incorrect subjective information), there is reality (e.g., the realistic aspect of a scientific modeling when it uses correct information). And the reverse direction also holds true.

In relation to the finiteness-transfiniteness principle, if there is finiteness (e.g., the finite number of separate points in some "discrete" scientific systems), there is transfiniteness (e.g., the transfinite number of different points in some "continuous" scientific systems). And the reverse direction also holds true.

In relation to the preciseness-vagueness principle, if there is preciseness (e.g., the precise identification of six factors to evaluate a scientific model aforecited), there is vagueness (e.g., the vagueness in the identification of the six factors, since it is not clear why there must be only six, not seven, eight, and so on). And the reverse direction also holds true.

In relation to the simpleness-complicatedness principle, if there is simpleness (e.g., for Thomas Kuhn, "a paradigm shift offers radical simplification" of scientific explanations), there is complicatedness (e.g., "Kuhn argues that as science progresses, explanations tend to become more complex, before a paradigm shift offers radical simplification"). And the reverse direction also holds true.

In relation to the openness-hiddenness principle, if there is openness (e.g., the open exploration of the possibility of doing scientific modeling without doing experiments), there is hiddenness (e.g., the hidden bias in scientific modeling, because of its "simplified reflections of reality"). And the reverse direction also holds true.

In relation to the denseness-emptiness principle, if there is denseness (e.g., the relatively denser concentration of "subjective" information in Bayesian statistics), there is emptiness (e.g., the relatively less dense, or more empty, concentration of "subjective" information in frequency-based statistics). (WK 2011ll) And the reverse direction also holds true.

In relation to the change-constancy principle, if there is change (e.g., the ever new ways to do science over time, as in the use of "scientific modeling"), there is constancy (e.g., the ever constant existence of different problems in doing science over time, as shown in the problems of scientific modeling pointed out by the critics). And the reverse direction also holds true.

In relation to the order-chaos principle, if there is order (e.g., the more or less orderly reality to be described by a scientific modeling which is "linear"), there is chaos (e.g,. the more or less chaotic reality to be described by a scientific modeling which is "nonlinear").

In relation to the slowness-quickness principle, if there is slowness (e.g., the relatively slower speed of a scientific model to solve computatonal problems if it adds "complexity" into the system to "improve…realism"), there is quickness (e.g., the relatively quicker speed of a scientific model to solve computational problems if it only requires "simplicity" in the system for expediency). And the reverse direction also holds true.

In relation to the expansion-contraction principle, if there is expansion (e.g., the relatively more developed ability of a scientist to deal with realism if he uses "complexity" in a system), there is contraction (e.g., the relatively less developed ability of a scientist to deal with expediency if he uses "complexity" in a system). And the reverse direction also holds true.

In relation to the theory-praxis principle, if there is theory (e.g., the theoretical construction of the basics of scientific modeling), there is praxis (e.g.,the practical application of the basics of scientific modeling to the field of "Modelling and Simulation" for the understanding of "Defence capability development process"). (WK 2011jj) And the reverse direction also holds true.

In relation to the convention-novelty principle, if there is convention (e.g., the conventional wisdom of doing science by experiments), there is novelty (e.g., the alternative novel challenge to the conventional wisdom about doing experiments in science, by the new idea of doing science by scientific modeling instead). And the reverse direction also holds true.

In relation to the evolution-transformation principle, if there is evolution (e.g., the natural evolution of humans to observe, act, and understand reality in the state of nature in a primitive way), there is transformation (e.g., the technical transformation of human ability to understand nature by the invention of scientific modeling). And the reverse direction also holds true.

In relation to the symmetry-asymmetry principle, if there is symmetry (e.g., the existence of different ways to do science), there is asymmetry (e.g., the popularity of doing science by experiments in the older days—but the popularity of doing science by modeling nowadays, especially with the advance in computers). And the reverse direction also holds true.

In relation to the softness-hardness principle, if there is softness (e.g., the appreciation of scientific modeling, as shown in its use of "Modelling and Simulation" for the understanding of "Defence capability development process"), there is hardness (e.g., the critique of scientific modeling, as shown in the problems pointed out by the critics as cited). And the reverse direction also holds true.

In relation to the seriousness-playfulness principle, if there is seriousness (e.g., the serious business of a scientist in modeling to "represent empirical objects, phenomena, and physical processes in a logical and objective way"), there is playfulness (e.g., the playful part of a scientist in modeling, when he plays with different factors like "simplicity," "aesthetic appeal," "user friendliness," etc. to evaluate a model).

In relation to the regression-progression principle, if there is regression (e.g., the regression made by scientific modeling, as shown in the limits pointed out by the critics), there is progression (e.g., the progress made by scientific modeling, as shown in the useful applications aforecited). And the reverse direction also holds true.

In relation to the same-difference principle, if there is similarity in outcome (e.g., the contribution to the molding and control of beliefs, values, and behaviors, regardless of whether this be done by way of scientific modeling, or by way of scientific experiment), there is difference in outcome (e.g., the contribution to the molding and control of beliefs, values, and behaviors by way of scientific modeling for a relatively more simplified worldview—but the contribution to the molding and control of beliefs, values, and behaviors by way of scientific experiment for a relatively more realistic worldview). And the reverse direction also holds true.

And in relation to the stability-reaction principle, if there is stability (e.g., the relatively more stable condition of a paradigm when, for Kuhn, it can offer "explanations" which "tend to become more complex" as science progresses), there is reaction (e.g., a paradigm can lead to further changes for simplification, as "Kuhn argues that as science progresses,…a paradigm shift [towards a new paradigm] offers radical simplification"). And the reverse direction also holds true.

Quantitative Research and Society

Quantitative research, when examined from the perspective of society, can illuminate powerful aspects of its possibility and desirability, and this can be shown by way of two case studies, namely, (2.4.1) quantitative research, and the problem of portfolio analytics, and (2.4.2) quantitative research, and the dispute over social statistics—to be addressed in what follows, in that order.

Quantitative Research, and the Problem of Portfolio Analytics

One good case study here to evaluate the possibility and desirability of quantitative research from the perspective of society is to examine quantitative research and the problem of portfolio analytics.

The term "analytics" refers to "the application of computer technology, operational research, and statistics to solve problems in business and industry....Business managers may choose to make decisions based on past experiences or rules of thumb, or there might be other qualitative aspects to decision making; but unless there are data involved in the process, it would not be considered analytics." (WK 2011nn)

Alhough "analytics closely resembles statistical analysis and data mining," it "tends to be based on modeling involving extensive computation." (WK 2011nn)

A good illustration of analytics in business and industry (or "business analytics" in short) concerns the mathematical modeling of finance, known as "portfolio analytics." (WK 2011nn)

The Promise of Portfolio Analytics

Portfolio analytics is based "modern portfolio theory (MPT)" which was introduced by Harry Markowitz in a 1952 article titled "Portfolio Selection" and in a 1959 book with the same title. (WK 2011qq)

The main logic behind MPT is that "the assets in an investment portfolio should not be selected individually, each on their own merits. Rather, it is important to consider how each asset changes in price relative to how every other asset in the portfolio changes in price. Investing is a tradeoff between risk and expected return. In general, assets with higher expected returns are riskier. For a given amount of risk, MPT describes how to select a portfolio with the highest possible expected return. Or, for a given expected return, MPT explains how to select a portfolio with the lowest possible risk." (WK 2011qq; E. Elton 1997)

MPT is therefore a form of diversification based on mathematical modeling; for instance, "MPT models an asset's return as a normally distributed function…, defines risk as the standard deviation of return, and models a portfolio as a weighted combination of assets so that the return of a portfolio is the weighted combination of the assets' returns. By combining different assets whose returns are not perfectly positively correlated, MPT seeks to reduce the total variance of the portfolio return." (WK 2011qq)

Applications of Portfolio Analytics

Portfolio analytics can be applied to both financial and non-financial fields, as shown in the examples below: (WK 2011qq; S. Chandra 2007)

- In Finance—MPT suggests that "an investor can reduce portfolio risk simply by holding combinations of instruments which are not perfectly positively correlated (correlation coefficient $-1 \leq p_{ij} < 1$). In other words, investors can reduce their exposure to individual asset risk by holding a diversified portfolio of assets. Diversification may allow for the same portfolio expected return with reduced risk. If all the asset pairs have correlations of 0—they are perfectly uncorrelated—the portfolio's return variance is the sum over all assets of the square of the fraction held in the asset times the asset's return variance (and the portfolio standard deviation is the square root of this sum)."
- In Information Science—MPT models "the uncertainty and correlation between documents in information retrieval. Given a query, the aim is to maximize the overall relevance of a ranked list of documents and at the same time minimize the overall uncertainty of the ranked list."
- In Micro-Sociology—MPT suggests that "[w]hen the self attributes comprising the self-concept constitute a well-diversified portfolio, then psychological outcomes at the level of the individual such as mood and self-esteem should be more stable than when the self-concept is undiversified. This prediction has been confirmed in studies involving human subjects."

Surely, there can be other applications, so the examples above are illustrative (not exhaustive).

Criticisms of Portfolio Analytics

However, MPT has been subject to a lot of criticisms over the years. Consider, for illustration, a few criticisms below, which are organized on the basis of three categories: (WK 2011qq; B. Mandelbrot 2004)

- MPT Uses Questionable Assumptions
  - –Assumption: "Asset returns are (jointly) normally distributed random variables." Critique: "In fact, it is frequently observed that returns in equity and other markets are not normally distributed. Large swings (3 to 6 standard deviations from the mean) occur in the market far more frequently than the normal distribution assumption would predict."
  - –Assumption: "Correlations between assets are fixed and constant forever." Critique: "Correlations depend on systemic relationships between the underlying assets, and change when these relationships change....During times of financial crisis all assets tend to become positively correlated, because they all move (down) together. In other words, MPT breaks down precisely when investors are most in need of protection from risk."
  - –Assumption: "All investors aim to maximize economic utility (in other words, to make as much money as possible, regardless of any other considerations)." Critique: Investors have other concerns, and "this is a key assumption of the efficient market hypothesis, upon which MPT relies."
  - –Assumption: "All investors are rational and risk-averse." Critique: "This is another assumption of the efficient market hypothesis, but we now know from behavioral economics that market participants are not rational."
  - –Assumption: "All investors have access to the same information at the same time." Critique: "This also comes from the efficient market hypothesis. In fact, real markets contain information asymmetry, insider trading, and those who are simply better informed than others."
  - –Assumption: "Investors have an accurate conception of possible returns, i.e., the probability beliefs of investors match the true distribution of returns." Critique: "A different possibility is that investors' expectations are biased, causing market prices to be informationally inefficient."
  - –Assumption: "There are no taxes or transaction costs." Critique: "Real financial products are subject both to taxes and transaction

costs (such as broker fees), and taking these into account will alter the composition of the optimum portfolio."

–Assumption: "All investors are price takers, i.e., their actions do not influence prices." Critique: "In reality, sufficiently large sales or purchases of individual assets can shift market prices for that asset and others (via cross-elasticity of demand.) An investor may not even be able to assemble the theoretically optimal portfolio if the market moves too much while they are buying the required securities."

–Assumption: "Any investor can lend and borrow an unlimited amount at the risk free rate of interest." Critique: "In reality, every investor has a credit limit."

–Assumption: "All securities can be divided into parcels of any size." Critique: "In reality, fractional shares usually cannot be bought or sold, and some assets have minimum orders sizes."

- "MPT Does Not Really Model the Market"
  - –"The risk, return, and correlation measures used by MPT are based on expected values, which means that they are mathematical statements about the future....In practice investors must substitute predictions based on historical measurements of asset return and volatility for these values in the equations. Very often such expected values fail to take account of new circumstances which did not exist when the historical data were generated."
  - –"Variance is a symmetric measure that counts abnormally high returns as just as risky as abnormally low returns. Some would argue that, in reality, investors are only concerned about losses, and do not care about the dispersion or tightness of above-average returns. According to this view, our intuitive concept of risk is fundamentally asymmetric in nature."
  - –"Essentially, the mathematics of MPT view the markets as a collection of dice. By examining past market data we can develop hypotheses about how the dice are weighted, but this isn't helpful if the markets are actually dependent upon a much bigger and more complicated chaotic system—the world."
  - –"MPT does not account for the personal, environmental, strategic, or social dimensions of investment decisions. It only attempts to maximize risk-adjusted returns, without regard to other consequences. In a narrow sense, its complete reliance on asset prices makes it vulnerable to all the standard market failures such as those arising from information asymmetry, externalities, and public goods. It also rewards corporate fraud and dishonest

accounting. More broadly, a firm may have strategic or social goals that shape its investment decisions, and an individual investor might have personal goals. In either case, information other than historical returns is relevant."

- "MPT Does Not Take its Own Effect on Asset Prices into Account"
  —"Diversification eliminates non-systematic risk, but at the cost of increasing the systematic risk. Diversification forces the portfolio manager to invest in assets without analyzing their fundamentals, solely for the benefit of eliminating the portfolio's non-systematic risk....This artificially increased demand pushes up the price of assets that, when analyzed individually, would be of little fundamental value. The result is that the whole portfolio becomes more expensive and, as a result, the probability of a positive return decreases (i.e. the risk of the portfolio increases)."

Again, there can be other criticisms, so the examples above are illustrative (not exhaustive).

Beyond Portfolio Analytics

These criticisms of MPT do not mean to dismiss the mathematical modeling of finance but are to show us the two opposing sides of the debate, such that the possibility and desirability of quantitative research (from the perspective of society with quantitative research and the problem of portfolio analytics as a case study here) are not to the extent that the respective defenders would like us to believe.

In addition, the analysis of quantitative research and the problem of portfolio analytics reveal something insightful about the ontological principles in existential dialectics, and good examples include the formalness-informalness principle, the absoluteness-relativeness principle, the partiality-totality principle, the predictability-unpredictability principle, the explicability-inexplicability principle, the fiction-reality principle, the finiteness-transfiniteness principle, the preciseness-vagueness principle, the simpleness-complicatedness principle, the openness-hiddenness principle, the denseness-emptiness principle, the change-constancy principle, the order-chaos principle, the slowness-quickness principle, the expansion-contraction principle, the theory-praxis principle, the convention-novelty principle, the evolution-transformation principle, the symmetry-asymmetry principle, the softness-hardness principle, the seriousness-playfulness principle, the regression-progression principle, the same-difference principle, and the stability-reaction principle.

For instance, in relation to the formalness-informalness principle, if there is formalness (e.g., the formal logical requirement of soundness in a system of ideas, as in MPT), there is informalness (e.g., the non-formal existence of unsoundness in MPT, as shown in the questionable assumptions aforecited). And the reverse direction also holds true.

In relation to the absoluteness-relativeness principle, if there is absoluteness (e.g., the absolute view of MPT which "does not account for the personal, environmental, strategic, or social dimensions of investment decisions"), there is relativeness (e.g., what is right for MPT in regard to its failure to take into account "the personal, environmental, strategic, or social dimensions of investment decisions" is not necessarily so for its critics, who suggest their inclusion). And the reverse direction also holds true.

In relation to the partiality-totality principle, if there is partiality (e.g., the partial view of MPT to only "maximize risk-adjusted returns"), there is totality (e.g., the more holistic view of "the personal, environmental, strategic, or social dimensions of investment decisions," such that the whole is not the sum of its parts, that is, the possible emergence of "all the standard market failures such as those arising from information asymmetry, externalities, and public goods," which could not be well understood if only considering to "maximize risk-adjusted returns" alone within MPT). And the reverse direction also holds true.

In relation to the predictability-unpredictability principle, if there is predictability (e.g., the predictable tendency of those in MPT to measure "risk, return, and correlation" on the basis of "expected values, which means that they are mathematical statements about the future"), there is unpredictability (e.g., the more difficult task to predict exactly the extent to which these mathematical statements are accurate, since "very often such expected values fail to take account of new circumstances which did not exist when the historical data were generated"). And the reverse direction also holds true.

In relation to the explicability-inexplicability principle, if there is explicability (e.g., the explanation by those in MPT that diversification is desirable because it "eliminates non-systematic risk"), there is inexplicability (e.g., the more difficult task for those in MPT to explain why diversification is necessarily desirable, since it occurs "at the cost of increasing the systematic risk," because "[t]his artificially increased demand pushes up the price of assets…and, as a result,…the risk of the portfolio increases"). And the reverse direction also holds true.

In relation to the fiction-reality principle, if there is fiction (e.g., the fictional aspect of MPT, as shown in its questionable assumptions which are not really true), there is reality (e.g., the realistic aspect of MPT, as shown in some successful applications in finance and non-finance aforecited). And the reverse direction also holds true.

In relation to the finiteness-transfiniteness principle, if there is finiteness (e.g., the finite amount of "credit" that an investor can have in practice), there is transfiniteness (e.g., the transfinite amount of "credit" that an investor can have in principle, within MPT). And the reverse direction also holds true.

In relation to the preciseness-vagueness principle, if there is preciseness (e.g., the precise identification of three categories of criticisms against MPT as cited), there is vagueness (e.g., the vagueness in the identification of the three categories of criticisms against MPT as cited, since it is not clear why there must be only three, not two, four, five, and so on). And the reverse direction also holds true.

In relation to the simpleness-complicatedness principle, if there is simpleness (e.g., the relatively simple analysis of investment decisions by MPT to only "maximize risk-adjusted returns"), there is complicatedness (e.g., the relatively more complicated analysis of investment decisions by the critics of MPT to also consider "the personal, environmental, strategic, or social dimensions of investment decisions"). And the reverse direction also holds true.

In relation to the openness-hiddenness principle, if there is openness (e.g., the open exploration of investment decisions on the basis of diversification by MPT), there is hiddenness (e.g., the hidden bias in MPT, as shown in the problems pointed out by the critics aforecited). And the reverse direction also holds true.

In relation to the denseness-emptiness principle, if there is denseness (e.g., the relatively denser concentration of diversified assets if invested according to MPT), there is emptiness (e.g., the relatively less dense, or more empty, concentration of diversified assets if invested according to the critics of MPT). And the reverse direction also holds true.

In relation to the change-constancy principle, if there is change (e.g., the ever new ways to invest assets, as shown in the rise of MPT), there is constancy (e.g., the ever constant existence of different problems with investment decisions, as shown in the problems with MPT pointed out by the critics). And the reverse direction also holds true.

In relation to the order-chaos principle, if there is order (e.g., the more or less orderly markets for investment decisions if viewed within MPT), there is chaos (e.g,. the more or less chaotic world for investment decisions, if viewed by the critics of MPT, who believe that "the markets are actually dependent upon a much bigger and more complicated chaotic system—the world").

In relation to the slowness-quickness principle, if there is slowness (e.g., the relatively slower readiness of those who reject MPT to diversify their assets), there is quickness (e.g., the relatively quicker readiness of those who accept MPT to diversify their assets). And the reverse direction also holds true.

In relation to the expansion-contraction principle, if there is expansion (e.g., the relatively more developed ability of those in MPT to deal with investment decisions solely on how to "maximize risk-adjusted returns"), there is contraction (e.g., the relatively less developed ability of of those in MPT to deal with problems related to "personal, environmental, strategic, or social dimensions of investment decisions"). And the reverse direction also holds true.

In relation to the theory-praxis principle, if there is theory (e.g., the theoretical construction of MPT), there is praxis (e.g.,the practical application of MPT to the field of micro-sociology or social psychology). (WK 2011jj) And the reverse direction also holds true.

In relation to the convention-novelty principle, if there is convention (e.g., the conventional way of analyzing finance without the use of portfolio analytics), there is novelty (e.g., the alternative novel challenge to the conventional way of analyzing finance, by the new idea of portfolio analytics which makes good use of mathematical modeling, together with the use of "computer technology, operational research, and statistics"). (WK 2011nn & qq) And the reverse direction also holds true.

In relation to the evolution-transformation principle, if there is evolution (e.g., the natural evolution of humans to take risks in the state of nature), there is transformation (e.g., the technical transformation of human ability to take risks by the invention of portfolio analytics for the diversification of investment decisions). And the reverse direction also holds true.

In relation to the symmetry-asymmetry principle, if there is symmetry (e.g., the existence of different views about risk, be it focused on losses or returns), there is asymmetry (e.g., for the critics of MPT, "in reality, investors are only concerned about losses, and do not care about the dispersion or tightness of above-average returns," so "our intuitive concept of risk is fundamentally asymmetric in nature"). And the reverse direction also holds true.

In relation to the softness-hardness principle, if there is softness (e.g., the support of portfolio analytics, as shown in its applications in other fields like micro-sociology), there is hardness (e.g., the critique of portfolio analytics, as shown in the problems pointed out by the critics as cited). And the reverse direction also holds true.

In relation to the seriousness-playfulness principle, if there is seriousness (e.g., the serious business of those in MPT to understand investment decisions, especially on how to "maximize risk-adjusted returns"), there is playfulness (e.g., the playful part of those in MPT to mess with different questionable assumptions).

In relation to the regression-progression principle, if there is regression (e.g., the regression made by MPT, as shown in the problems pointed out by the critics), there is progression (e.g., the progress made by MPT, as shown in its applications to other fields aforecited). And the reverse direction also holds true.

In relation to the same-difference principle, if there is similarity in outcome (e.g., the contribution to the molding and control of beliefs, values, and behaviors, regardless of whether this be done by way of following MPT, or by way of following the alternative to MPT like behavioral economics), there is difference in outcome (e.g., the contribution to the molding and control of beliefs, values, and behaviors by way of following MPT for a relatively more simplified worldview—but the contribution to the molding and control of beliefs, values, and behaviors by way of following the alternative to MPT like behavioral economics for a relatively more realistic worldview). And the reverse direction also holds true.

And in relation to the stability-reaction principle, if there is stability (e.g., the relatively more stable condition of diversification when it "eliminates non-systematic risk" in the short run), there is reaction (e.g., diversification can lead to further changes over time, when it has its own effect on the larger market by "increasing the systematic risk," because "diversification forces the portfolio manager to invest in assets without analyzing their fundamentals, solely for the benefit of eliminating the portfolio's non-systematic risk" and "[t]his artificially increased demand pushes up the price of assets that, when analyzed individually, would be of little fundamental value," so "the result is that the whole portfolio becomes more expensive and, as a result, the probability of a positive return decreases" and "the risk of the portfolio increases"). And the reverse direction also holds true.

Quantitative Research, and the Dispute over Social Statistics

Another good case study here to evaluate the possibility and desirability of quantitative research from the perspective of society is to examine quantitative research and the dispute over social statistics.

The Varieties of Social Statistics

The term "social statistics" refers to "the use of statistical measurement systems to study human behavior in a social environment. This can be accomplished through polling a particular group of people, evaluating a particular subset of data obtained about a group of people, or by observation and statistical analysis of a set of data that relates to people and their behaviors." (WK 2011rr)

The use of statistical measurement systems has become "a key feature of contemporary social science. Statistics is and has been perhaps most important in economics and psychology but is also employed in political science, sociology and anthropology." (WK 2011rr)

However, "[t]here is...currently a heated debate regarding the questionable uses and value of statistical methods in social science, especially in political science, with many statisticians questioning the policy conclusions of political partisans who overestimate the interpretive power that non-robust statistical methods such as simple and multiple linear regression allow. Indeed, an important mantra that social scientists cite, but often forget, is that 'correlation does not imply causation.'" (WK 2011rr)

Since there can be many different ways to do statistics in social science, consider, for illustration, three major ones, namely, (a) survey sampling, (b) cluster analysis, and (c) factor analysis—to be analyzed below, respectively.

The Case of Survey Sampling

The first case of social statistics to be illustrated here concerns "survey sampling," which refers to "the process of selecting a sample of elements from a target population in order to conduct a survey." (WK 2011ss)

In contrast, unlike a sampling, "a survey that measures the entire target population is called a census." (WK 2011ss)

There are two types of survey sampling, namely, (a) probability sampling and (b) non-probability sampling, as described below.

(a) Probabliity Sampling

In probability sampling (also known as "scientific" or "random" sampling), "each member of the target population has a known and non-zero probability of inclusion in the sample. A survey based on a probability sample can in theory produce statistical measurements of the target population" which have two important features, as shown below: (WK 2011ss; L. Kish 1965)

- "unbiased, the expected value of the sample mean is equal to the population mean $E(\bar{y})=\mu$"
- "have a measurable sampling error, which can be expressed as a confidence interval, or margin of error"

The task of propability sampling can be further distinguished in terms of two sub-tasks.

The first sub-task is "selection procedure," which refers to "a randomized process for selecting units from the sample frame" (that is, "constructing a list of the target population"). (WK 2011ss)

And the second sub-task is "selection method or mode," which refers to "contacting selected units to and enabling them complete the survey." (WK 2011ss; R. Groves 2009)

However, probability sampling has two major problems, as shown below: (WK 2011ss)

- Complex and Expensive: "For some target populations, this process may be easy, for example, sampling the employees of a company by using payroll list. However, in large, disorganized populations, simply constructing a suitable sample frame is often a complex and expensive task."
- Biased: Bias in probability sampling can come from three sources.
  - –"Non-Response Bias"—"When individuals or households selected in the survey sample cannot or will not complete the survey, there is the potential for bias to result from this non-response. Nonresponse bias occurs when the observed value deviates from the population parameter due to differences between respondents and nonrespondents."
  - –"Coverage Bias"—"Coverage bias can occur when population members do not appear in the sample frame (undercoverage). Coverage bias occurs when the observed value deviates from the population parameter due to differences between covered and non-covered units. Telephone surveys suffer from a well known

source of coverage bias because they cannot include households without telephones."

–"Selection Bias"—"Selection bias occurs when some units have a differing probability of selection that is unaccounted for by the researcher. For example, some households have multiple phone numbers making them more likely to be selected in a telephone survey than households with only one phone number."

Of course, this does not mean that probability sampling is useless; in fact, "the purpose of sampling is to reduce the cost and/or the amount of work that it would take to survey the entire target population." (WK 2011ss)

(b) Non-Probabliity Sampling

In non-probability sampling, however, a survey is selected by "finding a suitable collection of respondents to complete the survey. Some common examples of non-probability sampling" are shown below: (WK 2011ss)

- "Judgement Samples"—"A researcher decides which population members to include in the sample based on his or her judgement."
- "Snowball Samples"—"Often used when a target population is rare, members of the target population recruit other members of the population for the survey."
- "Quota Samples"—"The sample is designed to include a designated number of people with certain specified characteristics. For example, 100 coffee drinkers. This type of sampling is common in non-probability market research surveys."
- "Convenience Samples"—"The sample is composed of whatever persons can be most easily accessed to fill out the survey."

However, non-probability sampling has its own problems, and a good illustration is shown below: (WK 2011ss; H. Weisberg 2005)

- No External Validity—"Surveys that are not based on probability sampling have no way of measuring their bias or sampling error. Surveys based on non-probability samples are not externally valid. They can only be said to be representative of the people that have actually completed the survey."

Surely, there can be other problems, so the example above is illustrative (not exhaustive).

The Case of Cluster Analysis

The second case of social statistics to be illustrated here concerns "cluster analysis" (also known as "automatic classification," "numerical taxonomy," "botryology and typological analysis," etc.) which refers to "the assignment of a set of observations into subsets (called clusters), so that observations in the same cluster are similar in some sense." (WK 2011tt)

For instance, in educational research, cluster analysis was used in 2002 for "the project 'School Like Mine' to compare students' achievement in literacy and numeracy by the type of school they attended. 2707 majority and minority students in New Zealand are classified into different clusters according to school size, student ethnicity, region, size of civil jurisdiction and socioeconomic status for comparison. The cluster in this research is calculated across five dimensions, decile, region, size, minority and rurality. All schools are placed into one of twenty clusters… as a basis of student achievement comparison. The result shows that using the power of socioeconomic status to describe schools is…found inadequate….By clustering schools," the project found that "school types had no significant relation with performance of schools." (WK 2011tt; C. Huberty 2005)

However, cluster analysis has both advantages and disadvantages, as shown below.

(a) Advantages of Cluster Analysis

Cluster analysis has two main advantages, as shown below:

- "A Quick Review of Data"—"cluster analysis is a good way for quick review of data, especially if the objects are classified into many groups. In the 'Schools Like Mine' example, 23 clusters of schools with different properties were clearly clustered. It is easy for users to assign or nominate themselves into a cluster they would most like to compare with in a school cluster database because each cluster is clearly named with understandable terms." (WK 2011tt; C. Huberty 2005; N. Entwistle 1971)
- "A Simple Profile of Individuals"—"cluster analysis provides a simple profile of individuals. Given a number of analysis units, for example school size, student ethnicity, region, size of civil jurisdiction and social economic status in this example, each of which is described by a set of characteristics and attributes. Cluster Analysis also suggests how groups of units are determined such that units within groups are

> similar in some respect and unlike those from other groups." (WK 2011tt; J. Frisvad 2011)

But cluster analysis also have its own disadvantages.

(b) Disadvantages of Cluster Analysis

Cluster analysis have some main disadvantages, as shown below: (WK 2011tt; C. Huberty 2005)

- "Assignment Rigidity"—"an object can be assigned in one cluster only. For example in 'Schools Like Mine,' schools are automatically assigned into the first twenty-two clusters. However, if schools want to compare themselves with integrated schools, they will have to manually assign themselves into cluster twenty-three. Data-driven clustering may not represent reality, because once a school is assigned to a cluster, it cannot be assigned to another one. Some schools may have more than one significant property or fall on the edge of two clusters."
- "Detrimental Effects"—"clustering may have detrimental effects to teachers who work in low-decile schools, students who are educated in them, and parents who support them, by telling them the schools are classified as ineffective, when in fact many are doing well in some unique aspects that are not sufficiently illustrated by the clusters formed."

There can be other disadvantages too; for example, "in k-means clustering methods, it often requires several analysis before the number of clusters can be determined. It can be very sensitive to the choice of initial cluster centres." (WK 2011tt)

The Case of Factor Analysis

And the third case of social statistics to be illustrated here concerns "factor analysis," which refers to "a statistical method used to describe variability among observed variables in terms of a potentially lower number of unobserved variables called factors." (WK 2011uu)

In factor analysis, "it is possible, for example, that variations in three or four observed variables mainly reflect the variations in a single unobserved variable, or in a reduced number of unobserved variables. Factor analysis searches for such joint variations in response to unobserved latent variables. The observed variables are modeled as linear combinations of the potential factors, plus 'error' terms. The information gained about the interdependencies between observed variables can be used later to reduce the set of variables in a dataset." (WK 2011uu)

For illustration, suppose a researcher "proposes a theory that there are two kinds of intelligence, 'verbal intelligence' and 'mathematical intelligence,' neither of which is directly observed. Evidence for the theory is sought in the examination scores from each of 10 different academic fields of 1000 students. If each student is chosen randomly from a large population, then each student's 10 scores are random variables. The…theory may say that, for each of the 10 academic fields, the score averaged over the group of all students who share some common pair of values for verbal and mathematical 'intelligences' is some constant times their level of verbal intelligence plus another constant times their level of mathematical intelligence, i.e., it is a linear combination of those two 'factors.' The numbers for a particular subject, by which the two kinds of intelligence are multiplied to obtain the expected score, are posited by the theory to be the same for all intelligence level pairs, and are called 'factor loadings' for this subject," as shown below: (WK 2011uu)

(10×the student's verbal intelligence) +
(6 ×the student's mathematical intelligence)

So, "the numbers 10 and 6 are the factor loadings….Other academic subjects may have different factor loadings." (WK 2011uu)

Now, "two students having identical degrees of verbal intelligence and identical degrees of mathematical intelligence may have different aptitudes…because individual aptitudes differ from average aptitudes. That difference is called the 'error'—a statistical term that means the amount by which an individual differs from what is average for his or her levels of intelligence." (WK 2011uu)

In this example, "[t]he observable data that go into factor analysis would be 10 scores of each of the 1000 students, a total of 10,000 numbers. The factor loadings and levels of the two kinds of intelligence of each student must be inferred from the data." (WK 2011uu)

However, factor analysis has its own advantages and disadvantages, as shown below.

(a) Advantages of Factor Analysis

Factor analysis has some main advantages, as shown below: (WK 2011uu)

- "Both objective and subjective attributes can be used provided the subjective attributes can be converted into scores."

- "Factor Analysis can be used to identify hidden dimensions or constructs which may not be apparent from direct analysis."
- "It is easy and inexpensive to do."

But factor analysis also has its own disadvantages.

(b) Disadvantages of Factor Analysis

Factor analysis has some main disadvantages, as shown below: (WK 2011uu)

- "Usefulness depends on the researchers' ability to collect a sufficient set of…attributes. If important attributes are missed, the value of the procedure is reduced."
- "If sets of observed variables are highly similar to each other and distinct from other items, factor analysis will assign a single factor to them. This may make it harder to identify factors that capture more interesting relationships."
- "Naming the factors may require background knowledge or theory because multiple attributes can be highly correlated for no apparent reason."

Surely, there can be other disadvantages, so the examples above are illustrative (not exhaustive).

Beyond Social Statistics

This analysis of the disadvantages (and also advantages) of the three different versions of social statistics is not to reject their usefulness but are to show us the two opposing sides of the debate, such that the possibility and desirability of quantitative research (from the perspective of society with quantitative research and the dispute over social statistics as a case study here) are not to the extent that the respective defenders would like us to believe.

More fundamentally, the analysis of quantitative research and the dispute over social statistics can teach us something important about the ontological principles in existential dialectics, and good examples include the formalness-informalness principle, the absoluteness-relativeness principle, the predictability-unpredictability principle, the explicability-inexplicability principle, the fiction-reality principle, the preciseness-vagueness principle, the simpleness-complicatedness principle, the openness-hiddenness principle, the denseness-emptiness principle, the change-constancy principle, the slowness-quickness principle, the expansion-contraction principle, the theory-praxis principle, the convention-novelty principle, the evolution-transformation principle, the symmetry-asymmetry principle, the softness-hardness principle, the seriousness-playfulness principle, the regression-progression principle, the same-difference principle, and the stability-reaction principle.

For instance, in relation to the formalness-informalness principle, if there is formalness (e.g., the formal logical requirement of completeness and soundness in a system of ideas, as in social statistics), there is informalness (e.g., the non-formal existence of incompleteness and unsoundness in social statistics, as shown in the critique against social statistics in regard to different problems, such that "correlation does not imply causation," and so on). And the reverse direction also holds true.

In relation to the absoluteness-relativeness principle, if there is absoluteness (e.g., the absolute "assignment rigidity" in cluster analysis, such that "an object can be assigned in one cluster only"), there is relativeness (e.g., what is right for cluster analysis in regard to "assignment rigidity" is not necessarily so for its critics, who call for more flexibility). And the reverse direction also holds true.

In relation to the predictability-unpredictability principle, if there is predictability (e.g., the predictable tendency of non-probability sampling to produce surveys which "can only be said to be representative of the people that have actually completed" them), there is unpredictability (e.g., the more difficult task to predict exactly the extent to which non-probability sampling can at least represent the target population, since "surveys that are not based on probability sampling have no way of measuring their bias or sampling error"). And the reverse direction also holds true.

In relation to the explicability-inexplicability principle, if there is explicability (e.g., the explanation by the advocates of cluster analysis that it is desirable, because it provides "a quick review of data" and "a simple profile of individuals"), there is inexplicability (e.g., the lack of sufficient explanation by the advocates of why cluster analysis is necessarily desirable, since it also produces "detrimental effects" as cited by the critics). And the reverse direction also holds true.

In relation to the fiction-reality principle, if there is fiction (e.g., the fictional aspect of factor analysis when the researcher also uses "subjective attributes" which may not accurately reflect reality), there is reality (e.g., the realistic aspect of factor analysis when the researcher uses "objective attributes" which may more likely reflect reality). And the reverse direction also holds true.

In relation to the preciseness-vagueness principle, if there is preciseness (e.g., the precise identification of three advantages and three disadvantages of factor analysis as cited), there is vagueness (e.g., the vagueness in the identification of the three advantages and three disadvantages of factor analysis as cited, since it is not clear why there must be only three, not four, five, and so on). And the reverse direction also holds true.

In relation to the simpleness-complicatedness principle, if there is simpleness (e.g., the relatively simple task for probability sampling to construct "a suitable sample frame" in a small population like "sampling the employees of a company by using payroll list"), there is complicatedness (e.g., the relatively more complicated task for probability sampling to construct "a suitable sample frame" in "large, disorganized populations"). And the reverse direction also holds true.

In relation to the openness-hiddenness principle, if there is openness (e.g., the open exploration of correlation among multiple attributes in factor analysis), there is hiddenness (e.g., the hidden nature of some "multiple attributes" which "can be highly correlated for no apparent reason"). And the reverse direction also holds true.

In relation to the denseness-emptiness principle, if there is denseness (e.g., the relatively denser concentration of samples in non-probability sampling which have no external validity), there is emptiness (e.g., the relatively less dense, or more empty, concentration of samples in probability sampling which have no external validity). And the reverse direction also holds true.

In relation to the change-constancy principle, if there is change (e.g., the ever new ways to do social statistics, as shown in the three versions aforediscussed), there is constancy (e.g., the ever constant existence of different problems with social statistics, as shown in the problems of the three versions pointed out by the critics). And the reverse direction also holds true.

In relation to the slowness-quickness principle, if there is slowness (e.g., the relatively slower review of data if unstructured or disorganized), there is quickness (e.g., the relatively quicker review of data if organized in clusters as in cluster analysis). And the reverse direction also holds true.

In relation to the expansion-contraction principle, if there is expansion (e.g., the relatively more developed ability of those in factor analysis to discover "unobserved latent variables"), there is contraction (e.g., the relatively less developed ability of those in factor analysis to work on action research). And the reverse direction also holds true.

In relation to the theory-praxis principle, if there is theory (e.g., the theoretical construction of cluster analysis), there is praxis (e.g.,the practical application of cluster analysis to the field of educational research as shown in "the project 'School Like Mine'"). (WK 2011jj) And the reverse direction also holds true.

In relation to the convention-novelty principle, if there is convention (e.g., the conventional wisdom of doing statistics by probability sampling), there is novelty (e.g., the alternative novel challenge to the conventional wisdom about doing statistics by probability sampling, by the new idea of factor analysis for "unobserved latent variables"). And the reverse direction also holds true.

In relation to the evolution-transformation principle, if there is evolution (e.g., the natural evolution of humans to observe and measure things in the state of nature in a primitive way), there is transformation (e.g., the technical transformation of human ability to measure things by the invention of social statistics). And the reverse direction also holds true.

In relation to the symmetry-asymmetry principle, if there is symmetry (e.g., the existence of different versions of social statistics), there is asymmetry (e.g., the strength of factor analysis is to discover "unobserved latent variables"—but the strength of cluster analysis is to provide "a quick review of data"). And the reverse direction also holds true.

In relation to the softness-hardness principle, if there is softness (e.g., the appreciation of cluster analysis, as shown in identifying its ability to provide "a quick review of data"), there is hardness (e.g., the critique of cluster analysis, as shown in the problems pointed out by the critics as cited). And the reverse direction also holds true.

In relation to the seriousness-playfulness principle, if there is seriousness (e.g., the serious business of those in non-probability sampling to find "respondents to complete" an survey), there is playfulness (e.g., the playful part of those in non-probability sampling when they choose a sample "composed of whatever persons can be most easily accessed to fill out the survey"). And the reverse direction also holds true.

In relation to the regression-progression principle, if there is regression (e.g., the regression made by social statistics, as shown in the disadvantages and problems pointed out by the critics aforecited), there is progression (e.g., the progress made by social statistics, as shown in the advantages of the three versions aforecited). And the reverse direction also holds true.

In relation to the same-difference principle, if there is similarity in outcome (e.g., the contribution to the molding and control of beliefs, values, and behaviors, regardless of whether this be done by way of probability sampling, or by way of non-probability sampling), there is difference in outcome (e.g., the contribution to the molding and control of beliefs, values, and behaviors by way of probability sampling for a relatively more scientific mindset—but the contribution to the molding and control of beliefs, values, and behaviors by way of non-probability sampling for a relatively more subjective mindset). And the reverse direction also holds true.

And in relation to the stability-reaction principle, if there is stability (e.g., the relatively more stable condition of clusters in cluster analysis when each "object" is "assigned in one cluster only"), there is reaction (e.g., the clustering can lead to further changes later, when the analysis produces "detrimental effects to teachers who work in low-decile schools, students who are educated in them, and parents who support them, by telling them the schools are classified as ineffective, when in fact many are doing well in some unique aspects that are not sufficiently illustrated by the clusters formed," and this in turn may put pressue on the researchers to change the cluster analysis so as to make the results less offensive to some groups). And the reverse direction also holds true.

Quantitative Research and Culture

Quantitative research, when examined from the perspective of culture, can cast a bright light on its possibility and desirability, and this can be shown by way of two case studies, namely, (2.5.1) quantitative research and the debate on causality, and (2.5.2) quantitative research and diverse probability interpretations—to be addressed hereafter, in that order.

Quantitative Research, and the Debate on Causality

One major criticism against social statistics (in the previous section) is the "important mantra that social scientists cite," namely, that "correlation does not imply causation." (WK 2011rr)

Does quantitative research really show us causation, not just correlation, in the world at all? This then brings us to the enquiry concerning causality itself.

The word "causality" refers to "the relationship between an event (the *cause*) and a second event (the *effect*), where the second event is understood as a consequence of the first. Though the causes and effects are typically related to changes or events, candidates include objects, processes, properties, variables, facts, and states of affairs." (WK 2011vv)

However, "characterizing the causal relationship can be the subject of much debate." (WK 2011vv)

Characteristics of Causality

The characteristics of causality have been subject to much debate over the centuries in the history of philosophy. Consider, for illustration, two debatable issues about causality below.

(a) What are the Types of Causality?

Aristotle started the debate by distinguishing "four causes, or four explanations, that each answer the question 'why?' in different ways," as shown below: (WK 2011vv; G. Cuizon 2007; PP 2011)

- "Material Cause"—"is the physical matter, the mass of 'raw material' of which something is 'made' (of which it consists)."
- "Formal Cause"—"tells us what, by analogy to the plans of an artisan, a thing is intended and planned to be."
- "Efficient Cause"—"is that external entity from which the change or the ending of the change first starts."
- "Final Cause"—"is that for the sake of which a thing exists, or is done—including both purposeful and instrumental actions. The final cause, or telos, is the purpose, or end, that something is supposed to serve."

For the critics, however, there are three main problems with Aristotle's distinction of the four types of causation.

Firstly, one major problem with Aristotle's distinction of the four types of causation is that "formal" and "final" causes do not exist at all, as the critics like Niccolò Machiavelli and Francis Bacon pointed out. (WK 2011vv)

Others suggested that causation is to be "distinguished into two types: Necessary and sufficient. A third type of causation, which requires neither necessity nor sufficiency in and of itself, but which contributes to the effect, is called a 'contributory cause.'" (WK 2011vv; S. Epp 2004)

On the one hand, in regard to "necessary causes"—"if *x* is a necessary cause of *y*, then the presence of *y* necessarily implies the presence of *x*. The presence of *x*, however, does not imply that *y* will occur." (WK 2011vv)

On the other hand, in regard to "sufficient causes"—"if *x* is a sufficient cause of *y*, then the presence of *x* necessarily implies the presence of *y*. However, another cause *z* may alternatively cause *y*. Thus the presence of *y* does not imply the presence of *x*." (WK 2011vv)

Secondly, another major problem with Aristotle's distinction of the four types of causation is that, for other critics, only "efficient causes" are plausible, and the other three are not. So, it is better to focus on two types of "efficient causality" instead, namely, "deterministic" vs. "chance." (WK 2011vv)

And thirdly, still another major problem with Aristotle's distinction of the four types of causation is that, for David Hume, there is no causality perceived in human experience, other than "regular sequences and correlations of natural phenomena," as "the natural sciences describe[e] *how* things happen rather than asking *why* they happen." (WK 2011vv)

(b) How Is Causality to be Evaluated?

And David Hume started the debate by proposing eight ways to evaluate causality (as "regular sequences and correlations"), as shown below: (WK 2011vv)

- "The cause and effect must be contiguous in space and time."
- "The cause must be prior to the effect."
- "There must be a constant union betwixt the cause and effect."
- "The same cause always produces the same effect, and the same effect never arises but from the same cause."
- "Where several different objects produce the same effect, it must be by means of some quality, which we discover to be common amongst them."
- "The difference in the effects of two resembling objects must proceed from that particular, in which they differ."
- "When any object encreases or diminishes with the encrease or diminution of its cause, 'tis to be regarded as a compounded effect, deriv'd from the union of the several different effects, which arise from the several different parts of the cause."
- An "object, which exists for any time in its full perfection without any effect, is not the sole cause of that effect, but requires to be assisted by some other principle, which may forward its influence and operation."

However, Hume's eight ways to evaluate causality is limited in their application.

For instance, "according to [John] Sowa (2000), citing Max Born in 1949, 'relativity and quantum mechanics have forced physicists to abandon these assumptions as exact statements of what happens at the most fundamental levels, but they remain valid at the level of human experience.'" (WK 2011vv; J. Sowa 2006)

Theories of Causality

Besides the characteristics of causality (which are subject to debate)—there are theories to account for causality. Consider, for illustration, some main ones below.

(a) Counterfactual Theories

The first type of theories to be introduced here in regard to causality concerns "counterfactual conditional." (WK 2011vv)

Theories like this focus on a particular kind of conditional known as "counterfactual conditional" (also known as "subjunctive conditional" or "remote conditional"), which is "a conditional (or 'if-then') statement indicating what would be the case if its antecedent were true. This is to be contrasted with an indicative conditional, which indicates what is (in fact) the case if its antecedent is (in fact) true." (WK 2011vv)

Counterfactual theories are more common in psychology; for instance, "[p]sychological research shows that people's thoughts about the causal relationships between events influences their judgments of the plausibility of counterfactual alternatives, and conversely, their counterfactual thinking about how a situation could have turned out differently changes their judgements of the causal role of events and agents." (WK 2011vv)

However, people's thoughts and judgments are often subjective and do not necessarily imply causality in the real world, which can mean different things to different people and are not always consistent. For instance, "their identification of the cause of an event, and their counterfactual thought about how the event could have turned out differently do not always coincide. People distinguish between various different sorts of causes, e.g., strong and weak causes. Research in the psychology of reasoning shows that people make different sorts of inferences from different sorts of causes." (WK 2011vv; R. Byrne 2005; G. Miller 1976)

(b) Theory of Probabilistic Causation

The second type of theories to be introduced here in regard to causality concerns "probabilistic causation" (as opposed to "deterministic causation").

In a "deterministic causation," the relation between *A* and *B* is such that, "if *A* causes *B*, then A must *always* be followed by *B*. In this sense, war does not cause deaths, nor does smoking cause cancer." (WK 2011vv)

So an alternative interpretation is about "probabilistic causation," which means that "*A* probabilistically causes *B* if *A*'s occurrence increases the probability of *B*." (WK 2011vv)

However, the problem here is how to interprete what this might mean. For instance, one interpretation is that this is to "reflect imperfect knowledge of a deterministic system." (WK 2011vv) Another interpretation is that "the causal system under study is inherently probabilistic, such as quantum mechanics." (WK 2011vv)

(c) Theory of Causal Calculus

The third type of theories to be introduced here in regard to causality concerns "causal calculus."

This theory applies to situations "when experiments are infeasible or illegal," such that "the derivation of cause-effect relationship from observational studies must rest on some qualitative theoretical assumptions, for example, that symptoms do not cause diseases, usually expressed in the form of missing arrows in causal graphs such as Bayesian Networks or path diagrams." (WK 2011vv)

The mathematical part of the theory is to rely on "the distinction between conditional probabilities, as in *P*(cancer|smoking), and interventional probabilities, as in *P*(cancer|do{smoking}). The former reads: 'the probability of finding cancer in a person known to smoke' while the latter reads: 'the probability of finding cancer in a person forced to smoke.' The former is a statistical notion that can be estimated directly in observational studies, while the latter is a causal notion (also called 'causal effect') which is what we estimate in a controlled randomized experiment." (WK 2011vv)

The aim of this theory is "to infer interventional probabilities from conditional probabilities in causal Bayesian Networks with unmeasured variables." (WK 2011vv; J. Pearl 2000)

But the problem here is that it does not really tell us what "causality" is, since the inference of "interventional probabilities" is derived from "conditional probabilities," which in turn only gives us a statistical relationship between two variables (like "cancer" and "smoking" in the example).

(d) Manipulation Theories

The fourth type of theories to be introduced here in regard to causality concerns "manipulation."

These theories associate "causality with manipulability. Under these theories, *x* causes *y* only in the case that one can change *x* in order to change *y*. This coincides with commonsense notions of causations, since often we ask causal questions in order to change some feature of the world. For instance, we are interested in knowing the causes of crime so that we might find ways of reducing it." (WK 2011vv; R. Collingwood 1940; D. Gasking 1955; P. Menzies 1993; G. von Wright 1971)

However, this view has been subject to two criticisms.

The first criticism is that "these accounts are circular. Attempting to reduce causal claims to manipulation requires that manipulation is more basic than causal interaction. But describing manipulations in non-causal terms has provided a substantial difficulty." (WK 2011vv)

And the second criticism is about "anthropocentrism," because, in these theories, "causality is some existing relationship in the world that we can harness for our desires. If causality is identified with our manipulation, then this…makes humans overly central to interactions in the world."

(e) Process Theories

And the fifth type of theories to be introduced here in regard to causality concerns "process."

In these theories, the focus is on "distinguishing between causal processes and non-causal processes" or "between a process and a pseudo-process. As an example, a ball moving through the air (a process) is contrasted with the motion of a shadow (a pseudo-process). The former is causal in nature while the latter is not." (WK 2011vv; W. Salmon 1984; B. Russell 1948)

W. Salmon, for instance, argued that "causal processes can be identified by their ability to transmit an alteration over space and time. An alteration of the ball (a mark by a pen, perhaps) is carried with it as the ball goes through the air. On the other hand an alteration of the shadow (insofar as it is possible) will not be transmitted by the shadow as it moves along." (WK 2011vv)

In this way, "[t]hese theorists claim that the important concept for understanding causality is not causal relationships or causal interactions, but rather identifying causal processes. The former notions can then be defined in terms of causal processes." (WK 2011vv)

But the problem here is that it is question-begging, because one must already have a pre-conception of what "causality" is, as opposed to "non-causality," before even being able to distinguish between a "causal" process and a "non-causal" one.

Beyond Correlation and Causality

This analysis of the failure of quantitative research to address the different problems about the characteristics and theories of causality should not be misconstrued as a total dismissal of their validity but serves to show us the two opposing sides of the debate, such that the possibility and desirability of quantitative research (from the perspective of culture with quantitative research and the debate on causality as a case study here) are not to the extent that the respective defenders would like us to believe.

Besides, the analysis of quantitative research and the debate on causality also shows us some valuable insights about the ontological principles in existential dialectics, and good examples include the formalness-informalness principle, the absoluteness-relativeness principle, the predictability-unpredictability principle, the explicability-inexplicability principle, the fiction-reality principle, the preciseness-vagueness principle, the simpleness-complicatedness principle, the openness-hiddenness principle, the denseness-emptiness principle, the change-constancy principle, the order-chaos principle, the slowness-quickness principle, the expansion-contraction principle, the theory-praxis principle, the convention-novelty principle, the evolution-transformation principle, the symmetry-asymmetry principle, the softness-hardness principle, the seriousness-playfulness principle, the regression-progression principle, the same-difference principle, and the stability-reaction principle.

For instance, in relation to the formalness-informalness principle, if there is formalness (e.g., the formal logical requirement of completeness in a system of ideas, as in manipulation theories), there is informalness (e.g., the non-formal existence of incompleteness in manipulation theories, as shown in the criticism that "these accounts are circular"). And the reverse direction also holds true.

In relation to the absoluteness-relativeness principle, if there is absoluteness (e.g., the absolute view of "deterministic causation," in that, "if *A* causes *B*, then A must *always* be followed by *B*"), there is relativeness (e.g., what is true for "deterministic causation" is not necessarily so for "probabilistic causation," which suggests instead that "*A* probabilistically causes *B* if *A*'s occurrence increases the probability of *B*"). And the reverse direction also holds true.

In relation to the predictability-unpredictability principle, if there is predictability (e.g., the predictable tendency, if viewed within manipulation theories, of those who "are interested in knowing the causes of crime" to "find ways of reducing it"), there is unpredictability (e.g., the more difficult task to predict exactly the extent to which crimes will be reduced on a particular occasion, by those interested in knowing their causes). And the reverse direction also holds true.

In relation to the explicability-inexplicability principle, if there is explicability (e.g., the explanation, if viewed within "deterministic causation," of "*A* causes *B*" to mean that "if *A* causes *B*, then A must *always* be followed by *B*"), there is inexplicability (e.g., the lack of sufficient explanation, if viewed within "deterministic causation," of why war "causes" deaths, why smoking "causes" cancer," etc.). And the reverse direction also holds true.

In relation to the fiction-reality principle, if there is fiction (e.g., the fictional aspect of Aristotle's types of causality, as shown in the rejection of "formal" and "final" causes by the critics like Niccolò Machiavelli and Francis Bacon), there is reality (e.g., the realistic aspect of Aristotle's types of causality, as shown in the acceptance of "efficient" causation by some modern epistemic realists). And the reverse direction also holds true.

In relation to the preciseness-vagueness principle, if there is preciseness (e.g., the precise identification of four types of causality by Aristotle), there is vagueness (e.g., the vagueness in the identification of the four types of causality by Aristotle, since it is not clear why there must be only four, not one, two, three, five, and so on). And the reverse direction also holds true.

In relation to the simpleness-complicatedness principle, if there is simpleness (e.g., the relatively simple analysis of the different theories of causality by their respective advocates), there is complicatedness (e.g., the relatively more complicated analysis of the different theories of causality, by challenging the claims and assumptions, as shown in the problems pointed out by the critics as cited). And the reverse direction also holds true.

In relation to the openness-hiddenness principle, if there is openness (e.g., the open exploration of eight different ways to evaluate causality by Hume), there is hiddenness (e.g., the hidden bias in Hume's eight ways to evaluate causality, since, for the critic Max Born, "relativity and quantum mechanics have forced physicists to abandon these assumptions as exact statements of what happens at the most fundamental levels"). And the reverse direction also holds true.

In relation to the denseness-emptiness principle, if there is denseness (e.g., the relatively denser concentration of concerns with counterfactual conditionals, by those in counterfactual theories), there is emptiness (e.g., the relatively less dense, or more empty, concentration of concerns with counterfactual conditionals, by those in theories of deterministic causation). And the reverse direction also holds true.

In relation to the change-constancy principle, if there is change (e.g., the ever new ways to understand causality, as shown in the different theories aforediscussed), there is constancy (e.g., the ever constant existence of different problems with the different ways to understand causality, as shown in the problems with the different theories of causality aforediscussed). And the reverse direction also holds true.

In relation to the order-chaos principle, if there is order (e.g., the more or less orderly macro-world governed by "regular sequences and correlations of natural phenomena," as pointed out by David Hume), there is chaos (e.g,. the more or less chaotic micro-world of quantum mechanics, as pointed out by Max Born).

In relation to the slowness-quickness principle, if there is slowness (e.g., the relatively slower speed for somone to have cancer if he never smokes nor suffers from second-handed smoking and has lived a reasonable healthy lifestyle), there is quickness (e.g., the relatively quicker speed for somone to have cancer if he has smoked a lot for decades). And the reverse direction also holds true.

In relation to the expansion-contraction principle, if there is expansion (e.g., the relatively more developed ability of someone like Aristotle to interprete the world according to the four types of causation), there is contraction (e.g., the relatively less developed ability of someone like Aristotle to interprete the world according to the quantum mechanical view of causation by Max Born). And the reverse direction also holds true.

In relation to the theory-praxis principle, if there is theory (e.g., the theoretical construction of causality in philosophy), there is praxis (e.g.,the practical application of the theoretical construction of causality in philosophy to the field of psychology, as shown in manipulation theories). (WK 2011jj) And the reverse direction also holds true.

In relation to the convention-novelty principle, if there is convention (e.g., the conventional view about the four types of causality in ancient Greece), there is novelty (e.g., the alternative novel challenge to the conventional view about the four types of causality in ancient Greece, by the new idea of Hume about "regular sequences and correlations of natural phenomena," as "the natural sciences describe[e] *how* things happen rather than asking *why* they happen). And the reverse direction also holds true.

In relation to the evolution-transformation principle, if there is evolution (e.g., the natural evolution of humans to experience relationships among different "events..., objects, processes, properties, variables, facts, and states of affairs" in the state of nature), there is transformation (e.g., the technical transformation of human ability to understand causality by the invention of different theories of causality). And the reverse direction also holds true.

In relation to the symmetry-asymmetry principle, if there is symmetry (e.g., the existence of different views about causality), there is asymmetry (e.g., the popularity of Aristotle's idea of the four types of causality in ancient Greece—but the popularity of Hume's idea of regularity in the modern West). And the reverse direction also holds true.

In relation to the softness-hardness principle, if there is softness (e.g., the appreciation of Aristotle's idea of the four types of casuality, as shown in the acceptance of "efficient" causality by some modern epistemic realists), there is hardness (e.g., the critique of Aristotle's idea of the four types of casuality, as shown in the modern rejection of "formal" and "final" causality by Machiavelli and Bacon). And the reverse direction also holds true.

In relation to the seriousness-playfulness principle, if there is seriousness (e.g., the serious business of those in "counterfactual theories" to understand counterfactual scenarios), there is playfulness (e.g., the playful part of those in "counterfactual theories" to play with different "counterfactual alternatives"). And the reverse direction also holds true.

In relation to the regression-progression principle, if there is regression (e.g., the regression made by different theories of causality, as shown in the problems pointed out by the critics), there is progression (e.g., the progress made by different theories of causality, as shown in the applications by the practitioners aforecited). And the reverse direction also holds true.

In relation to the same-difference principle, if there is similarity in outcome (e.g., the contribution to the molding and control of beliefs, values, and behaviors, regardless of whether this be done by the belief in "deterministic causation" as in Christian religion or by the belief in "probabilistic causation" as in quantum mechanics), there is difference in outcome (e.g., the contribution to the molding and control of beliefs, values, and behaviors by the belief in "deterministic causation" for a relatively more certain worldview—but the contribution to the molding and control of beliefs, values, and behaviors by way of the belief in "probabilistic causation" for a relatively more indeterministic worldview). And the reverse direction also holds true.

And in relation to the stability-reaction principle, if there is stability (e.g., the relatively more stable condition of an individual's deep-rooted interest in "knowing the causes of crime," for instance), there is reaction (e.g., this individual's interest in "knowing the causes of crime" can lead to further changes, when he "might find ways of reducing it," in accordance to manipulation theories). And the reverse direction also holds true.

Quantitative Research, and Diverse Probability Interpretations

The issue of causality (in the previous section) raises another important question, this time, about probability (i.e., the tendency for something to occur), although it does not answer the question of causality itself (e.g., whether that something is "caused" by, not just "correlated" with, something else to occur).

Thus, another way to evaluate the possibility and desirability of quantitative research from the perspective of culture concerns quantitative research and probability interpretations in a given philosophical tradition.

However, the word "probability" is tricky enough, since it "has been used in a variety of ways since it was first coined in relation to games of chance. Does probability measure the real, physical tendency of something to occur, or is it just a measure of how strongly one believes it will occur?" (WK 2011ww)

To understand the multiple interpretations of "probability," there are "two broad categories of probability interpretations" to be considered here, namely, (a) "physical" probability interpretations and (b) "evidential" probability interpretations—to be addressed hereafter, in that order.

Physical Probability Interpretations

The first broad category of probability interpretations to be considered here concerns "physical" probabiltiies (also called "objective" or "frequency" probabilities) which refer to "random physical systems such as roulette wheels, rolling dice and radioactive atoms. In such systems, a given type of event (such as the dice yielding a six) tends to occur at a persistent rate, or 'relative frequency,' in a long run of trials." (WK 2011ww)

Thus, "physical" probabilities "either explain, or are invoked to explain, these stable frequencies. Thus talk about physical probability makes sense only when dealing with well defined random experiments." (WK 2011ww)

More specifically, "physical" probabilities can be divided into two main types for analysis, namely, (i) "frequentist accounts (such as those of Venn, Reichenbach and von Mises)" and "propensity accounts (such as those of [Karl] Popper, [Richard] Miller, [Ronald] Giere and Fetzer)." (WK 2011ww)

(a) Frequentist Interpretation

For the frequentist theorists, "the probability of an event is its relative frequency over time, i.e., its relative frequency of occurrence after repeating a process a large number of times under similar conditions. This is also known as aleatory probability. The events are assumed to be governed by some random physical phenomena, which are either phenomena that are predictable, in principle, with sufficient information; or phenomena which are essentially unpredictable. Examples of the first kind include tossing dice or spinning a roulette wheel; an example of the second kind is radioactive decay. In the case of tossing a fair coin, frequentists say that the probability of getting a heads is 1/2, not because there are two equally likely outcomes but because repeated series of large numbers of trials demonstrate that the empirical frequency converges to the limit 1/2 as the number of trials goes to infinity." (WK 2011ww)

So, if the number of trials, *n*, approaches infinity, then, the probability *P* for event *A* is, as shown below: (WK 2011ww)

$$\text{if } \lim_{n\to\infty} \frac{n_a}{n} = p, \quad P(\mathcal{A}) = p$$

However, the frequentist interpretation has its own problems. Consider three main problems below.

Firstly, one main problem with the frequentist interpretation is that "[i]t is of course impossible to actually perform an infinity of repetitions of a random experiment to determine the probability of an event. But if only a finite number of repetitions of the process are performed, different relative frequencies will appear in different series of trials. If these relative frequencies are to define the probability, the probability will be slightly different every time it is measured. But the real probability should be the same every time." (WK 2011ww)

Secondly, another main problem with the frequentist interpretation is that it is circuluar. The reason is that, "[i]f we acknowledge the fact that we only can measure a probability with some error of measurement attached, we still get into problems as the error of measurement can only be expressed as a probability, the very concept we are trying to define. This renders even the frequency definition circular." (WK 2011ww)

And thirdly, still another main problem with the frequentist interpretation is that it is often applied by those interested in "statistical hypothesis testing," which, however, has been subject to a lot of criticisms over the years, as shown in the following examples: (WK 2011dd)

- "The test is a flawed application of probability theory"

–"While the data can be unlikely given the null hypothesis, the alternative hypothesis can be even more unlikely. (Nobody can be that lucky. vs. Clairvoyance is impossible.)"

- "The test result is a function of sample size."
- "The test result is uninformative."
- "Statistical significance does not imply practical significance."
- "Using statistical significance as a criterion for publication results in problems collectively known as publication bias."

–"Published Type I errors are difficult to correct."

–"Meta-studies are biased by the invisibility [non-publication] of tests which failed to reach significance."

–"Type II errors (false negatives) are common."

Of course, these examples are solely illustrative (not exhaustive).

(b) Propensity Interpretation

For the propensity theorists, "probability" can be interpreted as "a physical propensity, or disposition, or tendency of a given type of physical situation to yield an outcome of a certain kind or to yield a long run relative frequency of such an outcome. This kind of objective probability is sometimes called 'chance.'" (WK 2011ww)

In this sense, "propensities, or chances, are not relative frequencies, but purported causes of the observed stable relative frequencies. Propensities are invoked to explain why repeating a certain kind of experiment will generate a given outcome type at a persistent rate, which are known as propensities or chances." (WK 2011ww)

On the other hand, "frequentists are unable to take this approach, since relative frequencies do not exist for single tosses of a coin, but only for large ensembles or collectives. In contrast, a propensitist is able to use the law of large numbers to explain the behaviour of long-run frequencies. This law, which is a consequence of the axioms of probability, says that if (for example) a coin is tossed repeatedly many times, in such a way that its probability of landing heads is the same on each toss, and the outcomes are probabilistically independent, then the relative frequency of heads will (with high probability) be close to the probability of heads on each single toss. This law allows that stable long-run frequencies are a manifestation of invariant single-case probabilities. In addition to explaining the emergence of stable relative frequencies, the idea of propensity is motivated[by the desire to make sense of single-case probability attributions in quantum mechanics, such as the probability of decay of a particular atom at a particular time." (WK 2011ww)

However, the propensity interpretation has its own problems. Consider two main problems below.

Firstly, one main problem with the propensity interpretation is that "[t]he main challenge facing propensity theories is to say exactly what propensity means…[a]nd then, of course, to show that propensity thus defined has the required properties….At present, unfortunately, none of the well-recognised accounts of propensity comes close to meeting this challenge." (WK 2011ww)

And secondly, another main problem with the propensity interpretation is that it is not clear what "theoretical" role probability plays in science at all, because some "propensity theorists (e.g. Ronald Giere) do not explicitly define propensities at all, but rather see propensity as defined by the theoretical role it plays in science." (WK 2011ww)

These theorists argue, for instance, that "physical magnitudes such as electrical charge cannot be explicitly defined either, in terms of more basic things, but only in terms of what they do (such as attracting and repelling other electrical charges). In a similar way, propensity is whatever fills the various roles that physical probability plays in science." (WK 2011ww)

But the deeper questios here are, "What roles does physical probability play in science? What are its properties?" (WK 2011ww)

Evidential Probability Interpretations

And the second broad category of probability interpretations to be considered here concerns "evidential" probabiltiies (also called "Bayesian" probabilities or "subjectivist" probabilities) which "can be assigned to any statement whatsoever, even when no random process is involved, as a way to represent its subjective plausibility, or the degree to which the statement is supported by the available evidence. On most accounts, evidential probabilities are considered to be degrees of belief, defined in terms of dispositions to gamble at certain odds." (WK 2011ww)

More specifically, "evidential" probabilities can be divided into four main types for analysis, namely, (i) the "classical" interpretation (e.g., P. Laplace), (ii) the "subjective" interpretation (e.g., de Finetti and L. Savage), (iii) the "epistemic or inductive" interpretation (e.g., F. Ramsey, D. Cox) and (iv) the "logical" interpretation (e.g., J. Keynes and R. Carnap)." (WK 2011ww)

(a) Classical Interpretation

Pierre-Simon Laplace first gave a "mathematical rigour in the field of probability," which "is now known as the classical definition. Developed from studies of games of chance (such as rolling dice) it states that probability is shared equally between all the possible outcomes, provided these outcomes can be deemed equally likely." (WK 2011ww; SEP 2006)

As Laplace (1951) put it, "the theory of chance consists in reducing all the events of the same kind to a certain number of cases equally possible, that is to say, to such as we may be equally undecided about in regard to their existence, and in determining the number of cases favorable to the event whose probability is sought. The ratio of this number to that of all the cases possible is the measure of this probability, which is thus simply a fraction whose numerator is the number of favorable cases and whose denominator is the number of all the cases possible." (WK 2011ww)

However, the classical interpretation has its own problems. For illustration, consider two main problems below.

Firstly, one main problem with the classical interpretation is that it is not clear how to accurately find out the favorable cases.

And secondly, another main problem with the classical interpretation is that it is also not clear how to accurately understand the subjective aspect involved, since it requires that "probability is shared equally between all the possible outcomes, provided these outcomes can be deemed equally likely."

(b) Subjective Interpretation

The subjectivists (also known as the "Bayesians") "give the notion of probability a subjective status by regarding it as a measure of the 'degree of belief' of the individual assessing the uncertainty of a particular situation....[S]ubjective probability is sometimes called credence, as opposed to the term chance for a propensity probability." (WK 2011ww)

For example, a subjectivist is "to assign a probability to the proposition that a proposed law of physics is true, and to determine how probable it is that a suspect committed a crime, based on the evidence presented." (WK 2011ww)

In this way, "[t]he use of Bayesian probability involves specifying a prior probability. This may be obtained from consideration of whether the required prior probability is greater or lesser than a reference probability associated with an urn model or a thought experiment. The issue is that for a given problem, multiple thought experiments could apply, and choosing one is a matter of judgement: different people may assign different prior probabilities, known as the reference class problem. The 'sunrise problem' provides an example." (WK 2011ww)

However, the subjecitivst interpretation has its own problems. The most obvious problem with the subjectivist interpretation is that it is too subjective.

In other words, "the use of Bayesian probability raises the philosophical debate as to whether it can contribute valid justifications of belief." (WK 2011ww; E. Jaynes 2003)

(c) Epistemic (Inductive) and Logical Interpretations

The last two types of "evidential" probabilities (namely, "epistemic" probability interpretation and "logical" probability interpretation) can be grouped together for analysis, since the differences between them are rather small.

Consider, for illustration, two distinctive differences between the two interpretations.

The first difference between the two interpretations concerns "the relation between probability and belief." (WK 2011ww)

On the one hand, for the logical theorists, "[l]ogical probabilities are conceived to be objective logical relations between propositions (or sentences), and hence not to depend in any way upon belief. They are degrees of (partial) entailment, or degrees of logical consequence, not degrees of belief...They do, nevertheless, dictate proper degrees of belief." (WK 2011ww)

On the other hand, the epistemic theorists like Frank P. Ramsey (1926) were "skeptical about the existence of such objective logical relations and argued that (evidential) probability is 'the logic of partial belief'....In other words, Ramsey held that epistemic probabilities simply are degrees of rational belief, rather than being logical relations that merely constrain degrees of rational belief." (WK 2011ww)

And the second difference between the two interpretations concerns "the *uniqueness* of evidential probability, relative to a given state of knowledge." (WK 2011ww)

On the one hand, the logical theorists like Rudolf Carnap "held...that logical principles always determine a unique logical probability for any statement, relative to any body of evidence." (WK 2011ww)

On the other hand, the epistemic theorists like Ramsey "thought that while degrees of belief are subject to some rational constraints (such as, but not limited to, the axioms of probability) these constraints usually do not determine a unique value. Rational people, in other words, may differ somewhat in their degrees of belief, even if they all have the same information." (WK 2011ww)

In the end, the disagreements between the last two types of "evidential" probabilities, that is, "epistemic or inductive" probability interpretation and "logical" probability interpretation, thus reveal some problems in each of them, such that they agree to disagree.

Beyond Probability Interpretations

All these different problems with different probability interpretations do not mean to reject their usefulness for quantitative research but are to show us the two opposing sides of the debate, such that the possibility and desirability of quantitative research (from the perspective of culture with quantitative research and diverse probability interpretations as a case study here) are not to the extent that the respective defenders would like us to believe.

More importantly, the analysis of quantitative research and diverse probability interpretations reveals something worthwhile about the ontological principles in existential dialectics, and good examples include the formalness-informalness principle, the absoluteness-relativeness principle, the predictability-unpredictability principle, the explicability-inexplicability principle, the fiction-reality principle, the finiteness-transfiniteness principle, the preciseness-vagueness principle, the simpleness-complicatedness principle, the openness-hiddenness principle, the denseness-emptiness principle, the change-constancy principle, the slowness-quickness principle, the expansion-contraction principle, the theory-praxis principle, the convention-novelty principle, the evolution-transformation principle, the symmetry-asymmetry principle, the softness-hardness principle, the seriousness-playfulness principle, the regression-progression principle, the same-difference principle, and the stability-reaction principle.

For instance, in relation to the formalness-informalness principle, if there is formalness (e.g., the formal logical requirement of completeness in a system of ideas, as in the theroy of frequentist probability), there is informalness (e.g., the non-formal existence of incompleteness in the theroy of frequentist probability, as shown in the problem of circularity aforediscussed).

In relation to the absoluteness-relativeness principle, if there is absoluteness (e.g., the absolute view of the logical theorists, in that "[l]ogical probabilities are conceived to be objective logical relations between propositions…, and hence not to depend in any way upon belief"), there is relativeness (e.g., what is right for the logical theorists is not necessarily so for the epistemic theorists like Ramsey who were "skeptical about the existence of such objective logical relations and argued that…evidential…probability is 'the logic of partial belief"). And the reverse direction also holds true.

In relation to the predictability-unpredictability principle, if there is predictability (e.g., the predictable tendency of those in frequentist interpretation to claim that, in the case of tossing a fair coin, "the probability of getting a heads is ½"), there is unpredictability (e.g., the more difficult task to predict exactly if a particular tossing of a coin in practice will turn out to be a head on a particular occasion, if viewed within the frequentist interpretation). And the reverse direction also holds true.

In relation to the explicability-inexplicability principle, if there is explicability (e.g., the explanation by those in propensity interpretation that "propensities, or chances, are not relative frequencies, but purported causes of the observed stable relative frequencies"), there is inexplicability (e.g., the lack of sufficient explanation by those in propensity interpretation of "exactly what propensity means," since "none of the well-recognised accounts of propensity comes close to meeting this challenge"). And the reverse direction also holds true.

In relation to the fiction-reality principle, if there is fiction (e.g., the fictional aspect of subjectivist interpretation, since it gives "the notion of probability a subjective status by regarding it as a measure of the 'degree of belief' of the individual assessing the uncertainty of a particular situation"), there is reality (e.g., the realistic aspect of subjectivist interpretation, when its assignment of a probability to a proposition is "based on the evidence presented"). And the reverse direction also holds true.

In relation to the finiteness-transfiniteness principle, if there is finiteness (e.g., "only a finite number of repetitions" is performed in the process of determining frequency probability), there is transfiniteness (e.g., according to the law of large numbers, "the empirical frequency converges to the limit 1/2 as the number of trials goes to infinity"). And the reverse direction also holds true.

In relation to the preciseness-vagueness principle, if there is preciseness (e.g., the precise classification of four types of "evidential" probabilities in the literature aforecited), there is vagueness (e.g., the vagueness in the classification of the four types of "evidential" probabilities in the literature aforecited, since it is not clear why there must be four in the classification, not three, five, six, and so on). And the reverse direction also holds true.

In relation to the simpleness-complicatedness principle, if there is simpleness (e.g., the relatively simple analysis of probability in the classical interpretation), there is complicatedness (e.g., the relatively more complicated analysis of probability nowadays, as shown in the diverse probability interpretations and the problems in each, as pointed out by the critics as cited). And the reverse direction also holds true.

In relation to the openness-hiddenness principle, if there is openness (e.g., the open exploration of probability and belief in subjectivist interpretation), there is hiddenness (e.g., the hidden bias of subjectivity in subjectivist interpretation, since "for a given problem, multiple thought experiments could apply, and choosing one is a matter of judgement: different people may assign different prior probabilities"). And the reverse direction also holds true.

In relation to the denseness-emptiness principle, if there is denseness (e.g., the relatively denser concentration of concerns with the "degree of belief" in subjectivist interpretation), there is emptiness (e.g., the relatively less dense, or more empty, concentration of concerns with the "degree of belief" in frequentist interpretation). And the reverse direction also holds true.

In relation to the change-constancy principle, if there is change (e.g., the ever new ways to study probability, as shown in the diversity of probability interpretations), there is constancy (e.g., the ever constant existence of different problems with different probability interpretations pointed out by the critics). And the reverse direction also holds true.

In relation to the slowness-quickness principle, if there is slowness (e.g., the relatively slower readiness of those in logical interpretation to argue that "[l]ogical probabilities are conceived to be objective logical relations between propositions…, and hence not to depend in any way upon belief"), there is quickness (e.g., the relatively faster readiness of those in epistemic interpretation to argue that "probability is 'the logic of partial belief'"). And the reverse direction also holds true.

In relation to the expansion-contraction principle, if there is expansion (e.g., the relatively more developed ability of those in logical interpretation to "determine a unique logical probability for any statement, relative to any body of evidence"), there is contraction (e.g., the relatively less developed ability of those in logical interpretation to assign "degrees of rational belief" to any statement concerning probability). And the reverse direction also holds true.

In relation to the theory-praxis principle, if there is theory (e.g., the theoretical construction of probability in mathematics), there is praxis (e.g.,the practical application of probability in mathemtics to the fields of both natural and social sciences, for hypothesis testing, for example). (WK 2011jj) And the reverse direction also holds true.

In relation to the convention-novelty principle, if there is convention (e.g., the conventional notion of probability like the classical interpretation by Laplace), there is novelty (e.g., the alternative novel challenge to the classical interpretation by Laplace, by the different new interpretations later on, as shown in the diverse probability interpretations aforediscussed). And the reverse direction also holds true.

In relation to the evolution-transformation principle, if there is evolution (e.g., the natural evolution of humans to count things in the state of nature in a primitive way), there is transformation (e.g., the technical transformation of human ability to measure things by the invention of probability theory and its diverse interpretations for practical applications). And the reverse direction also holds true.

In relation to the symmetry-asymmetry principle, if there is symmetry (e.g., the existence of different ways to understand probability), there is asymmetry (e.g., propensity theorists focus more on propensities—but subjectivist theorists focus more on the degrees of belief). And the reverse direction also holds true.

In relation to the softness-hardness principle, if there is softness (e.g., the defense of logical interpretation by Carnap), there is hardness (e.g., the critique of logical interpretation by Ramsey). And the reverse direction also holds true.

In relation to the seriousness-playfulness principle, if there is seriousness (e.g., the serious business of those in subjectivist interpretation to understand the nature of probability), there is playfulness (e.g., the playful part of those in subjectivist interpretation, when they play with "multiple thought experiments" and different "judgements" in assigning different "prior probabilities"). And the reverse direction also holds true.

In relation to the regression-progression principle, if there is regression (e.g., the regression made by probability interpretations, as shown in the different problems for each type of probability interpretation aforediscussed), there is progression (e.g., the progress made by probability interpretations, as shown in their applications for hypothesis testing, for example). And the reverse direction also holds true.

In relation to the same-difference principle, if there is similarity in outcome (e.g., the contribution to the molding and control of beliefs, values, and behaviors, regardless of whether this be done by way of physical probability interpreations, or by way of evidential probability interpreations), there is difference in outcome (e.g., the contribution to the molding and control of beliefs, values, and behaviors by way of physical probability interpreations for a relatively more objective mindset—but the contribution to the molding and control of beliefs, values, and behaviors by way of evidential probability interpreations for a relatively more subjective mindset). And the reverse direction also holds true.

And in relation to the stability-reaction principle, if there is stability (e.g., the relatively more stable condition of a posterior probability in subjectivist interpretation, that is, the degree of belief about a hypothesis when the outcome of an experiment is known), there is reaction (e.g., the probability in subjectivist interpretation can be subject to further changes later on, when new experiments are carried out with different new evidences and outcomes). And the reverse direction also holds true.

The Narrowness of Quantitative Research

This comprehensive analysis of the possibility and desirability of quantitative research, from the perspectives of the mind, nature, society, and culture, has the scholarly value to show us the different ways in which quantitative research is both possible and desirable, but not to the extent that the spokespersons from each side would like us to believe.

Quantitative research thus has its preciseness, just as it has its narrowness too, as the other side of the same coin.

But this is only one part of a larger story, since there is the other side of the story, which concerns qualitative research, the opposite of quantitative research. Since this chapter already deals with quantitative research, the next chapter will therefore address the issue of qualitative research.

As already pointed out earlier, these dual analyses should not be underestimated, because those on the side of quantitative research often downgrade qualitative research merely as the receiving end of data analysis without really appreciating it from the vantage point of qualitative research too.

With this in mind, let's now turn to Chapter Three for the study of qualitative research and its limitation—for which we now turn to chapter three.

Table 2.1. Quantitative Research and its Inadequacy

- **Quantitative Research and the Mind**
  —Ex: quantitative research, and the challenge of action science
  —Ex: quantitative research, and the controversy of cognitive biases

- **Quantitative Research and Nature**
  —Ex: quantitative research, confirmation bias, and paradigm shifts
  —Ex: quantitative research, and the limits of scientific modeling

- **Quantitative Research and Society**
  —Ex: quantitative research, and the problem of portfolio analytics
  —Ex: quantitative research, and the dispute over social statistics

- **Quantitative Research and Culture**
  —Ex: quantitative research, and the debate about causality
  —Ex: quantitative research, and diverse probability interpretations

Notes: The examples in the categories are solely illustrative (not exhaustive), and the comparison is relative (not absolute), nor are they necessarily mutually exclusive. And some can be easily re-classified elsewhere. As generalities, they allow exceptions.
Source: A summary of Ch.2 of *FPHDA*

Table 2.2. Quantitative Research, and the Challenge of Action Science

- **The General Framework of Action Science**
  - —Ex: "input" (for "unfreezing")
  - —Ex: "transformation" (for "changing")
  - —Ex: "output" (for "refreezing")

- **Different Versions of Action Science**
  - —Ex: Chris Argyris' action science
  - —Ex: John Heron and Peter Reason's cooperative inquiry
  - —Ex: Paulo Freire's participatory action research
  - —Ex: William Torbert's developmental action inquiry
  - —Ex: J. Whitehead's living theory/J. McNiff's action research

- **Problems with Action Science**
  - —Firstly, one major criticism of action science is that the tension among its conflicting goals remains unresolved.
  - —Secondly, another major criticism of action science is that its innovative combination of "problem solving actions" and "data-driven collaborative analysis or research" remains to be reconciled. In other words, the problem here is that "problem solving actions" are more praxis-oriented but "data-driven collaborative analysis or research" is more reflection-oriented . Although the two types of research do not have to be mutually exclusive, it is not clear how or to what extent their integration is achieved in the context of action science.
  - —And thirdly, still another major criticism of action science is that it is not clear why the focus on action theorizing is necessarily more desirable than the conventional attention to reflective knowledge instead. The point here does not mean to favor reflective knowledge but to question the convincingness of the main claim in action science.

Notes: The examples in the categories are solely illustrative (not exhaustive), and the comparison is relative (not absolute), nor are they necessarily mutually exclusive. And some can be easily re-classified elsewhere. As generalities, they allow exceptions.
Source: From *Sec. 2.2.1* of *FPHDA*. See book for citations.

Table 2.3. Quantitative Research, and the Limits of Scientific Modeling

---

- **Basics of Scientific Modeling**
  —Ex: language
  —Ex: simulation
  —Ex: structure
  —Ex: systems
  —Ex: the process of generating a model
  —Ex: the process of evaluating a model
  —Ex: visualization

- **Limits of Scientific Modeling**
  —The first limit of scientific modeling is simplification, in that the use of assumptions then means that "[a]ll models are in simulacra, that is, simplified reflections of reality…." (WK 2011jj)
  —The second limit of scientific modeling is subjectivity, in that scientific modeling can also use subjective information, especially when necessarily information for modeling is not available.
  —The third limit of scientific modeling is extrapolation, which refers to "the same question for events or data points outside the observed data." (WK 2011kk)
  —And the fourth limit of scientific modeling is causality or cause-effect relationship (not just correlation). As the old saying goes, "correlation does not imply causation." (WK 2011mm)

Notes: The examples in the categories are solely illustrative (not exhaustive), and the comparison is relative (not absolute), nor are they necessarily mutually exclusive. And some can be easily re-classified elsewhere. As generalities, they allow exceptions.
Source: From *Sec. 2.3.2* of *FPHDA*. See book for citations.

• Part Three •

Qualitative Research

CHAPTER 3
QUALITATIVE RESEARCH AND ITS LIMITATION

> The image of the world around us, which we carry in our head, is just a model. Nobody in his head imagines all the world, government or country. He has only selected concepts, and relationships between them, and uses those to represent the real system.
> —Jay W. Forrester (WK 2011xx)

The Broadness of Qualitative Research

As already explained in *Sec. 1.2* and in *Sec. 2.1*—data analysis involves both quantitative and qualitative research.

This relationship between quantitative and qualitative research is not mutually independent, however, because both can be exploited simulteanously. For instance, "[q]uantitative methods might be used with a global qualitative frame. Qualitative methods might be used to understand the meaning of the numbers produced by quantitative methods. Using quantitative methods, it is possible to give precise and testable expression to qualitative ideas. This combination of quantitative and qualitative data gathering is often referred to as mixed-methods research." (WK 2011d)

Thus, any study of quantitative research can be relevant to qualitative reserch too, as one cannot be completely separated from the other—and vice versa.

The issues concerning quantitative research were already addressed in Chapter Two, so what is left concerns qualitative research, which is to be analyzed in this chapter (Chapter Three)—with the caveat that the two cannot be completely separated.

With this dialectic treatment in mind—a good way to examine qualitative research and its limitation is by way of the evaluation of the extent to which it is in fact both possible and desirabe. And this can be done by way of a comprehensive analysis of qualitative research in the context of data analysis from the four perspectives of the mind, nature, society, and culture (in accordance to my sophisticated methodological holism, as explained in *Sec. 1.8*).

In other words, this chapter is organized in four main sections, in relation to (3.2) qualitative research and the mind, (3.3) qualitative research and nature, (3.4) qualitative research and society, and (3.5) qualitative research and culture—to be addressed hereafter, in that order (and summarized in *Table 3.1*).

Qualitative Research and the Mind

Qualitatve research, when examined from the perspective of the mind, can cast an illuminating light about its possibility and desirability, and this can be shown by way of two case studies, namely, (3.2.1) qualitative research, and the role of phenomenology, and (3.2.2) qualitative research, interpretation, and psychology—to be addressed in what follows, respectively.

Qualitative Research, and the Role of Phenomenology

A good stating point to evaluate the possibility and desirability of qualitative research from the perspective of the mind is to qualitative research and the role of phenomenology.

As already introduced in *Sec. 1.4*, the term "phenomenology" refers to "a broad…movement emphasizing the study of conscious experience. It was founded in the early years of the 20th century by Edmund Husserl [1970], expanded together with a circle of his followers at the universities of Göttingen and Munich in Germany…." (WK 2011v)

For Husserl, "phenomenology…is primarily concerned with the systematic reflection on and analysis of the structures of consciousness, and the phenomena which appear in acts of consciousness…. Phenomenology is a method of philosophical inquiry that rejects the rationalist bias that has dominated Western thought since Plato in favor of a method of reflective attentiveness that discloses the individual's 'lived experience.'" (WK 2011v)

Rooted in the skeptic tradition called "epoché," Husserl's phenomenology "entails the suspension of judgment while relying on the intuitive grasp of knowledge, free of presuppositions and intellectualizing. Sometimes depicted as the 'science of experience,' the phenomenological method is rooted in intentionality, Husserl's theory of consciousness (developed from Brentano). Intentionality represents an alternative to the representational theory of consciousness which holds that reality cannot be grasped directly because it is available only through perceptions of reality which are representations of it in the mind." (WK 2011v)

When put in the context of data analysis, phenomenology is to question the objective reality that data analysts tries to interpret and instead proposes a methodological alternative which is more based on a subjective, intuitive grasp of reality, "free of presuppositions and intellectualizing." (WK 2011v)

Main Concerns of Phenomenology

Phenomenology can be summarized in some main concerns, as shown below (and summarized in *Table 3.2*).

(a) Systematic Reflection

The first concern of phenomenology to be introduced here is about "systematic reflection."

In other words, " phenomenology attempts to create conditions for the objective study of topics usually regarded as subjective: consciousness and the content of conscious experiences such as judgments, perceptions, and emotions. Although phenomenology seeks to be scientific, it does not attempt to study consciousness from the perspective of clinical psychology or neurology. Instead, it seeks through systematic reflection to determine the essential properties and structures of consciousness and conscious experience." (WK 2011v)

Precisely here, Husserl was influenced by "his teachers, the philosophers and psychologists Franz Brentano and Carl Stumpf. An important element of phenomenology that Husserl borrowed from Brentano was intentionality (often described as 'aboutness'), the notion that consciousness is always consciousness of something. The object of consciousness is called the intentional object, and this object is constituted for consciousness in many different ways, through for instance perception, memory, retention and protention, signification, etc. Throughout these different intentionalities, though they have different structures and different ways of being 'about' the object, an object is still constituted as the same identical object; consciousness is directed at the same intentional object in direct perception as it is in the immediately following retention of this object and the eventual remembering of it." (WK 2011v)

(b) Anti-Reductionism

The second concern of phenomenology to be introduced here is about "anti-reductionism."

Phenomenology is “anti-reductionistic” in a fundamental way (although “many of the phenomenological methods involve various reductions”), because “the reductions are mere tools to better understand and describe the workings of consciousness, not to reduce any phenomenon to these descriptions.” (WK 2011v)

Thus, “when a reference is made to a thing's essence or idea, or when one details the constitution of an identical coherent thing by describing what one 'really' sees as being only these sides and aspects, these surfaces, it does not mean that the thing is only and exclusively what is described here: The ultimate goal of these reductions is to understand how these different aspects are constituted into the actual thing as experienced by the person experiencing it. Phenomenology is a direct reaction to the psychologism and physicalism of Husserl's time.” (WK 2011v)

(c) Bracketing

And the third concern of phenomenology to be introduced here is about “the suspension of belief” (or “bracketing”).

Maurice Natanson (1973: 63) once commented that “[t]he radicality of the phenomenological method is both continuous and discontinuous with philosophy’s general effort to subject experience to fundamental, critical scrutiny: to take nothing for granted and to show the warranty for what we claim to know.” (WK 2011v)

In practice, this means “an unusual combination of discipline and detachment to suspend, or bracket, theoretical explanations and second-hand information while determining one's 'naive' experience of the matter. The phenomenological method serves to momentarily erase the world of speculation by returning the subject to his or her primordial experience of the matter, whether the object of inquiry is a feeling, an idea, or a perception.” (WK 2011v)

This bracketing is a “suspension of belief in what we ordinarily take for granted or infer by conjecture,” which “diminishes the power of what we customarily embrace as objective reality. According to [R.] Safranski [1998], '[Husserl and his followers’] great ambition was to disregard anything that had until then been thought or said about consciousness or the world [while] on the lookout for a new way of letting the things [they investigated] approach them, without covering them up with what they already knew.'” (WK 2011v)

In the end, this “bracketing,” if going far enough, can help consciousness reach the stage of “pure consciousness” in “transncendental ego” (as opposed to “empirical ego”). (WK 2011v)

Main Components of Phenomenology

Besides these main concerns, phenomenology also has some important components to be understood. Consider, for illustration, some major components below (as summarized in *Table 3.2*).

(a) Intentionality

The first component of phenomenology to be discussed here concerns "intentionality."

The term "intentionality" refers to "the notion that consciousness is always consciousness *of* something. The word itself should not be confused with the 'ordinary' use of the word intentional, but should rather be taken as playing on the etymological roots of the word. Originally, intention referred to a 'stretching out' ('in tension,' lat. *intendere*), and in this context it refers to consciousness *'stretching out'* towards its object (although one should be careful with this image, seeing as there is not some consciousness first that, subsequently, stretches out to its object. Rather, consciousness *occurs* as the simultaneity of a conscious act and its object.) Intentionality is often summed up as 'aboutness.'" (WK 2011v; E. Husserl 1970; M. Natanson 1973)

The content of this "aboutness," however, is not important, because "the object of consciousness doesn't have to be a *physical* object apprehended in perception: it can just as well be a fantasy or a memory. Consequently, these 'structures' of consciousness, i.e., perception, memory, fantasy, etc., are called *intentionalities*." (WK 2011v)

Historically, "the term intentionality originated with the Scholastics in the medieval period and was resurrected by Brentano who in turn influenced Husserl's conception of phenomenology, who refined the term and made it the cornerstone of his theory of consciousness. The meaning of the term is complex and depends entirely on how it is conceived by a given philosopher. The term should not be confused with 'intention' or the psychoanalytic conception of unconscious 'motive' or 'gain.'" (WK 2011v)

(b) Intuition

The second component of phenomenology to be discussed here concerns "intuition."

The term "intuition" refers to "those cases where the intentional object is directly present to the intentionality at play; if the intention is 'filled' by the direct apprehension of the object, you have an intuited object. Having a cup of coffee in front of you, for instance, seeing it, feeling it, or even imagining it—these are all filled intentions, and the object is then *intuited.* The same goes for the apprehension of mathematical formulae or a number. If you do not have the object as referred to directly, the object is not intuited, but still intended, but then *emptily*. Examples of empty intentions can be signitive intentions—intentions that only *imply* or *refer* to their objects." (WK 2011v)

(c) Evidence

The third component of phenomenology to be discussed here concerns "evidence."

The word "evidence" here differs from the usage in everday life. For instance, "[i]n everyday language, we use the word evidence to signify a special sort of relation between a state of affairs and a proposition: State A is evidence for the proposition 'A is true.' In phenomenology, however, the concept of evidence is meant to signify the 'subjective achievement of truth.'" (WK 2011v; R. Sokolowski 2000)

Of course, "[t]his is not an attempt to reduce the objective sort of evidence to subjective 'opinion,' but rather an attempt to describe the structure of having something present in intuition with the addition of having it present as i*ntelligible*: ' Evidence is the successful presentation of an intelligible object, the successful presentation of something whose truth becomes manifest in the evidencing itself.'" (WK 2011v; R. Sokolowski 2000)

(d) Noesis and Noema

The fourth component of phenomenology to be discussed here concerns "noesis and noema."

These two terms are derived from "the Greek *nous* (mind), designate respectively the real content, noesis, and the ideal content, noema, of an intentional act (an act of consciousness)." (WK 2011v)

In regard to the "real" content—"the Noesis is the part of the act which gives it a particular sense or character (as in judging or perceiving something, loving or hating it, accepting or rejecting it, and so on). This is real in the sense that it is actually part of what takes place in the consciousness (or psyche) of the subject of the act." (WK 2011v)

In regard to the "ideal" content—"the full Noema is a complex ideal structure comprising at least a noematic sense and a noematic core. The correct interpretation of what Husserl meant by the Noema has long been controversial, but the noematic sense is generally understood as the ideal meaning of the act and the noematic core as the act's referent or object *as it is meant in the act*. One element of controversy is whether this noematic object is the same as the actual object of the act (assuming it exists) or is some kind of ideal object." (WK 2011v; D. Smith 2007)

(d) Empathy and Intersubjectivity

The fourth component of phenomenology to be discussed here concerns "empathy and intersubjectivity."

The word "empathy" refers to "the experience of another human body *as* another. While we often identify others with their physical bodies, this type of phenomenology requires that we focus on the subjectivity of the other, as well as our intersubjective engagement with them." (WK 2011v)

For Husserl, "this was done by a sort of apperception built on the experiences of your own lived-body. The lived body is your own body as experienced by yourself, *as* yourself. Your own body manifests itself to you mainly as your possibilities of acting in the world. It is what lets you reach out and grab something, for instance, but it also, and more importantly, allows for the possibility of changing your point of view. This helps you differentiate one thing from another by the experience of moving around it, seeing new aspects of it (often referred to as making the absent present and the present absent), and still retaining the notion that this is the same thing that you saw other aspects of just a moment ago (it is identical). Your body is also experienced as a duality, both as object (you can touch your own hand) and as your own subjectivity (you experience being touched)." (WK 2011v)

Then, "[t]he experience of your own body as your own subjectivity is then applied to the experience of another's body, which, through apperception, is constituted as another subjectivity. You can thus recognise the Other's intentions, emotions, etc. This experience of empathy is important in the phenomenological account of intersubjectivity. In phenomenology, intersubjectivity constitutes objectivity (i.e., what you experience as objective is experienced as being intersubjectively available—available to all other subjects. This does not imply that objectivity is reduced to subjectivity nor does it imply a relativist position, cf. for instance intersubjective verifiability)." (WK 2011v)

Thus, empathy and intersubjectivity are closely related, because "[i]n the experience of intersubjectivity, one also experiences oneself as being a subject among other subjects, and one experiences oneself as existing objectively *for* these Others; one experiences oneself as the noema of Others' noeses, or as a subject in another's empathic experience. As such, one experiences oneself as objectively existing subjectivity. Intersubjectivity is also a part in the constitution of one's lifeworld, especially as 'homeworld.'" (WK 2011v)

(5) Lifeworld

And the fifth component of phenomenology to be discussed here concerns "lifeworld."

The term "lifeworld" (or "German: *Lebenswelt*) refers to "the 'world' each one of us lives in. One could call it the 'background' or 'horizon' of all experience, and it is that on which each object stands out as itself (as different) and with the meaning it can only hold for us. The lifeworld is both personal and intersubjective (it is then called a 'homeworld'), and, as such, it does not enclose each one of us in a solus ipse." (WK 2011v)

Criticisms of Phenomenology

However, Husserl's phenomenology has its own problems. Consider, for illustration, some illustrative criticisms below (as summarized in *Table 3.2*).

Firstly, one major criticism of Husserl's phenomenology concerns what Martin Heidegger called its "subjectivist tendencies." (WK 2011v)

In other words, "[w]hereas Husserl conceived humans as having been constituted by states of consciousness, Heidegger countered that consciousness is peripheral to the primacy of one's existence (i.e., the mode of being of Dasein) which cannot be reduced to one's consciousness of it. From this angle, one's state of mind is an 'effect' rather than a determinant of existence, including those aspects of existence that one is not conscious of. By shifting the center of gravity from consciousness (psychology) to existence (ontology), Heidegger altered the subsequent direction of phenomenology, making it at once both personal and mysterious." (WK 2011v)

Graham Harman also made a comparable point, albeit in a different way, when he criticised phenomenology "to be deficient in that it constantly subordinates the independent life of objects to our (human) access to them," and proposed a "radical break with the traditional use of terms such as intentionality" in phenomenology. (WK 2011v)

Secondly, another major criticism of Husserl's phenomenology concerns its holy grail of "transcendental ego" (as opposed to "empirical ego").

The reason is that "Husserl…wanted to exclude any hypothesis on the existence of external objects," so he "introduced the method of phenomenological reduction to eliminate them. What was left over was the pure transcendental ego, as opposed to the concrete empirical ego. Now (transcendental) phenomenology is the study of the essential structures that are left in pure consciousness: this amounts in practice to the study of the noemata and the relations among them. The philosopher Theodor

Adorno criticised Husserl's concept of phenomenological epistemology in his metacritique *Against Epistemology*, which is anti-foundationalist in its stance." (WK 2011v)

In fact, some existentialists made a comparable critique of Husserl's phenomenology by their "rejection of the transcendental ego. Merleau-Ponty objects to the ego's transcendence of the world, which for Husserl leaves the world spread out and completely transparent before the conscious." (WK 2011vv)

This "transcendental" search in Husserl's phenomenology (for its version of "transcendental ego," as opposed to "empirical ego") led Stephen Hicks (2004) to once aptly "describe...phenomenology as an outgrowth of European Kantianism" in modern times," as he wrote: "Neo-Kantianism evolved during the nineteenth century, and by the twentieth century two main forms had emerged. One form was Structuralism, of which Ferdinand de Saussure was a prominent exponent, representing the broadly rationalist wing of Kantianism. The other was Phenomenology, of which Edmund Husserl was a prominent representative, representing the broadly empiricist wing of Kantianism." (WK 2011v)

And thirdly, still another major criticism of Husserl's phenomenology concerns its neglect of the study of "unconsciousness."

This criticism has much to do with a consequence of Heidegger's critique of Husserl's phenomenology. In other words, "[b]y shifting the center of gravity from consciousness (psychology) to existence (ontology), Heidegger altered the subsequent direction of phenomenology, making it at once both personal and mysterious. One of the consequences of Heidegger's modification of Husserl's conception of phenomenology was its increased relevance to psychoanalysis. Whereas Husserl gave priority to a depiction of consciousness that was fundamentally alien to the psychoanalytic conception of the unconscious, Heidegger offered a way to conceptualize experience that could accommodate those aspects of one's existence that lie on the periphery of sentient awareness." (WK 2011v; M. Natanson 1973; R. Safranski 1998)

Beyond Phenomenology

These criticisms of phenomenology, however, do not mean to dismiss its usefulness for qualitative research (like the technique of bracketing) but are to show us the two opposing sides of the debate, such that the possibility and desirability of qualitative research (from the perspective of the mind with qualitative research and the role of phenomenology as a case study here) are not to the extent that the respective defenders would like us to believe.

In addition, the analysis of qualitative research and the role of phenomenology can teach us something valuable about the ontological principles in existential dialectics, and good examples include the formalness-informalness principle, the absoluteness-relativeness principle, the predictability-unpredictability principle, the explicability-inexplicability principle, the fiction-reality principle, the preciseness-vagueness principle, the simpleness-complicatedness principle, the openness-hiddenness principle, the denseness-emptiness principle, the change-constancy principle, the slowness-quickness principle, the expansion-contraction principle, the theory-praxis principle, the convention-novelty principle, the evolution-transformation principle, the symmetry-asymmetry principle, the softness-hardness principle, the seriousness-playfulness principle, the regression-progression principle, the same-difference principle, and the stability-reaction principle.

For instance, in relation to the formalness-informalness principle, if there is formalness (e.g., the formal logical requirement of completeness and soundness in a system of ideas, as in Husserl's phenomenology), there is informalness (e.g., the non-formal existence of incompletness and unsoundness in Husserl's phenomenology, as shown in the rejection of "transcendental ego" by its critics as non-existent). And the reverse direction also holds true.

In relation to the absoluteness-relativeness principle, if there is absoluteness (e.g., the absolute concern of Husserl with "consciousness"), there is relativeness (e.g., what is right for Husserl in regard to his focus on "consciousness" is not necessarily so for Heidegger who "countered that consciousness is peripheral to the primacy of one's existence," or "the mode of being of Dasein"). And the reverse direction also holds true.

In relation to the predictability-unpredictability principle, if there is predictability (e.g., the predictable tendency of those in Husserl's phenomenology to focus on "consciousness"), there is unpredictability (e.g., the more difficult task to predict exactly the extent to which those in Husserl's phenomenology will reach "pure consciousness" in their version of "transcendental ego"). And the reverse direction also holds true.

In relation to the explicability-inexplicability principle, if there is explicability (e.g., the explanation by Husserl that "bracketing" is desirable because it can eventually help us reach "pure consciousness" in "transcendental ego"), there is inexplicability (e.g., the lack of sufficient explanation by Husserl of why "bracketing" is necessarily desirable because, for Heidegger, "consciousness is peripheral to the primacy of one's existence," that is, "the mode of being of Dasein," which "cannot be reduced" to consciousness and is more important). And the reverse direction also holds true.

In relation to the fiction-reality principle, if there is fiction (e.g., the fictional aspect of Husserl's phenomenology, as shown in the rejection of "transcendental ego" by its critics as non-existent), there is reality (e.g., the realistic aspect of Husserl's phenomenology, as shown in its "general effort to subject experience to fundamental, critical scrutiny" as pointed out by Maurice Natanson). And the reverse direction also holds true.

In relation to the preciseness-vagueness principle, if there is preciseness (e.g., the precise identification of five major compoents in Husserl's phenomenology for discussion in the literature aforecited), there is vagueness (e.g., the vagueness in the identification of the five major compoents in Husserl's phenomenology for discussion in the literature aforecited, since it is not clear why there must be only five, not four, six, seven, and so on). And the reverse direction also holds true.

In relation to the simpleness-complicatedness principle, if there is simpleness (e.g., the relatively simple experience of the world in everyday life, according to Husserl's phenomenology), there is complicatedness (e.g., the relatively more complicated analysis of one's experience of the world by the use of "bracketing" in Husserl's phenomenology, such that it subjects "experience to fundamental, critical scrutiny: to take nothing for granted, as pointed out by Maurice Natanson). And the reverse direction also holds true.

In relation to the openness-hiddenness principle, if there is openness (e.g., the open exploration of the possibility of "pure consciousness" in Husserl's phenomenology), there is hiddenness (e.g., the hidden bias in the search for "pure consciousness" in Husserl's phenomenology, because of its hidden Kantian influence in regard to the holy grail of "transcendental ego," which, for the critics, does not exist). And the reverse direction also holds true.

In relation to the denseness-emptiness principle, if there is denseness (e.g., the relatively denser concentration of concerns with "bracketing" by those in Husserl's phenomenology), there is emptiness (e.g., the relatively less dense, or more empty, concentration of concerns with "bracketing" by those in Plato's rationalism). And the reverse direction also holds true.

In relation to the change-constancy principle, if there is change (e.g., the ever new ways to understand human experience with reality, as shown in Husserl's phenomenology), there is constancy (e.g., the ever constant existence of different problems with the new ways to understand human experience with reality, as shown in the problems of Husserl's phenomenology pointed out by the critics). And the reverse direction also holds true.

In relation to the slowness-quickness principle, if there is slowness (e.g., the relatively slower readiness of those in Plato's rationalism to subject "experience to fundamental, critical scrutiny: to take nothing for granted"), there is quickness (e.g., the relatively quicker readiness of those in Husserl's phenomenology to subject "experience to fundamental, critical scrutiny: to take nothing for granted"). And the reverse direction also holds true.

In relation to the expansion-contraction principle, if there is expansion (e.g., the relatively more developed ability of those in Husserl's phenomenology to deal with "subjectivist tendencies"), there is contraction (e.g., the relatively less developed ability of those in Husserl's phenomenology to deal with "the mode of being of Dasein"). And the reverse direction also holds true.

In relation to the theory-praxis principle, if there is theory (e.g., the theoretical construction of Heidgger's version of phenomenology in his critique of Husserl's), there is praxis (e.g.,the practical application of Heidgger's version of phenomenology to the field of psychoanalysis, because Heidegger criticized Husserl for failing to take into account other domains of reality that "one is not conscious of," such as unconsciousness, which is "both personal and mysterious"). (WK 2011v; M. Natanson 1973; R. Safranski 1998) And the reverse direction also holds true.

In relation to the convention-novelty principle, if there is convention (e.g., the conventional view of phenomenology dominated by Husserl's version), there is novelty (e.g., the alternative novel challenge to the conventional view of phenomenology dominated by Husserl's version, by the new idea of phenomenology proposed by Heideigger, with "its increased relevance to psychoanalysis" in relation to unconsciousness). (WK 2011v; M. Natanson 1973; R. Safranski 1998) And the reverse direction also holds true.

In relation to the evolution-transformation principle, if there is evolution (e.g., the natural evolution of humans to experience consciousness in the state of nature), there is transformation (e.g., the technical transformation of human ability to understand consciousness by the invention of phenomenology like the one by Husserl). And the reverse direction also holds true.

In relation to the symmetry-asymmetry principle, if there is symmetry (e.g., the existence of different schools of philosophy), there is asymmetry (e.g., the popularity of Plato's rationalism in antiquity—but the popularity of Husserl's phenomenology in the 20th century). And the reverse direction also holds true.

In relation to the softness-hardness principle, if there is softness (e.g., the appreciation of Husserl's phenomenology by its followers), there is hardness (e.g., the critique of Husserl's phenomenology by its critics). And the reverse direction also holds true.

In relation to the seriousness-playfulness principle, if there is seriousness (e.g., the serious business of phenomenologists to study "consciousness"), there is playfulness (e.g., the playful part of phenomenologists, when different scholars give it a different interpretation, like the ones by Max Scheler, Martin Heidegger, Alfred Schütz, Graham Harman, and so on). (WK 2011v) And the reverse direction also holds true.

In relation to the regression-progression principle, if there is regression (e.g., the regression made by Husserl's phenomenology, as shown in the problems pointed out by the critics), there is progression (e.g., the progress made by Husserl's phenomenology, as shown in its influence in the 20th century). And the reverse direction also holds true.

In relation to the same-difference principle, if there is similarity in outcome (e.g., the contribution to the molding and control of beliefs, values, and behaviors, regardless of whether this be done by way of Husserl's phenomenology, or by way of Plato's rationalism), there is difference in outcome (e.g., the contribution to the molding and control of beliefs, values, and behaviors by way of Husserl's phenomenology for a relatively more subjectivist lifeworld—but the contribution to the molding and control of beliefs, values, and behaviors by way of Plato's rationalism for a relatively more rationalist lifeworld). And the reverse direction also holds true.

And in relation to the stability-reaction principle, if there is stability (e.g., the relatively more stable condition of one's subjectivist tendencies for bracketing, if one follows Husserl's phenomenology), there is reaction (e.g., one's subjectivist tendencies for bracketing can lead to further changes later on, because, if going far enough in the process of interacting with others and the world, they can eventually lead one closer and closer to "pure consciousness," after myraid cycles of reduction, according to Husserl's phenomenology). And the reverse direction also holds true

Qualitative Research, Interpretation, and Psychology

The discussion of subjective consciousness in phenomenology (in the previous section) also raises an interesting question about the nature of interpretation in the mind (when dealing with its subjective experience of reality).

In this sense, another way to evaluate the possibility and desirability of qualitative research from the perspective of the mind is to explore qualitative research, interpretation, and psychology.

Interpretation and Humanistic Psychology

Psychology can help us understand the nature of interpretation in relation to the mind, and a good example is "humanistic psychology."

The term "humanistic psychology" refers to a branch of psychology "which rose to prominence in the mid-20th century, drawing on the work of early pioneers like Carl Rogers and the philosophies of existentialism and phenomenology. It adopts a holistic approach to human existence through investigations of meaning, values, freedom, tragedy, personal responsibility, human potential, spirituality, and self-actualization." (WK 2011yy; A. Colman 2009; C. Aanstoos 2000)

Humanistic psychology becomes involved in the nature of interpretation, because its foundation was grounded on "a holistic approach to human existence through investigations of meaning." (WK 2011yy)

This is so, even though the nature of interpretation is usually the main concern of hermeneutics in the humanities (not in psychology), but things started to change in recent times. For instance, "psychologists have recently become interested in hermeneutics, especially as alternatives to Cognitivism. Hubert Dreyfus' critique of conventional artificial intelligence has been influential not only in AI but in psychology, and psychologists are increasingly interested in hermeneutic approaches to meaning and interpretation as discussed by philosophers such as Heidegger (cf Embodied cognition) and the later Wittgenstein (cf discursive psychology). Hermeneutics is…influential in Humanistic Psychology." (WK 2011ff; D. Rennie 2007)

This involvement of humanistic psychology in hermeneutics (for the study of meaning and interpretation) is important for qualiative research in data analysis, even though it does not totally ignore quantitative research either. In other words, humanistic psychology "is not opposed to quantitative methods, but… [a] favors letting the methods be derived from the subject matter and not uncritically adopting the methods of natural science, and [b] advocates for methodological pluralism." (WK 2011yy)

Therefore, "much of the subject matter of [humanistic] psychology lends itself to qualitative approaches (e.g., the lived experience of grief) , and quantitative methods are mainly appropriate when something can be counted without leveling the phenomena (e.g., the length of time spent crying)." (WK 2011yy; A. Giorgi 2009)

Hermeneutic Circle for Psychological Study

A good illustration of the involvement of humanistic psychology in hermeneutics (for the study of meaning and interpretation) concerns "the hermeneutic circle."

Although the study of "hermeneutic circle" had been until recently the exclusive domain in the humanities (as in philosophy, theology, and the arts), it has recently been applied to other fields (like humanistic pyschology in the current context).

In any event, the term "hermeneutic circle" originally refers to "the process of understanding a text hermeneutically. It refers to the idea that one's understanding of the text as a whole is established by reference to the individual parts and one's understanding of each individual part by reference to the whole. Neither the whole text nor any individual part can be understood without reference to one another, and hence, it is a circle. However, this circular character of interpretation does not make it impossible to interpret a text; rather, it stresses that the meaning of a text must be found within its cultural, historical, and literary context." (WK 2011zz)

Consider, for illustration, a few versions of "the hermeneutic circle" as understood by some known scholars in the field.

(a) Friedrich Schleiermacher

Friedrich Schleiermacher (1768–1834) gave hermeneutics a psychological touch by expanding "the nature of understanding in relation not just to the problem of deciphering sacred texts, but to all human texts and modes of communication. The interpretation of a text must proceed by framing the content asserted in terms of the overall organization of the work. He distinguishes between grammatical interpretation and psychological interpretation. The former studies how a work is composed from general ideas, the latter considers the peculiar combinations that characterize the work as a whole." (WK 2011ff)

For Schleiermacher, "every problem of interpretation is a problem of understanding. He even defined hermeneutics as the art of avoiding misunderstanding. He provides a solution to avoidance of misunderstanding: knowledge of grammatical and psychological laws in trying to understand the text and the writer. There arose in his time a fundamental shift from understanding not only the exact words and their objective meaning to individuality of the speaker or author." (WK 2011ff; M. Forster,2002; B. Gjesdal 2007)

With this psychological touch, Schleiermacher stressed "the importance of the interpreter in the process of interpretation. Schleiermacher's hermeneutics focuses on the importance of the interpreter understanding the text as a necessary stage to interpreting it. Understanding, for Schleiermacher, does not simply come from reading the text, but involves knowledge of the historical context of the text and the psychology of the author." (WK 2011ff; H. Gadamer 1975)

(b) Wilhelm Dilthey

Wilhelm Dilthey divided the nature of interpretation from the vantage point of an individual into three levels, as shown below: (WK 2011ff)

- "Experience"—This first level "means to feel the situation or thing personally. Dilthey suggests that we can always grasp the meaning of unknown thinking when we try to experience it."
- "Expression"—This second level "converts experience into meaning because when we express something, it is no more a private and personal thing but an appeal to somebody outside of oneself. Every saying is an expression. Dilthey suggests that we can always return to an expression, especially its written form, and this practice has the same objective value as an experiment in sciences. The possibility of returning makes scientific analysis possible and therefore humanities may be labeled as science. Moreover, Dilthey assumes that expression may be 'saying' more than the speaker intended because the expression brings forward meanings that the individual conscience may not fully understand."
- "Comprehension"—This third level "is a dimension which contains both comprehension and incomprehension. Incomprehension means more or less wrong understanding. Dilthey presumes that comprehension produces coexistence: He who understands, understand others; he who does not understand stays alone."

Surely, more can be said about the three levels, which are summarized here for the purpose of introduction.

(c) Martin Heidegger

Martin Heidegger deepened hermeneutics through the study of "the experiences of the author" by shifting "the focus…to existential understanding, which was treated more as a direct, non-mediated, thus in a sense more authentic way of being in the world than simply as a way of knowing." (WK 2011ff)

Thus, for Heidegger, "such texts, and the people who produce them, cannot be studied using the same scientific methods as the natural sciences....Moreover,...such texts are conventionalized expressions of the experience of the author; thus, the interpretation of such texts will reveal something about the social context in which they were formed, but, more significantly, provide the reader with a means to share the experiences of the author. The reciprocity between text and context is part of what Heidegger called the hermeneutic circle. Among the key thinkers who elaborated this approach is the sociologist Max Weber." (WK 2011ff)

In other words, "Heidegger [1962] developed the concept of the Hermeneutic Circle to envision a whole in terms of a reality that was situated in the detailed experience of everyday existence by an individual (the parts)." (WK 2011zz)

(d) Hans-Georg Gadamer

Hans-Georg Gadamer gave hermeneutics a different treat by focusing on the "prejudice" of an interpreter in a given "tradition." (WK 2011ff)

In other words, "Gadamer asserts that methodical contemplation is opposite to experience and reflection. We can reach the truth only by understanding or even mastering our experience. According to Gadamer, experience isn't fixed but rather changing and always indicating new perspectives. The most important thing is to unfold what constitutes individual comprehension. Gadamer points out in this context that prejudice is a (nonfixed) reflection of that unfolding comprehension, and is not per se without value. Being alien to a particular tradition is a condition of understanding. Gadamer points out that we can never step outside of our tradition; all we can do is try to understand it. This further elaborates the idea of the hermeneutic circle." (WK 2011ff)

Problems with the Hermeneutic Circle in Psychology

But the hermeneutic circle in the context of interpretation and humanistic psycholoogy has its own problems. Consider, for illustration, some main criticisms below (as summarized in *Table 3.3*).

Firstly, one main criticism against the hermeneutic circle is that the "circle" in hermenutics remains unclear.

For instance, "Judith N. Shklar [1986] points out the ambiguity in the meaning and function of the 'circle' as a metaphor for understanding. It seems to imply a center, but it is unclear whether the interpreter he/herself stands there, or whether, on the contrary, some 'organizing principle and illuminating principle apart from him [is] there waiting to be discovered.'" (WK 2011zz)

Secondly, another main criticism against the hermeneutic circle is that it may not make much sense.

For instance, for Shklar (1986: 657-8), "the hermeneutic circle makes sense only if there is a known and closed whole, which can be understood in terms of its own parts and which has as its core God, who is its anchor and creator. Only the Bible really meets these conditions. It is the only possibly wholly self-sufficient text." (WK 2011zz)

Thirdly, still another main criticism against the hermeneutic circle is that it is too rigid.

For instance, "Gadamer and others assume a fixed role for tradition (individual and disciplinary/academic) in the process of any hermeneutic understanding, while it is more accurate to say that interpreters have multiple and sometimes conflicting cultural attachments, yet this do not prevent intercultural and/or interdisciplinary dialogue." (WK 2011zz)

Jürgen Habermas also "criticized the conservatism of previous hermeneutics, especially Gadamer, because the focus on tradition seemed to undermine possibilities for social criticism and transformation." (WK 2011ff)

And fourthly, one additional main criticism against the hermeneutic circle is that it still does not explain anything.

For instance, Shklar (1986) "warns that, at least in social science, interpretation is not a substitute for explanation." (WK 2011zz)

Beyond Interpretation and Psychology

These criticisms of the hermeneutic circle in the context of interpretation and humanistic psychology do not imply a dismissal of its usefulness but serve to show us the two opposing sides of the debate, such that the possibility and desirability of qualitative research (from the perspective of the mind with qualitative research, interpretation, and psychology as a case study here) are not to the extent that the respective defenders would like us to believe.

Furthermore, the analysis of qualitative research, interpretation, and psychology reveals something fascinating about the ontological principles in existential dialectics, and good examples include the formalness-informalness principle, the absoluteness-relativeness principle, the partiality-totality principle, the predictability-unpredictability principle, the explicability-inexplicability principle, the fiction-reality principle, the preciseness-vagueness principle, the simpleness-complicatedness principle, the openness-hiddenness principle, the denseness-emptiness principle, the change-constancy principle, the slowness-quickness principle, the expansion-contraction principle, the theory-praxis principle, the convention-novelty principle, the evolution-transformation principle, the symmetry-asymmetry principle, the softness-hardness principle, the seriousness-playfulness principle, the regression-progression principle, the same-difference principle, and the stability-reaction principle.

For instance, in relation to the formalness-informalness principle, if there is formalness (e.g., the formal logical requirement of completeness in a system of ideas, as in the hermeneutic circle), there is informalness (e.g., the non-formal existence of incompleteness in the hermeneutic circle, as shown in the criticism against Gadamer's version, since, for the critics, the focus on the rigid prejudice of tradition is question-begging or circular, since it can be used against the hermeneutic circle itself). And the reverse direction also holds true.

In relation to the absoluteness-relativeness principle, if there is absoluteness (e.g., the absolute view of Gadamer in regard to the "prejudice" of an interpreter in a given tradition), there is relativeness (e.g., what is right for Gadamer in regard to the "prejudice" of an interpreter in a given tradition is not necessarily so for Shklar, who argued that, although "interpreters have multiple and sometimes conflicting cultural attachments, yet this do not prevent intercultural and/or interdisciplinary dialogue," so as to go beyond "prejudice"). And the reverse direction also holds true.

In relation to the partiality-totality principle, if there is partiality (e.g., the partial view of the parts in the hermeneutic circle), there is totality (e.g., the holistic view of the whole in the hermeneutic circle, such that the whole cannot be reduced to the sum of its parts alone). And the reverse direction also holds true.

In relation to the predictability-unpredictability principle, if there is predictability (e.g., the predictable tendency of those in Schleiermacher's hermeneutics to focus on the grammatical interpretation and psychological interpretation of a text by an author), there is unpredictability (e.g., the more difficult task to predict exactly the extent to which a particular interpreter will succeed in interpreting correctly "the peculiar combinations that characterize the work as a whole" by an author in relation to psychological interpretation, since there can be multiple conflicting interpretations). And the reverse direction also holds true.

In relation to the explicability-inexplicability principle, if there is explicability (e.g., the explanation by Gadamer that interpretation is valuable because "[w]e can reach the truth only by understanding or even mastering our experience," even though "we can never step outside of our tradition," so "all we can do is try to understand it"), there is inexplicability (e.g., the lack of sufficient explanation by Gadamer of why "we can never step outside of our tradition," since, for the critics like Habermas, "the focus on tradition seemed to undermine possibilities for social criticism and transformation"). And the reverse direction also holds true.

In relation to the fiction-reality principle, if there is fiction (e.g., the fictional aspect of Dilthey's hermeneutic circle, when "incomprehension" produces "wrong" understanding), there is reality (e.g., the realistic aspect of Dilthey's hermeneutic circle, when "comprehension" produces "correct" understanding). And the reverse direction also holds true.

In relation to the preciseness-vagueness principle, if there is preciseness (e.g., the precise identification by Dilthey of three levels of interpretation from the vantage point of an individual), there is vagueness (e.g., the vagueness in the identification by Dilthey of the three levels of interpretation from the vantage point of an individual, since it is not clear why there must be only three, not two, four, five, and so on). And the reverse direction also holds true.

In relation to the simpleness-complicatedness principle, if there is simpleness (e.g., the relatively simple analysis of the hermeneutic circle by its advocates), there is complicatedness (e.g., the relatively more complicated analysis of the hermeneutic circle, by challenging the claims and assumptions, as shown in the different problems pointed out by the critics as cited). And the reverse direction also holds true.

In relation to the openness-hiddenness principle, if there is openness (e.g., the open exploration of a new way of understanding in the hermeneutic circle, for Gadamer), there is hiddenness (e.g., the hidden bias in the hermeneutic circle, because of the "prejudice" of an interpreter in a given tradition, for Gadamer). And the reverse direction also holds true.

In relation to the denseness-emptiness principle, if there is denseness (e.g., the relatively denser concentration of concerns with "prejudice" in Gadamer's hermeneutics), there is emptiness (e.g., the relatively less dense, or more empty, concentration of concerns with "prejudice" in Schleiermacher's hermeneutics). And the reverse direction also holds true.

In relation to the change-constancy principle, if there is change (e.g., the ever new ways to understand the nature of interpretation, as shown in the different scholars on the hermeneutic circle), there is constancy (e.g., the ever constant existence of different problems with the different new ways to understand the nature of interpretation, as shown in the different problems with the different versions of the hermeneutic circle pointed out by the critics). And the reverse direction also holds true.

In relation to the slowness-quickness principle, if there is slowness (e.g., the relatively slower readiness of those in the hermeneutic circle for psychological study to engage in quantitative analysis for statistical hypthothesis testing), there is quickness (e.g., the relatively quicker readiness of those in the hermeneutic circle for psychological study to engage in qualitative analysis for meaning and interpretation). And the reverse direction also holds true.

In relation to the expansion-contraction principle, if there is expansion (e.g., the relatively more developed ability of those in Gadamer's version of the hermeneutic circle to deal with the "prejudice" of an individual in a given tradition), there is contraction (e.g., the relatively less developed ability of those in Gadamer's version of the hermeneutic circle to deal with "the detailed experience of everyday existence by an individual" as advocated by those in Heidegger's version). And the reverse direction also holds true.

In relation to the theory-praxis principle, if there is theory (e.g., the theoretical construction of hermeneutics in philosophy), there is praxis (e.g.,the practical application of hermeneutics in philosophy to the field of humanisitc psychology in the 20th century). (WK 2011jj) And the reverse direction also holds true.

In relation to the convention-novelty principle, if there is convention (e.g., the conventional version of the hermeneutic circle as first proposed by Schleiermacher to give it a psychological touch), there is novelty (e.g., the alternative novel challenge to the conventional version of the hermeneutic circle as first proposed by Schleiermacher, by the new idea of Habermas to give it a practical turn for social change). And the reverse direction also holds true.

In relation to the evolution-transformation principle, if there is evolution (e.g., the natural evolution of humans to understand things in the state of nature in a primitive way), there is transformation (e.g., the technical transformation of human ability to understand things by the invention of the hermeneutic circle for humanistic psychology). And the reverse direction also holds true.

In relation to the symmetry-asymmetry principle, if there is symmetry (e.g., the existence of different versions of the hermeneutic circle), there is asymmetry (e.g., the popularity of Schleiermacher's version of the hermeneutic circle in the older days of the 19th century—but the popularity of Habermas's version of the hermeneutic circle in our time). And the reverse direction also holds true.

In relation to the softness-hardness principle, if there is softness (e.g., the defense of the hermeneutic circle by its supporters like Luis Schokel), there is hardness (e.g., the critique of the hermeneutic circle by its critics like Judith Shklar aforecited). (WK 2011zz; L. Schokel 1998) And the reverse direction also holds true.

In relation to the seriousness-playfulness principle, if there is seriousness (e.g., the serious business of those in the hermeneutic circle to study the nature of meaning and interpretation), there is playfulness (e.g., the playful part of those in the hermeneutic circle, since "interpretation must oscillate between particular and general, part and whole" in a circle, as pointed out by Heidegger, so an interpreter can play with different possible interpretations). (WK 2011zz) And the reverse direction also holds true.

In relation to the regression-progression principle, if there is regression (e.g., the regression made by the hermeneutic circle, as shown in the problems pointed out by the critics aforecited), there is progression (e.g., the progress made by the hermeneutic circle, as shown in its influence on humanistic psychology in our time aforediscussed). And the reverse direction also holds true.

In relation to the same-difference principle, if there is similarity in outcome (e.g., the contribution to the molding and control of beliefs, values, and behaviors, regardless of whether this be done by way of Gadamer's hermeneutics, or by way of Habermas's hermeneutics), there is difference in outcome (e.g., the contribution to the molding and control of beliefs, values, and behaviors by way of Gadamer's hermeneutics for a relatively more conservative lifeworld—but the contribution to the molding and control of beliefs, values, and behaviors by way of Habermas's hermeneutics for a relatively more liberal lifeworld). And the reverse direction also holds true.

And in relation to the stability-reaction principle, if there is stability (e.g., the relatively more stable condition of an individual's understanding, when it is within the boundary of "prejudice" in a given "tradition," according to Gadamer), there is reaction (e.g., an individual's understanding can lead to further changes over time, since, for Gadamer, "experience isn't fixed but rather changing and always indicating new perspectives" in the endless process of the hermeneutic circle). And the reverse direction also holds true.

Qualitative Research and Nature

Qualitative research, when examined from the perspective of nature, can show us in a refreshing way its possibility and desirability, and this can be shown by way of two case studies, namely, (3.3.1) qualitative research, and the promise of applied evolution, and (3.3.2) qualitative research, and the modeling of genetic regulatory networks—to be addressed in what follows, in that order.

Qualitative Research, and the Promise of Applied Evolution

One good way to evaluate the possibility and desirability of qualitative research from the perspective of nature is to explore qualitative research and applied evolution.

The Promise of Applied Evolution

The words "applied evolution" above refer to the use of "concepts and models" in evolutionary theory for different applications in different fields, especially in the sciences, as J. Bull (2001) had written a book about it. (WK 2011y)

For instance, "[i]n computer science, simulations of evolution using evolutionary algorithms and artificial life started in the 1960s, and was extended with simulation of artificial selection. Artificial evolution became a widely recognised optimisation method as a result of the work of Ingo Rechenberg in the 1960s. He used evolution strategies to solve complex engineering problems." (WK 2011y; A. Fraser 1958)

And "genetic algorithms in particular became popular through the writing of John Holland. Practical applications also include automatic evolution of computer programs. Evolutionary algorithms are now used to solve multi-dimensional problems...and also to optimise the design of systems." (WK 2011y; J. Holland 1975; J. Koza 1992; M. Jamshidi 2003)

Since qualitative research, as already introduced in *Sec. 1.2*, is useful for exploration, a good question here is, How effective is the application of evolutionary theory to other fields for idea generation?

An Example of Applied Evolution

Consider, for illustration, an example of applied evolution in the state of nature, namely, "genetic algorithm." (WK 2011aaa)

The term "genetic algorithm" refers to "a search heuristic that mimics the process of natural evolution. This heuristic is routinely used to generate useful solutions to optimization and search problems. Genetic algorithms belong to the larger class of evolutionary algorithms (EA), which generate solutions to optimization problems using techniques inspired by natural evolution, such as inheritance, mutation, selection, and crossover." (WK 2011aaa)

In a genetic algorithm, "a population of strings (called chromosomes or the genotype of the genome), which encode candidate solutions (called individuals, creatures, or phenotypes) to an optimization problem, evolves toward better solutions. Traditionally, solutions are represented in binary as strings of 0s and 1s, but other encodings are also possible. The evolution usually starts from a population of randomly generated individuals and happens in generations. In each generation, the fitness of every individual in the population is evaluated, multiple individuals are stochastically selected from the current population (based on their fitness), and modified (recombined and possibly randomly mutated) to form a new population. The new population is then used in the next iteration of the algorithm. Commonly, the algorithm terminates when either a maximum number of generations has been produced, or a satisfactory fitness level has been reached for the population. If the algorithm has terminated due to a maximum number of generations, a satisfactory solution may or may not have been reached." (WK 2011aaa)

In other words, a genetic algorithm requires two essential domains, as shown below: (WK 2011aaa)

- "A genetic representation of the solution domain"
- "A fitness function to evaluate the solution domain"

And genetic algorithin is carried out according to a four-step methodology, to be addressed in what follows.

(a) The Methodology of Genetic Algorithm

Genetic Algorithm has a methodology which can be divided into four major steps, as explained below (and summarized in *Table 3.4*).

(a1) Initialization

In the first step of the methodology of genetic algorithm, "many individual solutions are randomly generated to form an initial population. The population size depends on the nature of the problem, but typically contains several hundreds or thousands of possible solutions. Traditionally, the population is generated randomly, covering the entire range of possible solutions (the search space). Occasionally, the solutions may be 'seeded' in areas where optimal solutions are likely to be found." (WK 2011aaa)

(a2) Selection

In the second step of the methodology of genetic algorithm, "a proportion of the existing population is selected to breed a new generation. Individual solutions are selected through a fitness-based process, where fitter solutions (as measured by a fitness function) are typically more likely to be selected. Certain selection methods rate the fitness of each solution and preferentially select the best solutions. Other methods rate only a random sample of the population, as this process may be very time-consuming." (WK 2011aaa)

(a3) Reproduction

In the third step of the methodology of genetic algorithm, the task here is "to generate a second generation population of solutions from those selected through genetic operators: crossover (also called recombination), and/or mutation." (WK 2011aaa)

Over here, "a pair of 'parent' solutions is selected for breeding from the pool selected previously. By producing a 'child' solution using the above methods of crossover and mutation, a new solution is created which typically shares many of the characteristics of its 'parents.' New parents are selected for each new child, and the process continues until a new population of solutions of appropriate size is generated. Although reproduction methods that are based on the use of two parents are more 'biology inspired,' some research suggests more than two 'parents' are better to be used to reproduce a good quality chromosome." (WK 2011aaa; A. Eiben 1994; C. Ting 2005)

It it important to note that "[t]hese processes ultimately result in the next generation population of chromosomes that is different from the initial generation. Generally the average fitness will have increased by this procedure for the population, since only the best organisms from the first generation are selected for breeding, along with a small proportion of less fit solutions, for reasons already mentioned above." (WK 2011aaa)

Of course, "[a]lthough Crossover and Mutation are known as the main genetic operators, it is possible to use other operators such as regrouping, colonization-extinction, or migration in genetic algorithms." (WK 2011aaa; C. Ting 2005)

(a4) Termination

And in the final step of the methodology of genetic algorithm, the process will eventually end, upon the following possible conditions: (WK 2011aaa)

- "A solution is found that satisfies minimum criteria"
- "Fixed number of generations reached"
- "Allocated budget (computation time/money) reached"
- "The highest ranking solution's fitness is reaching or has reached a plateau such that successive iterations no longer produce better results"
- "Manual inspection"

- "Combinations of the above"

In the end, the search heuristic of genetic algorithm is embedded within itself a hidden assumption, namley, "the building block hypothesis (BBH)," which states that "a genetic algorithm performs adaptation by implicitly and efficiently implementing this heuristic." (WK 2011aaa)

Criticisms of Applied Evolution

However, applied evolution has its own problems, and the example of the search heuristic of genetic algorithm illustrates them clearly, as shown in the following criticisms (and summarized in *Table 3.4*).

Firstly, one major criticism of genetic algorithm is that "the building block hypothesis (BBH)" is theoretically unjustified and empirically inaccurate.

In other words, "the building block hypothesis has been sharply criticized on the grounds that it lacks theoretical justification, and experimental results have been published that draw the veracity of this hypothesis into question." (WK 2011aaa; G. Syswerda 1989; D. Fogel 2000)

Secondly, another major criticism of genetic algorithm is that it can be computatiionally prohibitive for complex problems.

For instance, "repeated fitness function evaluation for complex problems is often the most prohibitive and limiting segment of artificial evolutionary algorithms. Finding the optimal solution to complex high dimensional, multimodal problems often requires very expensive fitness function evaluations. In real world problems such as structural optimization problems, one single function evaluation may require several hours to several days of complete simulation. Typical optimization methods can not deal with such types of problem. In this case, it may be necessary to forgo an exact evaluation and use an approximated fitness that is computationally efficient." (WK 2011aaa)

Thirdly, still another major criticism of genetic algorithm is that it often reaches "local optima" (not "global optimum") solutions instead.

In other words, "[i]n many problems, GAs may have a tendency to converge towards local optima or even arbitrary points rather than the global optimum of the problem. This means that it does not 'know how' to sacrifice short-term fitness to gain longer-term fitness. The likelihood of this occurring depends on the shape of the fitness landscape: certain problems may provide an easy ascent towards a global optimum, others may make it easier for the function to find the local optima. This problem may be alleviated by using a different fitness function, increasing the rate of mutation, or by using selection techniques that maintain a diverse population of solutions, although the No Free Lunch theorem proves that there is no general solution to this problem." (WK 2011aaa; D. Wolpert 1995)

Fourthly, one additional major criticism of genetic algorithm is that it is less useful for those problems with "a single right/wrong measure."

For instance, "GAs cannot effectively solve problems in which the only fitness measure is a single right/wrong measure (like decision problems), as there is no way to converge on the solution (no hill to climb)." (WK 2011aaa)

And fifthly, another additional major criticism of genetic algorithm is that alternative optimization algorithms can be better genetic algorithm for some problems.

In other words, "for specific optimization problems and problem instances, other optimization algorithms may find better solutions than genetic algorithms (given the same amount of computation time). Alternative and complementary algorithms include evolution strategies, evolutionary programming, simulated annealing, Gaussian adaptation, hill climbing, and swarm intelligence (e.g.: ant colony optimization, particle swarm optimization) and methods based on integer linear programming. The question of which, if any, problems are suited to genetic algorithms (in the sense that such algorithms are better than others) is open and controversial." (WK 2011aaa)

Beyond Applied Evolution

These criticisms of the methodology of genetic algorithm as a case study of applied evolution should not be misunderstood as a rejection of its usefulness but serve to show us the two opposing sides of the debate, such that the possibility and desirability of qualitative research (from the perspective of nature with qualitative research and the promise of applied evolution as a case study here) are not to the extent that the respective defenders would like us to believe.

Indeed, the analysis of qualitative research and the promise of applied evolution also teaches us something valuable about the ontological principles in existential dialectics, and good examples include the formalness-informalness principle, the absoluteness-relativeness principle, the partiality-totality principle, the predictability-unpredictability principle, the explicability-inexplicability principle, the fiction-reality principle, the preciseness-vagueness principle, the simpleness-complicatedness principle, the openness-hiddenness principle, the denseness-emptiness principle, the change-constancy principle, the order-chaos principle, the slowness-quickness principle, the expansion-contraction principle, the theory-praxis principle, the convention-novelty principle, the evolution-transformation principle, the symmetry-asymmetry principle, the softness-hardness principle, the seriousness-playfulness principle, the regression-progression principle, the same-difference principle, and the stability-reaction principle.

For instance, in relation to the formalness-informalness principle, if there is formalness (e.g., the formal logical requirement of completeness and soundness in a system of ideas, as in genetic algorithm), there is informalness (e.g., the non-formal existence of incompleteness and unsoundness in genetic algorithm, as shown in the criticism by the critics that "the building block hypothesis has been sharply criticized on the grounds that it lacks theoretical justification, and experimental results have been published that draw the veracity of this hypothesis into question"). And the reverse direction also holds true.

In relation to the absoluteness-relativeness principle, if there is absoluteness (e.g., the absolute view in genetic algorithm to mimic "the process of natural evolution…to generate useful solutions to optimization and search problems"), there is relativeness (e.g., what is right for those in genetic algorithm in regard to its particular methodology is not necessarily so for the critics who offer alternative optimization algorithms like "simulated annealing, Gaussian adaptation, hill climbing, and swarm intelligence"). And the reverse direction also holds true.

In relation to the partiality-totality principle, if there is partiality (e.g., those cases of "local optima" in genetic algorithm), there is totality (e.g., those cases of "global optimum" in genetic algorithm, such that the whole is not the sum of its parts, that is, a "global optimum" is not simply the sum of "local optima"). And the reverse direction also holds true.

In relation to the predictability-unpredictability principle, if there is predictability (e.g., the predictable tendency of genetic algorithm to converge"[i]n many problems…towards local optima or even arbitrary points rather than the global optimum of the problem"), there is unpredictability (e.g., the more difficult task to predict exactly the extent to which "global optimum" will be achieved, because, according to the No Free Lunch theorem, "there is no general solution to this problem"). And the reverse direction also holds true.

In relation to the explicability-inexplicability principle, if there is explicability (e.g., the explanation by the advocates of genetic algorithms that they are desirable because "these algorithms frequently succeed at generating solutions of high fitness when applied to practical problems"), there is inexplicability (e.g., the lack of sufficient explanation by the advocates of genetic algorithms of why genetic algorithms can generate these solutions as claimed, because, for the critics, "their behavior is difficult to understand," since, "[i]n particular it is difficult to understand why these algorithms frequently succeed at generating solutions of high fitness when applied to practical problems" as claimed). And the reverse direction also holds true.

In relation to the fiction-reality principle, if there is fiction (e.g., the fictional aspect of genetic algorithm, as shown in the criticism by the critics that "it lacks theoretical justification, and experimental results have been published that draw the veracity of this hypothesis into question"), there is reality (e.g., the realistic aspect of genetic algorithm, when it "succeed[s] at generating solutions of high fitness when applied to practical problems"). And the reverse direction also holds true.

In relation to the preciseness-vagueness principle, if there is preciseness (e.g., the precise identification of four steps in the methodology of genetic algorithm aforecited), there is vagueness (e.g., the vagueness in the identification of the four steps in the methodology of genetic algorithm aforecited , since it is not clear why there must be only four, not three, five, six, and so on, as there is an alternative classification which has only three steps, for instance). (WK 2011aaa) And the reverse direction also holds true.

In relation to the simpleness-complicatedness principle, if there is simpleness (e.g., the relatively simple problems in genetic algorithm when they are computationally feasible), there is complicatedness (e.g., the relatively complex problems in genetic algorithm when they are computationally prohibitive). And the reverse direction also holds true.

In relation to the openness-hiddenness principle, if there is openness (e.g., the open exploration of using genetic algorithm to generate "solutions of high fitness when applied to practical problems"), there is hiddenness (e.g., the hidden nature of genetic algorithm, since, for the critics, "their behavior is difficult to understand," since, "[i]n particular it is difficult to understand why these algorithms frequently succeed at generating solutions of high fitness when applied to practical problems" as claimed). And the reverse direction also holds true.

In relation to the denseness-emptiness principle, if there is denseness (e.g., the relatively denser concentration of "local optima" in many problems toward which GAs tend to converge), there is emptiness (e.g., the relatively less dense, or more empty, concentration of "local optima" in some other problems to which GAs tend to converge, depending on "the shape of the fitness landscape," such that "certain problems may provide an easy ascent towards a global optimum" instead). And the reverse direction also holds true.

In relation to the change-constancy principle, if there is change (e.g., the ever new ways to achieve optimization, as shown in the use of GAs), there is constancy (e.g., the ever constant existence of different problems in the different ways to achieve optimization, as shown in the problems of GA's pointed out by the critics aforecited). And the reverse direction also holds true.

In relation to the order-chaos principle, if there is order (e.g., according to G. Syswerda [1989] in his critique of "the building block hypothesis" in general algorithm, "one and two-point crossover are much less disruptive" or relatively more orderly), there is chaos (e.g,. according to Syswerda, "uniform crossover is highly disruptive of short schemata" or relatively more chaotic). (WK 2011aaa) And the reverse direction also holds true.

In relation to the slowness-quickness principle, if there is slowness (e.g., the relatively slower speed of.GAs to solve problems, when they are complex and computationally prohibitive), there is quickness (e.g., the relatively faster speed of.GAs to solve problems, when they are simple and computationally feasible). And the reverse direction also holds true.

In relation to the expansion-contraction principle, if there is expansion (e.g., the relatively more developed ability of those in GAs to deal with optimization problems from the standpoint of evolutionary theory), there is contraction (e.g., the relatively less developed ability of those in GAs to deal with optimization problems from the standpoint of alternative optimizaiton methods like swarm intelligence). And the reverse direction also holds true.

In relation to the theory-praxis principle, if there is theory (e.g., the theoretical construction of natural evolution in biology), there is praxis (e.g., the practical application of natural evolution in biology to the field of computation, wth the use of GAs for applied evolution). (WK 2011jj) And the reverse direction also holds true.

In relation to the convention-novelty principle, if there is convention (e.g., the conventional wisdom of "natural evolution" as understood in evolution theory), there is novelty (e.g., the alternative novel challenge to the conventional wisdom about "natural evolution," by the new idea of "artificial evolution" as used in applied evolution like GAs). (WK 2011z) And the reverse direction also holds true.

In relation to the evolution-transformation principle, if there is evolution (e.g., the natural evolution of humans to understand reproduction by going through the process of reproduction in the state of nature), there is transformation (e.g., the technical transformation of human understanding of reproduction by the invention of evolutionary theory). And the reverse direction also holds true.

In relation to the symmetry-asymmetry principle, if there is symmetry (e.g., the existence of different ways to undrestand the change of things in nature over time), there is asymmetry (e.g., the popularity of creation theory in pre-modern times—but the popularity of evolutionary theory in modern times). And the reverse direction also holds true.

In relation to the softness-hardness principle, if there is softness (e.g., the defense of GAs, as shown by those who use to solve optimization problems), there is hardness (e.g., the critique of GAs, as shown in the problems pointed out by the critics as cited, especially by those who favor alternative methods). And the reverse direction also holds true.

In relation to the seriousness-playfulness principle, if there is seriousness (e.g., the serious business of those in GAs to achieve optimization), there is playfulness (e.g., the playful part of those in GAs, since there are many different fitness functions, selection techniques, termination conditions, and so on for them to choose from, or to play with, in the four steps of the methodology, and, according to the No Free Lunch theorem, "there is no general solution to this problem"). And the reverse direction also holds true.

In relation to the regression-progression principle, if there is regression (e.g., the regression made by GAs, as shown in the problems pointed out by the critics), there is progression (e.g., the progress made by GAs, as shown in applications by the advocates to solve optimization problems). And the reverse direction also holds true.

In relation to the same-difference principle, if there is similarity in outcome (e.g., the contribution to the molding and control of beliefs, values, and behaviors, regardless of whether this be done by way of applying evolutionary theory, or by way of applying creation theory), there is difference in outcome (e.g., the contribution to the molding and control of beliefs, values, and behaviors by way of applying evolutionary theory for a relatively more secular worldview—but the contribution to the molding and control of beliefs, values, and behaviors by way of applying creation theory for a relatively more sacred worldview). And the reverse direction also holds true.

And in relation to the stability-reaction principle, if there is stability (e.g., the relatively more stable condition of a GA, when it reaches termination in the final step of its methodology, in accordance to some specified conditions like fitness function, number of generations, computational time, etc.), there is reaction (e.g., a GA can lead to further changes later on, since the process can be repeated, after certain evaluation, for the next process of the search heursitic, with different specified new conditions). And the reverse direction also holds true.

Qualitative Research, and the Modeling of GRNs

Another good way to evaluate the possibility and desirability of qualitative research from the perspective of nature is to explore qualitative research and the modeling of GRNs.

The Nature of GRNs

In order to understanding the modeling of GRNs, it is necessary first to introduce what GRNs are.

GRN's stand for "gene regulatory networks" (also called "genetic regulatory networks"), which refer to "a collection of DNA segments in a cell which interact with each other (indirectly through their RNA and protein expression products) and with other substances in the cell, thereby governing the rates at which genes in the network are transcribed into mRNA." (WK 2011bbb)

Each of these mRNA molecules is "to make a specific protein (or set of proteins)" to do different things, as shown in the following three examples: (WK 2011bbb)

- "In some cases this protein will be structural, and will accumulate at the cell-wall or within the cell to give it particular structural properties."
- "In other cases the protein will be an enzyme; a micro-machine that catalyses a certain reaction, such as the breakdown of a food source or toxin."
- "Some proteins…serve only to activate other genes, and these are the transcription factors that are the main players in regulatory networks or cascades. By binding to the promoter region at the start of other genes they turn them on, initiating the production of another protein, and so on. Some transcription factors are inhibitory."

The structure of a typical GRN looks something like this: "The nodes of this network are proteins, their corresponding mRNAs, and protein/protein complexes. Nodes that are depicted as lying along vertical lines are associated with the cell/environment interfaces, while the others are free-floating and diffusible. Implied are genes, the DNA sequences which are transcribed into the mRNAs that translate into proteins. Edges between nodes represent individual molecular reactions, the protein/protein and protein/mRNA interactions through which the products of one gene affect those of another....These interactions can be inductive (the arrowheads), with an increase in the concentration of one leading to an increase in the other, or inhibitory (the filled circles), with an increase in one leading to a decrease in the other. A series of edges indicates a chain of such dependences, with cycles corresponding to feedback loops." (WK 2011bbb)

In GRNs, "[g]enes can be viewed as nodes in the network, with input being proteins such as transcription factors, and outputs being the level of gene expression. The node itself can also be viewed as a function which can be obtained by combining basic functions upon the inputs [like 'AND, OR, and NOT']....These functions have been interpreted as performing a kind of information processing within the cell, which determines cellular behavior. The basic drivers within cells are concentrations of some proteins, which determine both spatial (location within the cell or tissue) and temporal (cell cycle or developmental stage) coordinates of the cell, as a kind of 'cellular memory.'" (WK 2011bbb)

The Modeling of GRNs

This introduction to GRNs gives us a general idea of a GRN, but its understanding also depends on a specific model to be used. Consider, for illustration, the following three examples of modeling GRNs.

(a) Boolean Network

The first model of GRNs to be considered here is called "Boolean network."

In general, "a Boolean network can model a GRN together with its gene products (the outputs) and the substances from the environment that affect it (the inputs). Stuart Kauffman was amongst the first biologists to use the metaphor of Boolean networks to model genetic regulatory networks," as shown in the following four steps: (WK 2011bbb; S. Kauffman 1969 & 1993)

- "Each gene, each input, and each output is represented by a node in a directed graph in which there is an arrow from one node to another if and only if there is a causal link between the two nodes."
- "Each node in the graph can be in one of two states: on or off."
- "For a gene, 'on' corresponds to the gene being expressed; for inputs and outputs, 'on' corresponds to the substance being present."

- "Time is viewed as proceeding in discrete steps. At each step, the new state of a node is a Boolean function of the prior states of the nodes with arrows pointing towards it."

The validity of a model is to be tested by "comparing simulation results with time series observations," for instance. (WK 2011bbb)

(b) Continuous Networks

The second model of GRNs to be considered here is called "continuous networks."

In a way, "continuous network models of GRNs are an extension of the boolean networks….Nodes still represent genes, and connections between them [mean] regulatory influences on gene expression. Genes in biological systems display a continuous range of activity levels and it has been argued that using a continuous representation captures several properties of gene regulatory networks not present in the Boolean model." (WK 2011bbb; J. Vohradsky 2001)

Many of these continuous networks are "similar to an artificial neural network, as inputs to a node are summed up and the result serves as input to a sigmoid function,…but proteins do often control gene expression in a synergistic, i.e. non-linear, way. However there is now a continuous network model that allows grouping of inputs to a node thus realizing another level of regulation. This model is formally closer to a higher order recurrent neural network. The same model has also been used to mimic the evolution of cellular differentiation and even multicellular morphogenesis." (WK 2011bbb; N. Geard 2005; M. Schilstra 2002; J. Knabe 2006, 2006a, & 2008)

(c) Stochastic Gene Networks

And the third model of GRNs to be considered here is called "stochastic gene networks."

This modeling was motivated by some "recent experimental results [which] have demonstrated that gene expression is a stochastic process. Thus, many authors are now using the stochastic formalism….Works on single gene expression and small synthetic genetic networks, such as the genetic toggle switch of Tim Gardner and Jim Collins, provided additional experimental data on the phenotypic variability and the stochastic nature of gene expression." (WK 2100bbb; M. Elowitz 2000 & 2002; W. Blake 2003; A. Arkin 1998; J. Raser 2005; T. Gardner 2000)

However, "[s]ince some processes, such as gene transcription, involve many reactions and could not be correctly modeled as an instantaneous reaction in a single step, it was proposed to model these reactions as single step multiple delayed reactions in order to account for the time it takes for the entire process to be complete. From here, a set of reactions were proposed that allow generating GRNs. These are then simulated using a modified version of the Gillespie algorithm, that can simulate multiple time delayed reactions (chemical reactions where each of the products is provided a time delay that determines when it will be released in the system as a 'finished product')." (WK 2011bbb; M. Roussel 2006; A. Ribeiro 2006)

For instance, "a recent work proposed a simulator (SGNSim, Stochastic Gene Networks Simulator), that can model GRNs where transcription and translation are modeled as multiple time delayed events and its dynamics is driven by a stochastic simulation algorithm (SSA) able to deal with multiple time delayed events. The time delays can be drawn from several distributions and the reaction rates from complex functions or from physical parameters." (WK 2011bbb; A. Ribeiro 2007)

Problems with the Modeling of GRNs

But the modeling of GRNs has its own problems. Consider, for illustration, some main criticisms below.

Firstly, one major criticism of the modeling of GRNs is that it is often simplistic and does not accurately represent reality.

For instance, the critics pointed out that "[t]he approaches used to model gene regulatory networks have been constrained to be interpretable and, as a result, are generally simplified versions of the network. For example, Boolean networks have been used due to their simplicity and ability to handle noisy data but lose data information by having a binary representation of the genes. Also, artificial neural networks omit using a hidden layer so that they can be interpreted, losing the ability to model higher order correlations in the data." (WK 2011bbb)

Secondly, another major criticism of the modeling of GRNs is that it still does not help us understand some intriguing probblems.

For instance, "[s]ometimes, a 'self-sustaining feedback loop' ensures that a cell maintains its identity and passes it on. Less understood is the mechanism of epigenetics by which chromatin modification may provide cellular memory by blocking or allowing transcription." (WK 2011bbb)

Thirdly, still another major criticism of the modeling of GRNs is that it is often too abstract, such that its validity can only be inferred from knowledge outside the model, if available.

For instance, the critics pointed out that "the network structure is an abstraction of the system's chemical dynamics, describing the manifold ways in which one substance affects all the others to which it is connected.

In practice, such GRNs are inferred from the biological literature on a given system and represent a distillation of the collective knowledge about a set of related biochemical reactions." (WK 2011bbb)

And fourthly, one additional major criticism of the modeling of GRNs is that it is still in its infancy of development, so that much is still needed to be explored and understood.

For instance, for the critics, "[t]he gene networks are only beginning to be understood, and it is a next step for biology to attempt to deduce the functions for each gene 'node,' to help understand the behavior of the system in increasing levels of complexity, from gene to signaling pathway, cell or tissue level." (WK 2011bbb)

Beyond the Modeling of GRNs

These criticisms of the modeling of GRNs do not suggest its total uselessness but are to reveal the two opposing sides of the debate, such that the possibility and desirability of qualitative research (from the perspective of nature with qualitative research and the modeling of GRNs as a case study here) are not to the extent that the respective defenders would like us to believe.

In addition, the analysis of qualitative research and the modeling of GRNs can cast a bright light on the ontological principles in existential dialectics, and good examples include the formalness-informalness principle, the absoluteness-relativeness principle, the predictability-unpredictability principle, the explicability-inexplicability principle, the fiction-reality principle, the finiteness-transfiniteness principle, the preciseness-vagueness principle, the simpleness-complicatedness principle, the openness-hiddenness principle, the denseness-emptiness principle, the change-constancy principle, the order-chaos principle, the slowness-quickness principle, the expansion-contraction principle, the theory-praxis principle, the convention-novelty principle, the evolution-transformation principle, the symmetry-asymmetry principle, the softness-hardness principle, the seriousness-playfulness principle, the regression-progression principle, the same-difference principle, and the stability-reaction principle.

For instance, in relation to the formalness-informalness principle, if there is formalness (e.g., the formal logical requirement of completeness and soundness in a system of ideas, as in the modeling of GRNs), there is informalness (e.g., the non-formal existence of unsoundness and incompleteness in the modeling of GRNs, as shown in the criticism that it is often too abstract, with little empirical relevance, and that its validity can only be inferred from knowledge outside the model, if available). And the reverse direction also holds true.

In relation to the absoluteness-relativeness principle, if there is absoluteness (e.g., the absolute focus on discrete steps in boolean networks), there is relativeness (e.g., what is right for "boolean networks" in regard to its simple non-continuous representation of the range of activity levels in GRNs is not necessarily so for "continuous networks" which allow continuous representation). And the reverse direction also holds true.

In relation to the predictability-unpredictability principle, if there is predictability (e.g., the predictable tendency of a model of GRNs to represent a network in terms of "nodes," with "input being proteins such as transcription factors, and outputs being the level of gene expression"), there is unpredictability (e.g., the more difficult task to predict exactly the extent to which each "node" will function in increasing levels of complexity, because the modeling of GRNs is still in its infnacy and "[t]he gene networks are only beginning to be understood, and it is a next step for biology to attempt to deduce the functions for each gene 'node,' to help understand the behavior of the system in increasing levels of complexity, from gene to signaling pathway, cell or tissue level"). And the reverse direction also holds true.

In relation to the explicability-inexplicability principle, if there is explicability (e.g., the explanation of the process of a GRN on the basis that "[s]ometimes, a 'self-sustaining feedback loop' ensures that a cell maintains its identity and passes it on"), there is inexplicability (e.g., the lack of sufficient explanation about "the mechanism of epigenetics by which chromatin modification may provide cellular memory by blocking or allowing transcription" in a GRN, which is still "less understood"). And the reverse direction also holds true.

In relation to the fiction-reality principle, if there is fiction (e.g., the fictional aspect of the modeling of GRNs, when it is simplistic and does not accurately represent reality, as pointed out by the critics), there is reality (e.g., the realistic aspect of the modeling of GRNs, when it can help us understand better more simple cases, as shown by Boolean networks, which "have been used due to their simplicity and ability to handle noisy data"). And the reverse direction also holds true.

In relation to the finiteness-transfiniteness principle, if there is finiteness (e.g., the finite number of "discrete steps" in the modeling of a Boolan network), there is transfiniteness (e.g., the transfinite number of "the continuous range of activity levels" in the modeling of "continuous networks"). And the reverse direction also holds true.

In relation to the preciseness-vagueness principle, if there is preciseness (e.g., the precise identification of four steps in the modeling of Boolean networks aforedescribed), there is vagueness (e.g., the vagueness in the identification of the four steps in the modeling of Boolean networks aforedescribed, since it is not clear why there must be only four, not three, five, six, and so on). And the reverse direction also holds true.

In relation to the simpleness-complicatedness principle, if there is simpleness (e.g., "[t]he approaches used to model gene regulatory networks have been constrained to be interpretable and, as a result, are generally simplified versions of the network"), there is complicatedness (e.g., "[u]sing a model that is not constrained to be interpretable, a more accurate model can be produced," but it becomes more complicated). And the reverse direction also holds true.

In relation to the openness-hiddenness principle, if there is openness (e.g., the open exploration of the nature of "artificial neural networks" by the modeling of GRNs), there is hiddenness (e.g., the hidden bias in the modeling of GRNs, since "artificial neural networks omit using a hidden layer so that they can be interpreted, losing the ability to model higher order correlations in the data"). And the reverse direction also holds true.

In relation to the denseness-emptiness principle, if there is denseness (e.g., the relatively denser concentration of "single step multiple delayed reactions" in "stochastic gene networks"), there is emptiness (e.g., the relatively less dense, or more empty, concentration of "single step multiple delayed reactions" in Boolean networks). And the reverse direction also holds true.

In relation to the change-constancy principle, if there is change (e.g., the ever new ways to do modeling, as shown in the three models of GRNs), there is constancy (e.g., the ever constant existence of different problems with doing modeling, as shown in the problems with the modeling of GRNs pointed out by the critics afore-described). And the reverse direction also holds true.

In relation to the order-chaos principle, if there is order (e.g., the more or less orderly reality as represented by "Boolean networks"), there is chaos (e.g,. the more or less chaotic reality as represented by "continuous networks" for "gene expression in a…non-linear, way" or, alternatively, by "stochastic gene networks" in a non-deterministic way). And the reverse direction also holds true.

In relation to the slowness-quickness principle, if there is slowness (e.g., the relatively slower readiness of those in the modeling of Boolean networks to model "gene expression" as "a stochastic process"), there is quickness (e.g., the relatively quicker readiness of those in the modeling of "stochastic gene networks" to model "gene expression" as "a stochastic process"). And the reverse direction also holds true.

In relation to the expansion-contraction principle, if there is expansion (e.g., the relatively more developed ability of those in the modeling of Boolean networks to deal with simple networks on the basis of the four-step methodology), there is contraction (e.g., the relatively less developed ability of those in the modeling of Boolean networks to deal with the "continuous representation" of gene expression as modeled by those in "continuous networks," which allow the latter to capture "several properties of gene regulatory networks not present in the Boolean model"). And the reverse direction also holds true.

In relation to the theory-praxis principle, if there is theory (e.g., the theoretical construction of GRNs in biology), there is praxis (e.g.,the practical application of GRNs in biology to the fields of mathematics and computing science for modeling). And the reverse direction also holds true.

In relation to the convention-novelty principle, if there is convention (e.g., the conventional way of modeling GRNs on the basis of "discret" steps as in Boolean networks), there is novelty (e.g., the alternative novel challenge to the conventional way of modeling GRNs on the basis of "discret" steps as in Boolean networks, by the new way of modeling on the basis of "continuous networks"). And the reverse direction also holds true.

In relation to the evolution-transformation principle, if there is evolution (e.g., the natural evolution of intelligent life like humans and animals to have gene regulatory networks in the state of nature), there is transformation (e.g., the technical transformation of intelligent life like humans to understand gene regulatory networks by the invention of the modeling of GRNs). And the reverse direction also holds true.

In relation to the symmetry-asymmetry principle, if there is symmetry (e.g., the existence of different models of representing reality in GRNs), there is asymmetry (e.g., Boolean networks model GRNs on the basis of a "discrete" representation—but continuous networks model GRNs on the basis of a "continuous" representation). And the reverse direction also holds true.

In relation to the softness-hardness principle, if there is softness (e.g., the appreciation of the modeling of GRNs, as shown by the use of the three models aforedescribed), there is hardness (e.g., the critique of the modeling of GRNs, as shown in the criticisms aforecited). And the reverse direction also holds true.

In relation to the seriousness-playfulness principle, if there is seriousness (e.g., the serious business of those in GRNs, when they work hard to discover the complexity of gene expressions), there is playfulness (e.g., the playful part of those in GRNs, when they play with different assumptions and constraints to obtain some desired results, as shown in the case of Boolean networks which "have been used due to their simplicity and ability to handle noisy data but lose data information by having a binary representation of the genes"). And the reverse direction also holds true.

In relation to the regression-progression principle, if there is regression (e.g., the regression made by the modeling of GRNs, as shown in the problems pointed out by the critics), there is progression (e.g., the progress made by the modeling of GRNs, as shown in their uses to help us better undersand gene expressions aforedescribed). And the reverse direction also holds true.

In relation to the same-difference principle, if there is similarity in outcome (e.g., the contribution to the molding and control of beliefs, values, and behaviors, regardless of whether this be done by using Boolean networks, or by using stochastic gene networks), there is difference in outcome (e.g., the contribution to the molding and control of beliefs, values, and behaviors by using Boolean networks for a relatively more deterministic worldview—but the contribution to the molding and control of beliefs, values, and behaviors by using stochastic gene networks for a relatively more probabilistic worldview). And the reverse direction also holds true.

And in relation to the stability-reaction principle, if there is stability (e.g., the relatively more stable condition of a chemical reaction in a GRN, when it finishes the entire process after a time delay, according to "stochastic gene networks"), there is reaction (e.g., a chemical reaction can lead to further reactions, because a gene expression can take "many reactions and could not be correctly modeled as an instantaneous reaction in a single step," as it interacts with different entities in a cell and others in the larger environment, according to "stochastic gene networks"). And the reverse direction also holds true.

Qualitative Research and Society

Qualitative research, when examined from the perspective of society, can reveal some powerful aspects of its possibility and desirability, and this can be shown by way of two case studies, namely, (3.4.1) qualitative research, and the critique of critical theory,, and (3.4.2) qualitative research, phronesis, and science wars—to be addressed in what follows.

Qualitative Research, and the Critique of Critical Theory

A good way to evaluate the possibility and desirability of qualitative research from the perspective of society is to explore qualitative research and the disagreement on critical theory.

Two Versions of Critical Theory

As already pointed out in *Sec. 1.4*—the term "critical theory" refers to "an examination and critique of society and culture, drawing from knowledge across the social sciences and humanities. The term has two different meanings with different origins and histories: one originating in sociology and the other in literary criticism. This has led to the very literal use of 'critical theory' as an umbrella term to describe any theory founded upon critique." (WK 2011aa)

Jürgen Habermas (1968) in *Knowledge and Human Interests* made an epistemological distinction between these two branches of critical theory, as shown below: (WK 2011aa)

- Literary Critical Theory—"is ultimately a form of hermeneutics, i.e. knowledge via interpretation to understand the meaning of human texts and symbolic expressions—including the interpretation of texts which are themselves implicitly or explicitly the interpretation of other texts."
- Social Critical Theory—"is, in contrast, a form of self-reflective knowledge involving both understanding and theoretical explanation to reduce entrapment in systems of domination or dependence, obeying the emancipatory interest in expanding the scope of autonomy and reducing the scope of domination."

Critical theory is important to the field of data analysis, because it provides a qualitative approach which is essentially critical in orientation, by rejecting "the idea that a researcher's work is considered" as "objective" at all. (WK 2011aa)

Since this section focuses on the perspective of society, let's elaborate on the second version (social critical theory) below.

Main Aspects of Social Critical Theory

Social critical theory is distinctive in terms of two main aspects, namely, (a) the total view of society, and (b) the normative interest in social emancipation, as explained below (and summarized in *Table 3.5*).

(a) The Total View of Society

The first aspect of social critical theory to be considered here concerns the total view of society.

Max Horkheimer in his 1937 essay *Traditional and Critical Theory* defined "critical theory" for society as "a social theory oriented toward critiquing and changing society as a whole, in contrast to traditional theory oriented only to understanding or explaining it. Horkheimer wanted to distinguish critical theory as a radical, emancipatory form of Marxian theory, critiquing both the model of science put forward by logical positivism and what he and his colleagues saw as the covert positivism and authoritarianism of orthodox Marxism and Communism." (WK 2011aa)

More specifically, social critical theory so understood has two concepts in relation to the total view of society, as shown below: (WK 2011aa)

- "That critical social theory should be directed at the totality of society in its historical specificity (i.e. how it came to be configured at a specific point in time)"
- "That critical theory should improve understanding of society by integrating all the major social sciences, including geography, economics, sociology, history, political science, anthropology, and psychology"

Of course, more can be said about these two concepts, so the introduction here is a summary.

(b) The Normative Interest in Social Emancipation

And the second aspect of social critical theory to be considered here concerns the normative interest in social emancipation.

Let's start first with the word "critical" in social critical theory, which "derives from Kant's (18th-century) and Marx's (19th Century) use of the term 'critique,' as in Kant's *Critique of Pure Reason* and Marx's concept that his work *Das Kapital* (Capital) forms a 'critique of political economy.' For Kant's transcendental idealism, 'critique' means examining and establishing the limits of the validity of a faculty, type, or body of knowledge, especially through accounting for the limitations imposed by the fundamental, irreducible concepts in use in that knowledge system. Early on, Kant's notion associated critique with the disestablishment of false, unprovable, or dogmatic philosophical, social, and political beliefs, because Kant's critique of reason involved the critique of dogmatic theological and metaphysical ideas and was intertwined with the enhancement of ethical autonomy and the Enlightenment critique of superstition and irrational authority. Marx explicitly developed this notion into the critique of ideology and linked it with the practice of social revolution, as in...his *Theses on Feuerbach*, 'Philosophers have only interpreted the world in certain ways; the point is to change it.'" (WK 2011aa; K. Marx 2008)

However, both M. Horkheimer and T. Adorno (2002) questioned the optimism in Kantian and Marxist works in their *Dialectic of Enlightenment* (1947) and pointed out "a certain ambivalence concerning the ultimate source or foundation of social domination, an ambivalence which gave rise to the 'pessimism' of the new critical theory over the possibility of human emancipation and freedom. This ambivalence was rooted, of course, in the historical circumstances in which the work was originally produced, in particular, the rise of National Socialism, state capitalism, and mass culture as entirely new forms of social domination that could not be adequately explained within the terms of traditional Marxist sociology." (WK 2011aa; J. Habermas 1987)

The reason is that "state intervention in the economy had effectively abolished the tension in capitalism between the 'relations of production' and 'material productive forces of society,' a tension which, according to traditional critical theory, constituted the primary contradiction within capitalism. The market (as an 'unconscious' mechanism for the distribution of goods) and private property had been replaced by centralized planning and socialized ownership of the means of production." (WK 2011aa; T. Adorno 2002)

Unfortunately, "contrary to Marx's famous prediction in the *Preface to a Contribution to the Critique of Political Economy*, this shift did not lead to 'an era of social revolution,' but rather to fascism and totalitarianism. As such, critical theory was left, in Jürgen Habermas' words, without 'anything in reserve to which it might appeal; and when the forces of production enter into a baneful symbiosis with the relations of production that they were supposed to blow wide open, there is no longer any dynamism upon which critique could base its hope.' For Adorno and Horkheimer, this posed the problem of how to account for the apparent persistence of domination in the absence of the very contradiction that, according to traditional critical theory, was the source of domination itself." (WK 2011aa)

Later, Jürgen Habermas in *Knowledge and Human Interests* continued the debate by restating the need for social emancipation, as he made an epistemological distinction for the social sciences "by identifying critical knowledge as based on principles that differentiated it either from the natural sciences or the humanities, through its orientation to self-reflection and emancipation." (WK 2011aa)

Yet, "[t]ough unsatisfied with Adorno and Horkeimer's thought presented in *Dialectic of Enlightenment*, Habermas shares the view that, in the form of instrumental rationality, the era of modernity marks a move away from the liberation of enlightenment and toward a new form of enslavement." (WK 2011aa; W. Outhwaite 1988)

But the ideas of Habermas "regarding the relationship between modernity and rationalization are in this sense strongly influenced by Max Weber. Habermas dissolved further the elements of critical theory derived from Hegelian German Idealism, though his thought remains broadly Marxist in its epistemological approach. Perhaps his two most influential ideas are the concepts of the public sphere and communicative action; the latter arriving partly as a reaction to new post-structural or so-called 'post-modern' challenges to the discourse of modernity. Habermas engaged in regular correspondence with Richard Rorty and a strong sense of philosophical pragmatism may be felt in his theory; thought which frequently traverses the boundaries between sociology and philosophy." (WK 2011aa)

Commonalities between the Two Versions of Critical Theory

However, these main aspects of social critical theory do not mean that there is no commonality between social critical theory and literary critical theory.

On the contrary, there are two main aspects that both critical and literary social theories share, namely, (a) language and communication and (b) social construction, as described below (and summarized in *Table 3.5*).

(a) Language and Communication

The first aspects (that both critical and literary social theories share) concerns language and communication.

In the older days of the 1960s and 1970s, "language, symbolism, text, and meaning…[were] seen as the theoretical foundation for the humanities, through the influence of Ludwig Wittgenstein, Ferdinand de Saussure, George Herbert Mead, Noam Chomsky, Hans-Georg Gadamer,

Roland Barthes, Jacques Derrida and other thinkers in linguistic and analytic philosophy, structural linguistics, symbolic interactionism, hermeneutics, semiology, linguistically oriented psychoanalysis (Jacques Lacan, Alfred Lorenzer), and deconstruction." (WK 2011aa)

However, beginning in the 1970s and 1980s, "Jürgen Habermas redefined critical social theory as a theory of communication, i.e. communicative competence and communicative rationality on the one hand, distorted communication on the other, the two versions of critical theory began to overlap or intertwine to a much greater degree than before." (WK 2011aa)

(b) Social Construction

And the second aspects (that both critical and literary social theories share) concerns social construction.

This is another way to overlap the two versions of critical theory, in that "[b]oth versions of critical theory have focused on the processes by which human communication, culture, and political consciousness are created," as shown below: (WK 2011aa)

- "Whether it is through universal pragmatic principles through which mutual understanding is achieved (Habermas)"
- "The semiotic rules by which objects obtain symbolic meanings (Barthes)"
- "The psychological processes by which the phenomena of everyday consciousness are generated (psychoanalytic thinkers)"
- "The episteme that underlies our cognitive formations (Foucault)"

In addition, "[t]here is a common interest in the processes (often of a linguistic or symbolic kind) that give rise to observable phenomena and here there is some mutual influence among the different versions of critical theory. Ultimately this emphasis on production and construction goes back to the revolution wrought by Kant in philosophy, namely his focus in the *Critique of Pure Reason* on synthesis according to rules as the fundamental activity of the mind that creates the order of our experience." (WK 2011aa)

Problems with Critical Theory

However, for the critics, critical theory has its own major problems. Consider, for illustration, some main criticisms below (as summarized in *Table 3.5*).

Firstly, one major critique of critical theory in relation to the version in literary studies is that "literary critical theory" has no "normative" dimension, because "it is focused on interpretation and explanation rather than on social transformation." (WK 2011aa)

Secondly, another major critique of critical theory in relation to the version in the social sciences is that, while its emancipatory interest can be worthwhile, there is no consensual basis for everyone to agree upon, especially in regard to what "social transformation" is to be carried out in the human future. For instance, "who" is to say what is "good" or "right" for society to be transformed?

Thirdly, still another major critique of critical theory in relation to the version in the social sciences is that its idea about the distinctiveness of the social sciences is too rigid (or too "black or white").

As I already went to great lenghts to explain in *The Future of Post-Human Engineering* (2009: 64-65), "the Habermasian taxonomy is highly problematic for two major reasons. Firstly, the classificaiton of each human interest on the basis of a category of certain disciplines is too rigid, as it does not exhaust the possibilities of concerns in each field; even engineers, for instance, as will be clear in the rest of th[e]...book, are not as obsessed with 'technical interest' as Habermas would like to put them into that 'box' of classification, and the same can be said about the simplistic categorization of other interests in relation to other fields. Secondly, Habermas singled out 'emacipatory' interest as the most vital one for the long-term advancement of civilization, and this bias on his part reflects his ideological commitment to finish the 'unfinished project of the Enlightenment.'"

In addition, I also explained in the two volumes of *The Future of Human Civilization* (2000) and thus summarized in *FPHE* (2009: 64-65), in that "the fate of 'critical theory' is no more special or more desirable than other theories ever proposed in history, as they all are used and misused by the interests and powers that be since historical time. This is so, even though, in the end, critical reason constitutes one of the faculties (e.g., reasoning, emotions, imagination, and intuition) left over for human use after the death of 'God' in modernity, together with its aftermath in postmodernity."

Fourthly, one additional major critique of critical theory in relation to the two versions is that it is not broad enough in scope, as the focus of social critical theory is on the social sciences, whereas the focus of literary critical theory is on the humanities.

What is needed is a total consideration of all domains of knowledge, ranging from the social sciences through the formal sciences and natural sciences to the humanities, in relation to the mind, nature, society, and culture.

This is precisely what I have been proposing for what I originally called "the unified theory of everything" in the context of "existential dialectics.

And fifthly, another additional major critique of critical theory in relation to the two versions is that it is too obsessed with the "construction" of reality and thus fails to consider a way to transcend it without commiting the sin of "scientism."

But this is also something that my "unified theory of everything" in the context of "existential dialectics is supposed to achieve.

Beyond Critical Theory

These criticisms of critical theory do not mean, of course, a total dismissal of its usefulness but serve to show us the two opposing sides of the debate, such that the possibility and desirability of qualitative research (from the perspective of society with qualitative research and the critique of critical theory as a case study here) are not to the extent that the respective defenders would like us to believe.

Furthermore, the analysis of qualitative research and the critique of critical theory can reveal important insights about the ontological principles in existential dialectics, and good examples include the formalness-informalness principle, the absoluteness-relativeness principle, the partiality-totality principle, the predictability-unpredictability principle, the explicability-inexplicability principle, the fiction-reality principle, the preciseness-vagueness principle, the simpleness-complicatedness principle, the openness-hiddenness principle, the denseness-emptiness principle, the change-constancy principle, the order-chaos principle, the slowness-quickness principle, the expansion-contraction principle, the theory-praxis principle, the convention-novelty principle, the evolution-transformation principle, the symmetry-asymmetry principle, the softness-hardness principle, the seriousness-playfulness principle, the regression-progression principle, the same-difference principle, and the stability-reaction principle.

For instance, in relation to the formalness-informalness principle, if there is formalness (e.g., the formal logical requirement of consistency in a system of ideas, as in the history of critical theory), there is informalness (e.g., the non-formal existence of inconsistency in the history of critical theory, as shown in "a certain ambivalence concerning the ultimate source or foundation of social domination, an ambivalence which gave rise to the 'pessimism' of the new critical theory over the possibility of human emancipation and freedom" optimistically held by those in the older days like the Kantians and the Marxists). And the reverse direction also holds true.

In relation to the absoluteness-relativeness principle, if there is absoluteness (e.g., the absolute concern in social critical theory by Habermas with "emancipation"), there is relativeness (e.g., what is right for Habermas in regard to his concern with "emancipation" is not necessarily so for those in literary social theory with their focus on the analysis of texts, not radical social change). And the reverse direction also holds true.

In relation to the partiality-totality principle, if there is partiality (e.g., the partial view of each discipline in the social sciences, like economics, political science, etc.), there is totality (e.g., the more holistic view of the entirety of society, as in social critical theory, such that the whole is not the sum of its parts, that is, the interelations among different disciplines are not simply the sum of all disciplines). And the reverse direction also holds true.

In relation to the predictability-unpredictability principle, if there is predictability (e.g., the predictable tendency of social critical theory to undergo changes over time, as shown in the questioning of Marxist optimism by Horkheimer and Adorno), there is unpredictability (e.g., the more difficult task to predict exactly the extent to which the pessimism of Horkheimer and Adorno will last in the future history of social critical theory). And the reverse direction also holds true.

In relation to the explicability-inexplicability principle, if there is explicability (e.g., the explanation by those in social critical theory that "emancipation" is desirable because it aims to overthrow oppressive social domination), there is inexplicability (e.g., the lack of sufficient explanation by those in social critical theory of why "emancipation" is necessarily desirable, since the critics can then ask, "who" is to say what is "good" or "right" for society to be transformed?"). And the reverse direction also holds true.

In relation to the fiction-reality principle, if there is fiction (e.g., the fictional aspect of critical theory, insofar as it deals with reality as "constructed" in the analysis of texts), there is reality (e.g., the realistic aspect of critical theory, insofar as it deals with the real existence of oppressive "social dominaton"). And the reverse direction also holds true.

In relation to the preciseness-vagueness principle, if there is preciseness (e.g., the precise identification of four ways in which reality is "constructed" in critical theory), there is vagueness (e.g., the vagueness in the identification of the four ways in which reality is "constructed" in critical theory, since it is not clear why there must be only four, not five, six, and so on). And the reverse direction also holds true.

In relation to the simpleness-complicatedness principle, if there is simpleness (e.g., the relatively simple optimism about liberation in orthodox Marxism), there is complicatedness (e.g., the relatively more complicated analysis of liberation, by challenging the claims and assumptions, as shown in the questioning of optimism pointed out by Horkheimer and Adorno afore-discussed). And the reverse direction also holds true.

In relation to the openness-hiddenness principle, if there is openness (e.g., the open exploration of "social domination" and its liberation by those in orthodox Marxism), there is hiddenness (e.g., the hidden bias in orthodox Marxism, as shown in the questioning of optimism about liberation by Horkheimer and Adorno). And the reverse direction also holds true.

In relation to the denseness-emptiness principle, if there is denseness (e.g., the relatively denser concentration of concerns with the "construction" of reality in critical theory), there is emptiness (e.g., the relatively less dense, or more empty, concentration of concerns with the "construction" of reality in positivism). And the reverse direction also holds true.

In relation to the change-constancy principle, if there is change (e.g., the ever new ways to understand society, as shown in the rise of critical theory), there is constancy (e.g., the ever constant existence of different problems with the new ways to understand society, as shown in the problems of critical theory pointed out by the critics). And the reverse direction also holds true.

In relation to the order-chaos principle, if there is order (e.g., the more or less orderly state when institutionalized by the oppressive class for social domination, in accordance to the Marxists), there is chaos (e.g,. the more or less chaotic state when brought about by social revolutionaries, at the beginning stage, for social liberation, in accordance to the Marxists). And the reverse direction also holds true.

In relation to the slowness-quickness principle, if there is slowness (e.g., the relatively slower speed of the market, if modified by "centralized planning and socialized ownership," to ever lead towards a successful social revolution as originally conceived by Marx against state capitalism, to the point that it could not happen at all in practice, as pointed out by Horkheimer and Adorno, based on what they witnessed in the 1st half of the 20th century), there is quickness (e.g., the relatively quicker speed of the market, when "replaced by centralized planning and socialized ownership," to lead instead towards "fascism and totalitarianism," as this happened in the 1st half of the 20th century, as pointed out by Horkheimer and Adorno). And the reverse direction also holds true.

In relation to the expansion-contraction principle, if there is expansion (e.g., the relatively more developed ability of those in critical theory to deal with issues concerning "interpretation," "construction of reality," etc.), there is contraction (e.g., the relatively less developed ability of those in critical theory to deal with issues concerning "truth-seeking" as in positivism). And the reverse direction also holds true.

In relation to the theory-praxis principle, if there is theory (e.g., the theoretical construction of communiative competence in the fields like linguistics and communication studies), there is praxis (e.g., the practical application of communiative competence in the fields like linguistics and communication studies to the field of critical theory). And the reverse direction also holds true.

In relation to the convention-novelty principle, if there is convention (e.g., the conventional view in the 1960s and 1970s about "language, symbolism, text, and meaning…as the theoretical foundation for the humanities"), there is novelty (e.g., the alternative novel challenge to the conventional view in the 1960s and 1970s about "language, symbolism, text, and meaning…as the theoretical foundation for the humanities," by the new idea of Habermas to use the theoretical studies of communication to combine the two versions of critical theory). And the reverse direction also holds true.

In relation to the evolution-transformation principle, if there is evolution (e.g., the natural evolution of humans to communicate with each other in the state of nature in a primitive way), there is transformation (e.g., the technical transformation of human ability to communicate by the invention of communication studies like the one used by Habermas). And the reverse direction also holds true.

In relation to the symmetry-asymmetry principle, if there is symmetry (e.g., the existence of different views about language and communication), there is asymmetry (e.g., the popularity of the view in the 1960s and 1970s about "language, symbolism, text, and meaning…as the theoretical foundation for the humanities"—but the popularity of the view in the 1970s and 1980s about redefining critical theory "as a theory of communication"). And the reverse direction also holds true.

In relation to the softness-hardness principle, if there is softness (e.g., the appreciation of Marxism, as shown in its incorporation into social critical theory), there is hardness (e.g., the critique of Marxism, as shown in the problems pointed out by Horkheimer and Adorno as cited). And the reverse direction also holds true.

In relation to the seriousness-playfulness principle, if there is seriousness (e.g., the serious business of those in literary critical theory, when they work hard to understand "language, symbolism, text, and meaning"), there is playfulness (e.g., the playful part of those in literary critical theory, when they play with numerous possible diferent interpretations about the meanings of texts). And the reverse direction also holds true.

In relation to the regression-progression principle, if there is regression (e.g., the regression made by critical theory, as shown in the problems pointed out by the critics), there is progression (e.g., the progress made by critical theory, as shown in the works aforecited which help us understand society in a critical way, "drawing from knowledge across the social sciences and humanities"). And the reverse direction also holds true.

In relation to the same-difference principle, if there is similarity in outcome (e.g., the contribution to the molding and control of beliefs, values, and behaviors, regardless of whether this be done by way of social critical theory, or by way of literary critical theory), there is difference in outcome (e.g., the contribution to the molding and control of beliefs, values, and behaviors by way of social critical theory for a relatively more society-transforming lifeworld—but the contribution to the molding and control of beliefs, values, and behaviors by way of literary critical theory for a relatively more text-analyzing lifeworld). And the reverse direction also holds true.

And in relation to the stability-reaction principle, if there is stability (e.g., the relatively more stable condition of society, when it is subtlely imposed by the dominating class through a system of social domination like the use of "mass culture industry," "socialized ownership," etc., in accordance to social critical theorists), there is reaction (e.g., this oppressive condition of society can lead to further changes over time, as those who are oppressed will work for their social emancipation, in accordance to social critical theorists). And the reverse direction also holds true.

Qualitative Research, Phronesis, and Science Wars

Another good way to evaluate the possibility and desirability of qualitative research from the perspective of society is to analyze qualitative research, phronesis, and science wars.

Phronesis and Science Wars

The word "phronesis" refers to "practical thought" and "is usually translated 'practical wisdom,' sometimes as 'prudence.'" (WK 2011ccc)

Aristotle, for instance, distinguished "sophia" from "phronesis," as explained below: (WK 2011ccc)

- "Sophia"—is "usually translated" as "wisdom" and refers to "the ability to think well about the nature of the world, to discern why the world is the way it is (this is sometimes equated with science); *sophia* involves reason concerning universal truths."
- "Phronesis"—"is the capability to consider the mode of action in order to deliver change, especially to enhance the quality of life. Phronesis is concerned with particulars, because it is concerned with how to act in particular situations. One can learn the principles of action, but applying them in the real world, in situations one could not have foreseen, requires experience of the world. For example, if one knows that one should be honest, one might act in certain situations in ways that cause pain and offense; knowing how to apply honesty in balance with other considerations and in specific contexts requires experience."

Bent Flyvbjerg in his book *Making Social Science Matter* (2001) "argued that instead of trying to emulate the natural sciences, the social sciences should be practiced as phronesis. According to him, Phronetic social science focuses on four value-rational questions" as shown below: (WK 2011ccc)

- "Where are we going?"
- "Who gains and who loses, by which mechanisms of power?"
- "Is this development desirable?"
- "What should we do about it?"

Flyvbjerg's book makes two major claims, namely, (a) that "the social sciences have failed as science" because they are incorrectly modeled on the natural science, and (b) that. "in order to matter again the social sciences must model themselves after phronesis (as opposed to episteme, which is at the core of natural science)." (WK 2011ddd)

In fact, this challenge by Flyvbjerg constitutes a latest addition to the heated debate known as "the science wars" which "were a series of intellectual battles in the 1990s, between scientific realists and postmodernist critics, about the nature of scientific theory. The postmodernists questioned scientific objectivity, and undertook a wide-ranging critique of the scientific method and of scientific knowledge, across the gamut of the disciplines of cultural studies, cultural anthropology, feminist studies, comparative literature, media studies, and science and technology studies. The scientific realists countered that objective scientific knowledge is real, and accused postmodernist critics of having little understanding of the science they were criticising." (WK 2011eee)

The Case of Praxis Intervention

In order to understand better this latest episode of the science wars, let's consider a case of the use of phronesis in the social sciences, namely, "praxis intervention" (as described below and summarized in *Table 3.6*). (WK 2011fff)

The term "praxis intervention" is a form of "participatory action research," which "emphasizes working on the *praxis potential* (phronesis) of its participants....'Praxis potential' means the 'members' potential to reflexively work on their respective 'mentalities'; by 'participant' it is not just meant the clientele beneficiaries of the praxis intervention project alone, but also the organizers and experts participating in such a project. The praxis intervention aims at leading its members through a 'participant objectivation'. The praxis intervention method prioritizes unsettling the settled mentalities, especially where the settled mindsets prevalent in the social world or individuals is suspected to have sustained or contributed to their suffering or marginality." (WK 2011fff; M. Archer 2000; P. Madhu 2005)

(a) Reasons for Praxis Intervention

Praxis intervention is to turn "research, creative expression or technology development into a bottom up process. It democratizes making of art, science, technology and critical conscience. The praxis intervention method aims at provoking members to unsettle their settled mindsets and to have a fresh look at the world around and intervene. For instance, members may take a fresh critical look on the gender relations existing, if the praxis intervention method is applied to study gender relations. They would be unsettling their biographically and structurally ingrained perceptions of gender relations and freshly look at it. A gradual process by which members are helped to reflexively recognize the arbitrary and discriminating mindsets within themselves and the world around and working towards correcting it is praxis intervention. The praxis intervention method helps members to struggle against structurally ingrained discrimination." (WK 2011fff)

In contrast, conventional methods like "[q]uestionnaire based surveys, formal interviews, and even focus group discussions are not useful to help respondents to come out with genuine answers to the questions posed at them. Praxis Intervention as it helps groups of people probing their own conditions phase by phase through prolonged discussions, experiments and conscious explorations is capable of coming out with better quality data that could be useful for the group to challenge existing epistemic structures and work out their own well being. Praxis intervention is useful wherever reflexivity is a major component in a research project." (WK 2011fff)

(b) Reflexive Praxis in Praxis Intervention

Praxis intervention makes a distinction between two forms of praxis, that is, "reflexive" vs. "routine" (or "non-reflexive") praxis, with credit to the work of Karl Marx. (WK 2011fff; A. Gouldner 1980:32–33)

In other words, "[t]he reflexive praxis is understood as the moment in the dialectic change, and the non-reflexive one as the routinising mechanism operating within the ideologies as a reproductive or status quo maintaining. It is, for Marx, the non-reflexive habituating praxis, which leads to false consciousness and alienation." (WK 2011fff)

As M. Markoviç (1974:64) explained, "moments of praxis include creativity instead of sameness, autonomy instead of subordination, sociality instead of massification, rationality instead of blind reaction and intentionality rather than compliance." (WK 2011fff)

(c) Systematic Intervention in Praxis Interevention

Praxis intervention "is not just a method in which one interferes with the *habitus* of the persons with whom the action research is carried out but also it is intervention on the *praxis intervention practitioner's habitus*." (WK 2011fff)

In this way, "the practice of praxis intervention is also a systematic *participant objectivation*. Participatory objectivation is 'objectifying the act of objectification.' By 'objectifying the objectification' it is meant the researcher, while observing and objectifying, taking a similar critical distance towards the objectification itself. It is being sensitive to the immensely possible biases from the researcher's social coordinates, field and intellectual orientation and self-critically problematising them to reduce the impact of the biases." (WK 2011fff; P. Bourdieu 1992: 39-42 & 2003).

In other words, "[t]he praxis intervention method problematises the bias of the researcher and her clientile emerging from their social origins, class, gender coordinates; their position in the intellectual field and their respective social space; and also their 'intellectual bias,' the results of viewing the world as a spectacle." (WK 2011fff)

For Pierre Bourdieu (1992:253 & 1990:34), for instance, "within the sociological analysis, the *participant objectivation* is the essential but difficult exercise of all because it requires the break with the deepest and most conscious adherence and adhesions, those quite often give the object its very *interest* for those who study it—i.e., everything about their relation to the object they try to know that they least want to know....It is through the participant objectivation [that] the practical relation to practice is substituted with the observer's relation to practice....Through the practice of participant objectivation, Bourdieu aims to make the critical and political activity of social research the 'solvent of doxa.' Though the practice of reflexive participant objectivation, the practitioner re-looks the taken for granted assumptions in order to wake up from her epistemic sleep and helps her clients too to help them to wake up from theirs." (WK 2011fff)

(d) Three Phases in Praxis Intervention

Praxis intervention can take place in three main phases, as described below.

In the first phase, the task here is coordinate "the clientele (and experts as well) of the problem under focus. The problem under focus could be anything like existing status of marginality, gender relation, health condition or status of present technology etc. Once the problem could properly be oriented from as much angles as possible the clientele, experts and others could involve discussing appropriate method to probe the issue under discussion. They could also decide on the phases through which the collective probing to be undertaken." (WK 2011fff)

In the second phase, the task here is to allow "clientele explor[e]...the issue themselves through the methods collectively agreed. They could maintain a journal to record their experiences and reflexive probing." (WK 2011fff)

And in the third phase, the task here is encourage "the clientele [to] com[e]...together with experts and sharing their experience with their fellow clients and experts. After discussing all the observations and experiences the method of enquiry can be further fine-tuned for better understanding. Gradually the clientele could be guided to intervene on the issue concerned and document their experience. Their action and reflexion could be discussed and fine tuned for further interventions. The phases and intervention strategies could be decided as it could be appropriate for different problems." (WK 2011fff)

(e) Different Applications of Praxis Intervention

Praxis intervention can be applied to different situations. Consider a few examples below, for illustration.

The first application of praxis intervention to be illustrated here involves the field of education. For instance, "[e]ducational process can be reconstituted in accordance to the learning capabilities of the sets of students. Students going under the process of praxis intervention would listen to the expert opinions and explanations of the phenomenon under discussion and they would carry out experiments in the laboratory or in appropriate practical field relevant to the phenomenon under focus individually, collectively and collaboratively as they are guided by the expert teachers. They would bring back the results of experiments to the classroom setting and clarify with each other and also with the expert teachers. In this process syllabus can be indicative negotiable taking into consideration of the advancement of knowledge creation. In this process knowledge becomes creative appropriation rather than indoctrination.This may help students to gain great quantity of quality knowledge within short periods of time. The method could be effectively used in pre-primary to higher schools of education.The method could be used in development of professional skills such as engineering, medicine, nursing, law, agricultural science, social work, management, career guidance, expertise development, teaching and other such fields." (WK 2011fff)

The second application of praxis intervention to be illustrated here involves the field of medical services. For instance, "praxis intervention can be used in helping patients to undergo a participatory diagnostic and experience sharing research with the medical professionals and other experts on their health conditions and take necessary action. Group of patients suffering from similar diseases, their family members, and their significant others, medical practitioners and relevant experts can sit together and be guided through informed dialogue, systematic corrective measures, and experience sharing. In this process both the facilitating professionals and the patients can overcome their respective personal and professional bias." (WK 2011fff)

And the third application of praxis intervention to be illustrated here involves the field of social work. For instance, "[t]he method could be fruitful in working with the marginalized people as marginalisation is usually a historical phenomenon. This would avoid people losing self-respect and dignity under the conditions of marginalisation. The method could be applied in other conditions of marginalisation such as working with people discriminated on the basis of gender. It can also serve as a model for opportunity scanning. The praxis intervention can be used as a method to initiate and implement participatory project provided the project has sufficient flexibility inbuilt for effecting a change from its pre-designs. A project management from praxis perspective should not have full-fledged blue print before hand, rather the projects should be flexible enough to wait till the participants themselves research and come out with a project plan." (WK 2011fff)

Problems with Praxis Intervention

However, for the critics, praxis intervention has its own major problems. Consider, for illustration, some main criticisms below (as summarized in *Table 3.6*).

Firstly, one major critique of praxis intervention is that it is not as bias-free as claimed.

For instance, praxis intervention already makes good use of the Marxist theory of understanding the world of non-reflexive (routine) praxis as false consciousness based on the oppressive social structure of class domination. But for those who do not accept the Marxist framing of the world, this Marxist way of looking at the world is biased and ideological, so the entire foundation of praxis intervention becomes questionable too.

Secondly, another major critique of praxis intervention is that it is not as effective as claimed.

For instance, the phenoemnon of "cognitive biases" (as discussed in *Sec. 2.2.2*) already shows how pervasive and deep-rooted they can be in the mind, although there are different explanations like "bounded rationality" (by H. Simon and others), "heuristics" (by A. Tversky, Paul Slovic, Melissa Finucane, Ellen Peters, Donald G. MacGregor, and others), "adaptive bias" (by Martie G. Haselton, David M. Buss, and others), and so on—as indicated in *Sec. 1.4* earlier (and elaborated in *Sec. 3.2.1* of *FPHGAM*).

After all, as already mentioned in the previous section, both Adorno and Horkeimer questioned the "optimism" held by the Marxists in relation to eventual emancipation—and even "Habermas shares the view that, in the form of instrumental rationality, the era of modernity marks a move away from the liberation of enlightenment and toward a new form of enslavement." (WK 2011aa; W. Outhwaite 1988) So, the cycle of oppression and false consciousness, instead of ending, continues, with no end in sight.

Thirdly, still another major critique of praxis intervention is that it is not necessarily applicable in all situations.

For instance, for the critics, "[t]here are sections of people who cannot take care of themselves and hence require absolute external care. For example, persons suffering from progressive, irreversible diseases characterised by degeneration of the brain cells such as Alzheimer's disease would require complete external care. The praxis practice would be inappropriate for the people who do not need care." (WK 2011fff)

Fourthly, one additional major critique of praxis intervention is that it assumes some debatable values for the outcomes of praxis for all.

For instance, Markoviç (1974:64), an advocate of praxis intervention, argued that "moments of praxis include creativity instead of sameness, autonomy instead of subordination, sociality instead of massification, rationality instead of blind reaction and intentionality rather than compliance." (WK 2011fff)

But the problem here is that these values like "creativity," "autonomy," "sociality," "rationality," "intentionality," etc. are not necessarily desirable, since, for the critics, some individuals may just prefer "sameness," "subordination," "compliance," etc.

As I already went to great lengths to explain in *The Future of Post-Human Personality* (2010), *Beyond Ethics to Post-Ethics* (2011), and *The Future of Post-Human Creativity* (2009), different individuals have different preferences for different values.

And fifthly, another additional major critique of praxis intervention is that it commits the same mistake, though in the opposite direction, that its enemies in the science wars have.

For instance, in the science wars, scientific realists often reject subjectivity, praxis, and hermeneutics as not legitimate in scientific endeavor, because they regard subjectivity (as too subjective), praxis (as too particular), and hermeneutics (as too interpretive).

Now, those in praxis intervention run the risk of commiting the same mistake, though in the opposite direction, by rejecting the scientific method as too formal-theoretic and thus not relevant to praxis.

One of my main arguments in this book is to show that they are all important, but none is possible and desirable to the extent that its proponents would like us to believe.

Beyond Praxis Intervention

In other words, these criticisms against praxis intervention are not meant to reject its usefulness but are to show us the two opposing sides of the debate, such that the possibility and desirability of qualitative research (from the perspective of society with qualitative research, phronesis, and science wars as a case study here) are not to the extent that the respective defenders would like us to believe.

In the end, the analysis of qualitative research, phronesis, and science wars can also teach us something worthwhile about the ontological principles in existential dialectics, and good examples include the formalness-informalness principle, the absoluteness-relativeness principle, the partiality-totality principle, the predictability-unpredictability principle, the explicability-inexplicability principle, the fiction-reality principle, the preciseness-vagueness principle, the simpleness-complicatedness principle, the openness-hiddenness principle, the denseness-emptiness principle, the change-constancy principle, the order-chaos principle, the slowness-quickness principle, the expansion-contraction principle, the theory-praxis principle, the convention-novelty principle, the evolution-transformation principle, the symmetry-asymmetry principle, the softness-hardness principle, the seriousness-playfulness principle, the regression-progression principle, the same-difference principle, and the stability-reaction principle.

For instance, in relation to the formalness-informalness principle, if there is formalness (e.g., the formal logical requirement of completeness in a system of ideas, as in praxis intervention), there is informalness (e.g., the non-formal existence of incompleteness in praxis intervention, as shown in the criticism of inherent bias in praxis intervention, because of its Marxist framing in regard to false consciousness). And the reverse direction also holds true.

In relation to the absoluteness-relativeness principle, if there is absoluteness (e.g., the absolute view of those in praxis intervention to eliminate different biases of the participants through reflexive praxis), there is relativeness (e.g., what is feasible for those in praxis intervention in regard to their attempt to eliminate different biases is not necessarily so for the critics, who argue instead that false consciousness can be deep-rooted and that some individuals are simply not reflexive enough to overcome it). And the reverse direction also holds true.

In relation to the partiality-totality principle, if there is partiality (e.g., the partial view of those in non-reflextive praxis because of their false consciousness), there is totality (e.g., the more holistic view of those in reflexive praxis in regard to the larger social structure of class domination and the need to overcome false consciousness, such that the whole is not the sum of its parts, that is, the reflexive view of the need to end false consciousness through praxis is not merely the sum of all individual forms of false consciousness). And the reverse direction also holds true.

In relation to the predictability-unpredictability principle, if there is predictability (e.g., the predictable tendency of those in praxis intervention "to struggle against structurally ingrained discrimination"), there is unpredictability (e.g., the more difficult task to predict exactly the extent to which those in praxis intervention "to struggle against structurally ingrained discrimination" will suceed on a particular occasion, because, as Alasdair MacIntyre argued, there is the problem of "the unpredictability of human beings," so even "one unpredictable human being can have a world-changing impact"). (WK 2011ccc) And the reverse direction also holds true.

In relation to the explicability-inexplicability principle, if there is explicability (e.g., the explanation, by those in praxis intervention, that reflexiveness is desirable, because it leads to "creativity instead of sameness, autonomy instead of subordination, sociality instead of massification, rationality instead of blind reaction and intentionality rather than compliance"), there is inexplicability (e.g., the lack of sufficient explanation by those in praxis intervention of why these ends like "creativity," "autonomy," "sociality," "rationality," "intentionality," etc. are necessarily desirable, since, for the critics, some individuals may instead prefer "sameness," "subordination," "compliance," etc.). And the reverse direction also holds true.

In relation to the fiction-reality principle, if there is fiction (e.g., false consciousness in non-reflexive praxis), there is reality (e.g., reflexion to wake up from "epistemic sleep" in reflexive praxis). (WK 2011ccc) And the reverse direction also holds true.

In relation to the preciseness-vagueness principle, if there is preciseness (e.g., the precise classification of three phases in praxis intervention in the literature afore-cited), there is vagueness (e.g., the vagueness in the classification of the three phases in praxis intervention in the literature afore-cited, since it is not clear why there must be only three phases, not four, five, and so on). And the reverse direction also holds true.

In relation to the simpleness-complicatedness principle, if there is simpleness (e.g., the relatively simple analysis of praxis intervention by its advocates afore-described), there is complicatedness (e.g., the relatively more complicated analysis of praxis intervention, by challenging the claims and assumptions, as shown in the problems pointed out by the critics afore-discussed). And the reverse direction also holds true.

In relation to the openness-hiddenness principle, if there is openness (e.g., the open exploration of the values in praxis intervention, like "creativity," "autonomy," "sociality," "rationality," "intentionality," etc., as pointed out by Markoviç), there is hiddenness (e.g., the hidden bias in the advocation of these values by Markoviç since, for the critics, some individuals may just prefer "sameness," "subordination," "compliance," etc.). And the reverse direction also holds true.

In relation to the denseness-emptiness principle, if there is denseness (e.g., the relatively denser concentration of values like "creativity," "autonomy," "sociality," "rationality," "intentionality," etc. in those for praxis intervention), there is emptiness (e.g., the relatively less dense, or more empty, concentration of values like "sameness," "subordination," "compliance," etc. in those for praxis intervention). And the reverse direction also holds true.

In relation to the change-constancy principle, if there is change (e.g., the ever new ways to do praxis, as shown in the new method of praxis intervention), there is constancy (e.g., the ever constant existence of different problems with the new ways to do praxis, as shown in the problems with praxis intervention pointed out by the critics). And the reverse direction also holds true.

In relation to the order-chaos principle, if there is order (e.g., the more or less orderly nature of non-reflexive praxis, because of its orientation for conformity), there is chaos (e.g,. the more or less chaotic nature of reflexive praxis, because of its orientation for disruption or unconformity). And the reverse direction also holds true.

In relation to the slowness-quickness principle, if there is slowness (e.g., the relatively slower readiness of a "reflexive" individual to conform to settled mentalities), there is quickness (e.g., the relatively quicker readiness of a "routine" individual to conform to settled mentalities). And the reverse direction also holds true.

In relation to the expansion-contraction principle, if there is expansion (e.g., the relatively more developed ability of those in reflexive praxis to deal with the challenge of creativity and autonomy), there is contraction (e.g., the relatively less developed ability of those in reflexive praxis to deal with the challenge of conformity and dependency). And the reverse direction also holds true.

In relation to the theory-praxis principle, if there is theory (e.g., the theoretical construction of praxis intervention), there is praxis (e.g.,the practical application of praxis intervention to the .fields of education, medical services, and social work afore-described). And the reverse direction also holds true.

In relation to the convention-novelty principle, if there is convention (e.g., the conventional routines in "non-reflexive" praxis for conformity), there is novelty (e.g., the alternative novel challenge to the conventional routines in non-reflexive praxis, by the new form known as "reflexive" praxis for creativity and autonomy). And the reverse direction also holds true.

In relation to the evolution-transformation principle, if there is evolution (e.g., the natural evolution of humans to understand group dynamics by living in groups in the state of nature), there is transformation (e.g., the technical transformation of human ability to understand group dynamics by the invention of praxis intervention). And the reverse direction also holds true.

In relation to the symmetry-asymmetry principle, if there is symmetry (e.g., the existence of different forms of praxis), there is asymmetry (e.g., non-reflexive praxis reflects social domination for the status quo—but reflexive praxis reflexts social rebellion against the status quo). And the reverse direction also holds true.

In relation to the softness-hardness principle, if there is softness (e.g., the defense of praxis intervention by its advocates), there is hardness (e.g., the critique of praxis intervention by its critics). And the reverse direction also holds true.

In relation to the seriousness-playfulness principle, if there is seriousness (e.g., the serious business of those in praxis intervention to understand their mentalities better by waking up from their "epistemic sleep"), there is playfulness (e.g., the playful part of those in praxis intervention, when they play around with their settled mentalities to find out the hidden assumptions). And the reverse direction also holds true.

In relation to the regression-progression principle, if there is regression (e.g., the regression made by praxis intervention, as shown in the problems pointed out by the critics afore-described), there is progression (e.g., the progress made by praxis intervention, as shown in the applications by their advocates aforecited). And the reverse direction also holds true.

In relation to the same-difference principle, if there is similarity in outcome (e.g., the contribution to the molding and control of beliefs, values, and behaviors, regardless of whether this be done by way of reflexive praxis, or by way of routine praxis), there is difference in outcome (e.g., the contribution to the molding and control of beliefs, values, and behaviors by way of reflexive praxis for a relatively more rebellious lifeworld—but the contribution to the molding and control of beliefs, values, and behaviors by way of routine praxis for a relatively more conformist lifeworld). And the reverse direction also holds true.

And in relation to the stability-reaction principle, if there is stability (e.g., the relatively more stable condition of an individual's mentality, when it becomes settled in routine praxis), there is reaction (e.g., an individual's settled mentality can lead to further changes when it interacts with others and the environment, and then experiences different mentalities with different assumptions). And the reverse direction also holds true.

Qualitative Research and Culture

Qualitative research, when examined from the perspective of culture, can show us in an exotic way its possibility and desirability, and this can be shown by way of two case studies, namely, (3.5.1) qualitative research, and the lies in ethnography, and (3.5.2) qualitative research, and a tale of two paradigms in GT—to be addressed hereafter, in that order.

Qualitative Research, and the Lies in Ethnography

A good way to evaluate the possibility and desirability of qualitative research from the perspective of culture is to examine qualitative research, and the lies in ethnography.

The word "ethnography" is "from Greek *ἔθνος*, ethnos" to mean "folk/people" and "*γράφω*, grapho" to mean "to write"—and "is a scientific research strategy often used in the field of social sciences, particularly in anthropology and in some branches of sociology, also known as part of historical science that studies people, ethnic groups and other ethnic formations, their ethnogenesis, composition, resettlement, social welfare characteristics, as well as their material and spiritual culture." (WK 2011ggg)

Ethnographic Techniques

Ethnography is "often employed for gathering empirical data on human societies and cultures....Ethnography aims to describe the nature of those who are studied (i.e., to describe a people, an ethnos) through writing. In the biological sciences, this type of study might be called a field study' or a 'case report,' both of which are used as common synonyms for 'ethnography.'" (WK 2011ggg; M. Maynard 1994; N. Boaz 1997)

In general, ethnograpy uses the following four main techniques for data collection:

- "Participant Observation"— "is direct, first-hand observation of daily participation." (WK 2011ggg)

- "Interviewing"—"may include conversation with different levels of form and can involve small talk to long interviews. A particular approach to transcribing interview data might be genealogical method. This is a set of procedures by which ethnographers discover and record connections of kinship, descent and marriage using diagrams and symbols." (WK 2011ggg)
- "Questionnaires"—"can be used to aid the discovery of local beliefs and perceptions and in the case of longitudinal research, where there is continuous long-term study of an area or site, they can act as valid instrument for measuring changes in the individuals or groups studied." (WK 2011ggg)
- "Informants"—can be used to focus "attention on a community," because they "know well the activities of the community. These informants are typically asked to identify other informants who represent the community, often using chain sampling. This process is often effective in revealing common cultural common denominators connected to the topic being studied." (WK 2011ggg; G. Garson 2008)

Surely, there can be other techniques, so these examples are illustrative. A short way to summarize ethnographic methods is that they rely "greatly on up-close, personal experience. Participation, rather than just observation, is one of the keys to this process." (WK 2011ggg)

Applications of Ethnographic Techniques

These ethnographic techniques have been used in different fields, so some differences are bound to happen, as shown in the following examples.

(a) Anthropology

One important field in which ethnographic techniques have been used is "anthropology," or more precisely, "cultural and social anthropology," which "were developed around ethnographic research," with classic examples like *Argonauts of the Western Pacific* (1922) by Bronisław Malinowski, *Ethnologische Excursion in Johore* (1875) by Nicholas Miklouho-Maclay, *Coming of Age in Samoa* (1928) by Margaret Mead, and *The Nuer* (1940) by E. E. Evans-Pritchard. (WK 2011ggg)

The first main characteristic of ethnogaphic research in cultural and social anthropology is that "[t]he typical ethnography is a document written about a particular people, almost always based at least in part on emic views of where the culture begins and ends. Using language or community boundaries to bound the ethnography is common. Ethnographies are also sometimes called 'case studies.' Ethnographers study and interpret culture, its universalities and its variations through ethnographic study based on fieldwork. An ethnography is a specific kind of written observational science which provides an account of a particular culture, society, or community. The fieldwork usually involves spending a year or more in another society, living with the local people and learning about their ways of life. Ethnographers are participant observers. They take part in events they study because it helps with understanding local behavior and thought," with classic examples like Carol Stack's *All Our Kin*, Jean Briggs' "Never in Anger," Richard Lee's "Kalahari Hunter-Gatherers," and son. (WK 2011ggg; L. Chavez 1997; H. Bernard 1998)

The second main characteristic of ethnogaphic research in cultural and social anthropology is that "a typical ethnography attempts to be holistic and typically follows an outline to include a brief history of the culture in question, an analysis of the physical geography or terrain inhabited by the people under study, including climate, and often including what biological anthropologists call habitat. Folk notions of botany and zoology are presented as ethnobotany and ethnozoology alongside references from the formal sciences. Material culture, technology and means of subsistence are usually treated next, as they are typically bound up in physical geography and include descriptions of infrastructure. Kinship and social structure (including age grading, peer groups, gender, voluntary associations, clans, moieties, and so forth, if they exist) are typically included. Languages spoken, dialects and the history of language change are another group of standard topics. Practices of childrearing, acculturation and emic views on personality and values usually follow after sections on social structure. Rites, rituals, and other evidence of religion have long been an interest and are sometimes central to ethnographies, especially when conducted in public where visiting anthropologists can see them. As ethnography developed, anthropologists grew more interested in less tangible aspects of culture, such as values, worldview and what Clifford Geertz termed the 'ethos' of the culture." (WK 2011ggg; C. Ember 2006; K. Heider 2001)

And the third main characteristic of ethnogaphic research in cultural and social anthropology is that "there are several sub-genres of ethnography. Beginning in the 1950s and early 1960s, anthropologists began writing 'bio-confessional' ethnographies that intentionally exposed the nature of ethnographic research [with examples like 'Tristes Tropiques' by Claude Lévi-Strauss and *The High Valley* by Kenneth Read]....Later 'reflexive' ethnographies refined the technique to translate cultural differences by representing their effects on the ethnographer [with examples like 'Deep Play: Notes on a Balinese Cockfight' by Clifford Geertz and *Reflections on Fieldwork in Morocco* by Paul Rabinow]….In the 1980s, the rhetoric of ethnography was subjected to intense scrutiny within the discipline, under the general influence of literary theory and post-colonial/post-structuralist thought," with examples like those in "experimental" ethnographies like *Shamanism, Colonialism, and the Wild Man* by Michael Taussig and *Debating Muslims* by Michael F. J. Fischer and Mehdi Abedi. (WK 2011ggg)

(b) Sociology

Another important field in which ethnographic techniques have been used is "sociology."

Unlike anthropology, sociology is less focused on ethnography, "even though many sub-fields and theoretical perspectives within sociology use ethnographic methods," but "ethnography is not the *sine qua non* of the discipline, as it is in cultural anthropology." (WK 2011ggg)

In any event, "sociology is another field which prominently features ethnographies. Urban sociology and the Chicago School in particular are associated with ethnographic research, with some well-known early examples being *Street Corner Society* by William Foote Whyte and *Black Metropolis* by St. Clair Drake and Horace R. Cayton, Jr.. Some of the influence for this can be traced to the anthropologist Lloyd Warner who was on the Chicago sociology faculty, and to Robert Park's experience as a journalist. Symbolic interactionism developed from the same tradition and yielded several excellent sociological ethnographies, including *Shared Fantasy* by Gary Alan Fine, which documents the early history of fantasy role-playing games. Other important ethnographies in the discipline of sociology include Pierre Bourdieu's work on Algeria and France, Paul Willis's *Learning To Labour* on working class youth, and the work of Elijah Anderson, Mitchell Duneier, Loic Wacquant on black America…." (WK 2011ggg)

(c) Communication Studies

And still another important field in which ethnographic techniques have been used is "communicaton studies."

For instance, "[b]eginning in the 1960s and 1970s, ethnographic research methods began to be widely employed by communication scholars. Studies such as Gerry Philipsen's analysis of cultural communication strategies in a blue-collar, working class neighborhood on the south side of Chicago, *Speaking 'Like a Man' in Teamsterville*, paved the way for the expansion of ethnographic research in the study of communication." (WK 2011ggg)

More specifically, "scholars of communication studies use ethnographic research methods to analyze communication behaviors, seeking to answer the 'why and 'how come' questions of human communication. Often this type of research results in a case study or field study such as an analysis of speech patterns at a protest rally or the way firemen communicate during 'down time' at a fire station. Like anthropology scholars, communication scholars often immerse themselves, participate in and/or directly observe the particular social group being studied." (WK 2011ggg; R. Rubin 2005; V. Bentz 1998)

Problems with Ethnographic Techniques

However, ethnographic techniques have their own problems. Consider, for illustratrion, some main criticisms below (as summarized in *Table 3.7*).

Firstly, one main criticism of ethnogrphic techniques is that there is no consensus on how to evaluate them.

For instance, "[e]thnographic methodology is not usually evaluated in terms of philosophical standpoint (such as positivism and emotionalism). Ethnographic studies nonetheless need to be evaluated in some manner," but "there is no consensus on evaluation standards." (WK 2011ggg)

L. Richardson (2000) once identified "five criteria that ethnographers might find helpful," as shown below: (WK 2011ggg)

- "Substantive Contribution"—"Does the piece contribute to our understanding of social-life?"
- "Aesthetic Merit"—"Does this piece succeed aesthetically?"
- "Reflexivity"—"How did the author come to write this text…Is there adequate self-awareness and self-exposure for the reader to make judgments about the point of view?"
- "Impact"—"Does this affect me? Emotionally? Intellectually? Does it move me?"
- "Expresses a Reality"—"Does it seem 'true'—a credible account of a cultural, social, individual, or communal sense of the 'real'?"

But Richardson did not say that these criteria are for everyone, but only for those who are interested.

And secondly, another main criticism of ethnogrphic techniques is that ethnographers violate professional ethical standards more often than they want to openly admit.

For instance, Gary Alan Fine Fine in "Ten Lies of Ethnography" (1993) argued that "ethnographic researchers…are typically not as ethical as they claim or assume to be." (WK 2011ggg)

Fine wanted "to show that researchers often make idealized ethical claims and standards which in actuality are inherently based on partial truths and self-deceptions. Fine also acknowledges that many of these partial truths and self-deceptions are unavoidable….He claims, 'Ethnographers cannot help but lie, but in lying, we reveal truths that escape those who are not so bold.' Based on these assertions, Fine establishes three conceptual clusters in which ethnographic ethical dilemmas can be situated," as shown below: (WK 2011ggg)

- Lies on Classic Virtues
  - "The Kindly Ethnographer"—"Most ethnographers present themselves as being more sympathetic than they actually are."
  - "The Friendly Ethnographer"—"In actuality, when hated individuals are found within research, ethnographers often crop them out of the findings."
  - "The Honest Ethnographer"—"Ethnographers often conceal what they know in order to increase the likelihood of acceptance."
- Lies on Technical Skills
  - "The Precise Ethnographer"—"Ethnographers often…giving credit to those undeserving by not using precise words but rather loose interpretations and paraphrasing. Researchers take near-fictions and turn them into claims of fact."
  - "The Observant Ethnographer"—"In reality, an ethnographer will always miss some aspect because they are not omniscient. Everything is open to multiple interpretations and misunderstandings."
  - "The Unobtrusive Ethnographer"—"As a 'participant' in the scene, the researcher will always have an effect on the communication that occurs within the research site."
- Lies on the Ethnographic Self
  - "The Candid Ethnographer"—"There is an illusion that everything reported has actually happened because the researcher has been directly exposed to it."
  - "The Chaste Ethnographer"—"When ethnographers participate within the field, they invariably develop relationships with research subjects/participants. These relationships are sometimes not accounted for within the reporting of the ethnography despite

the fact that they seemingly would influence the research findings."

–"The Fair Ethnographer"—"[O]bjectivity is an illusion and that everything in ethnography is known from a perspective."

–"The Literary Ethnographer"—"Representation is a balancing act of determining what to 'show' through poetic/prosaic language and style versus what to 'tell' via straightforward, 'factual' reporting. The idiosyncratic skill of the ethnographer influences the face-value of the research."

Of course, Fine did not suggest that all ethnographers lie equally in regard to these dilemmas, but they are the ones which are the most problematic in ethnographic research.

Beyond Ethnography

Again, these criticisms of ethnographic techniques should not be misinterpreted as a total rejection of their usefulness but are to show us the two opposing sides of the debate, such that the possibility and desirability of qualitative research (from the perspective of culture with qualitative research, and the lies in ethnography as a case study here) are not to the extent that the respective defenders would like us to believe.

In addition, the analysis of qualitative research, ethnography, and culture can illuminate some valuable lessons about the ontological principles in existential dialectics, and good examples include the formalness-informalness principle, the absoluteness-relativeness principle, the predictability-unpredictability principle, the explicability-inexplicability principle, the fiction-reality principle, the preciseness-vagueness principle, the simpleness-complicatedness principle, the openness-hiddenness principle, the denseness-emptiness principle, the change-constancy principle, the slowness-quickness principle, the expansion-contraction principle, the theory-praxis principle, the convention-novelty principle, the evolution-transformation principle, the symmetry-asymmetry principle, the softness-hardness principle, the seriousness-playfulness principle, the regression-progression principle, the same-difference principle, and the stability-reaction principle.

For instance, in relation to the formalness-informalness principle, if there is formalness (e.g., the formal logical requirement of consistency in a system of ideas, as in ethnography), there is informalness (e.g., the non-formal existence of inconsistency in ethnography, as shown by the ethical violations in ethnographic research pointed out by Fine). And the reverse direction also holds true.

In relation to the absoluteness-relativeness principle, if there is absoluteness (e.g., the absolute concern with ethnography in cultural anthropology), there is relativeness (e.g., what is the right focus for those in cultural anthropology is not necessarily so for those in sociology, who also use ethnographic technqiues but are not so obsessed). And the reverse direction also holds true.

In relation to the predictability-unpredictability principle, if there is predictability (e.g., the predictable tendency of those in communication studies to use more and more ethnographic techniques nowadays), there is unpredictability (e.g., the more difficult task to predict exactly the extent to which the spread of ethnogrpahic techniques will be in communication studies in a particular era of future history). And the reverse direction also holds true.

In relation to the explicability-inexplicability principle, if there is explicability (e.g., the explanation by those in ethnographic research that honesty is an important professional ethical virtue so that research outcomes are not "skewed"), there is inexplicability (e.g., the lack of sufficient explanation by those in ethnographic research of why in actually they often lie, in spite of the official idealization of honesty as a professonal ethical standard, as pointed out by Fine). And the reverse direction also holds true.

In relation to the fiction-reality principle, if there is fiction (e.g., the fictional aspect of ethnographic research, when "researchers take near-fictions and turn them into claims of fact"), there is reality (e.g., the realistic aspect of ethnographic research, when researchers try their best to report as accurately as possible). And the reverse direction also holds true.

In relation to the preciseness-vagueness principle, if there is preciseness (e.g., the precise identification of "five criteria that ethnographers might find helpful" by Richardson), there is vagueness (e.g., the vagueness in the identification of the "five criteria that ethnographers might find helpful" by Richardson, since it is not clear why there must be only five, not six, seven, and so on). And the reverse direction also holds true.

In relation to the simpleness-complicatedness principle, if there is simpleness (e.g., the relatively simple view of idealized ethical claims by the practitioners in ethnography), there is complicatedness (e.g., the relatively more complicated view of idealized ethical claims by the practitioners in ethnography, by challenging the claims and assumptions, as shown in the lies pointed out by the critics like Fine). And the reverse direction also holds true.

In relation to the openness-hiddenness principle, if there is openness (e.g., the open exploration of different professional ethical standards to follow for ethnographic research), there is hiddenness (e.g., the hidden bias in holding these professional ethical stndards, as in actually researchers do not follow them, more often than they would like to admit in public, according to Fine). And the reverse direction also holds true.

In relation to the denseness-emptiness principle, if there is denseness (e.g., the relatively denser concentration of concerns with ethnographic technqiues in cultural anthropology), there is emptiness (e.g., the relatively less dense, or more empty, concentration of concerns with formal modeling in cultural anthropology). And the reverse direction also holds true.

In relation to the change-constancy principle, if there is change (e.g., the ever new ways to do qualitative research, as shown in the four ethnographic techniques afore-described), there is constancy (e.g., the ever constant existence of different problems with the different ways to do qualitative research, as shown in the problems of ethnographic techniques pointed out by the critics). And the reverse direction also holds true.

In relation to the slowness-quickness principle, if there is slowness (e.g., the relatively slower readiness of an ethnographic researcher to pubicly admit that he lies in different ways and different degrees, so as to defend both personal and occupationl reputation, according to Fine), there is quickness (e.g., the relatively quicker readiness of an ethnographic researcher to lie secretly in his ethnographic work in different ways and different degrees, according to Fine). And the reverse direction also holds true.

In relation to the expansion-contraction principle, if there is expansion (e.g., the relatively more developed ability of a cultural anthropologist to do ethnography), there is contraction (e.g., the relatively less developed ability of a cultural anthropologist to do computer simulation). And the reverse direction also holds true.

In relation to the theory-praxis principle, if there is theory (e.g., the theoretical construction of ethnographic methods), there is praxis (e.g.,the practical application of ethnographic methods to the field of communication studies, for example). And the reverse direction also holds true.

In relation to the convention-novelty principle, if there is convention (e.g., the conventional wisdom about ethical claims in the ethnographic profession), there is novelty (e.g., the alternative novel challenge to the conventional wisdom about ethical claims in the ethnographic profession, by Fine's new view that ethnographic researchers lie more often than they would like to admit in public). And the reverse direction also holds true.

In relation to the evolution-transformation principle, if there is evolution (e.g., the natural evolution of humans to observe and participate in the state of nature), there is transformation (e.g., the technical transformation of human ability to observe and participate by the invention of ethnographic techniques). And the reverse direction also holds true.

In relation to the symmetry-asymmetry principle, if there is symmetry (e.g., the existence of different ways to do research), there is asymmetry (e.g., the popularity of ethnography among cultural anthropologists—but the popularity of regression analysis among microeconomists). And the reverse direction also holds true.

In relation to the softness-hardness principle, if there is softness (e.g., the use of ethnography by those in cultural anthropology), there is hardness (e.g., the critique of ethnography by the critics like Fine). And the reverse direction also holds true.

In relation to the seriousness-playfulness principle, if there is seriousness (e.g., the serious business of those in cultural ethnographic research to study a particular people in a given culture), there is playfulness (e.g., the playful part of those in cultural ethnographic research, when they play with the data for the results that they like to report, according to Fine). And the reverse direction also holds true.

In relation to the regression-progression principle, if there is regression (e.g., the regression made by ethnographic research, as shown in the problems pointed out by the critics aforecited), there is progression (e.g., the progress made by ethnographic research, as shown in their applications to different fields afore-described). And the reverse direction also holds true.

In relation to the same-difference principle, if there is similarity in outcome (e.g., the contribution to the molding and control of beliefs, values, and behaviors, regardless of whether this be done by a devotion to ethnographic ethical standards, or by a devotion to phenomenological reductions), there is difference in outcome (e.g., the contribution to the molding and control of beliefs, values, and behaviors by a devotion to ethnographic ethical standards for a relatively more objective worldview—but the contribution to the molding and control of beliefs, values, and behaviors by a devotion to phenomenological reductions for a relatively more intuitive worldview). And the reverse direction also holds true.

And in relation to the stability-reaction principle, if there is stability (e.g., the relatively more stable condition of ethnographic practices, when they are formalized by ethical stndards for professional reputation), there is reaction (e.g., these ethnographic practices can lead to further changes over time, as they receive evaluative feedbacks in practice, which then call for professional changes in regard to the techniques involved, as proposed by Fine). And the reverse direction also holds true.

Qualitative Research, and a Tale of Two Paradigms in GT

Another good way to evaluate the possibility and desirability of qualitative research from the perspective of culture is to explore qualitative research and a tale of two paradigms in grounded theory.

The term "grounded theory" (GT) refers to "a systematic methodology…emphasizing generation of theory from data in the process of conducting research. It is mainly used for qualitative research, but is also applicable to other data." (WK 2011hhh; P. Martin 1986)

In an explorative way, grounded theory is "a research method that operates almost in a reverse fashion from traditional research and at first may appear to be in contradiction of the scientific method. Rather than beginning by researching and developing a hypothesis, the first step is data collection, through a variety of methods. From the data collected, the key points are marked with a series of codes, which are extracted from the text. The codes are grouped into similar concepts in order to make them more workable. From these concepts, categories are formed, which are the basis for the creation of a theory, or a reverse engineered hypothesis. This contradicts the traditional model of research, where the researcher chooses a theoretical framework, and only then applies this model to the studied phenomenon." (WK 2011hhh; G. Allan 2003)

Four Stages of Grounded Theory

The research methodology of grounded theory can be divided into four steps for implementation, as shown below (and summarized in *Table 3.8*): (WK 2011hhh)

- "Codes"—Stage one is for "identifying anchors that allow the key points of the data to be gathered."
- "Concepts"—Stage two is for collecting "codes of similar content that allows the data to be grouped."
- "Categories"—Stage three is for organizing "broad groups of similar concepts that are used to generate a *theory*."

- "Theory"—Stage four is for generating "a collection of explanations that explain the subject of the research."

Of course, the classification is general and does not say much about the specific differences in methodology among different grounded theorists.

The Split between Two Paradigms in Grounded Theory

Precisely here, there is a "split" on grounded theory in regard to its general methodology, especially among two main paradigms, namely, the "Straussian" (after Anselm Strauss) and the "Glaserian" (after Barney Glaser).

As U. Kelle (2005) thus summarized the main difference among the two paradigms, "the controversy between Glaser and Strauss boils down to the question of whether the researcher uses a well defined 'coding paradigm' and always looks systematically for 'causal conditions,' 'phenomena/context, intervening conditions, action strategies' and 'consequences' in the data"—"or whether theoretical codes are employed as they emerge in the same way as substantive codes emerge, but drawing on a huge fund of 'coding families.'" (WK 2011hhh)

(a) The Glaserian Paradigm

For Barney Glaser, GT follows "the dictum 'all is data'"—which means that "not only interview or observational data but also surveys or statistical analyses or 'whatever comes the researcher's way while studying a substantive area'...can be used in the comparative process as well as literature data from science or media or even fiction. Thus the method according to Glaser is not limited to the realm of qualitative research." (WK 2011hhh)

But, for Glaser (1998), GT does not do things that some other methodologies do, namely: (WK 2011hhh)

- that "GT does not aim for the 'truth' but to conceptualize what's going on by using empirical data. In a way GT resembles what many researchers do when retrospectively formulating new hypotheses to fit data."
- that "in GT, the researcher does not pretend to have formulated the hypotheses in advance since preformed hypotheses are prohibited."
- that GT does not use literature review, taping, and talking, because "GT...gives the researcher freedom to generate new concepts explaining human behavior. This freedom is optimal when the researcher refrains from taping interviews, doing a pre research

literature review, and talking about the research before it is written up."

- that in "most GT studies…statistical methods are not used, and figures not presented. The results of GT are not a reporting of statistically significant probabilities but a set of probability statements about the relationship between concepts, or an integrated set of conceptual hypotheses developed from empirical data."

Instead, for Glaser, "validity…in GT…instead should be judged by fit, relevance, workability, and modifiability," as explained below: (WK 2011hhh; B. Glaser 1967, 1978 & 1998).

- "Fit"—"has to do with how closely concepts fit with the incidents they are representing, and this is related to how thoroughly the constant comparison of incidents to concepts was done."
- "Relevance"—has to do whether or not "[a] relevant study deals with the real concern of participants, evokes 'grab' (captures the attention) and is not only of academic interest."
- "Workability"—has to do whether or not "[t]he theory works when it explains how the problem is being solved with much variation."
- "Modifiability"—has to do whether or not "[a] modifiable theory can be altered when new relevant data is compared to existing data."

In the end, "a GT is never right or wrong, it just has more or less fit, relevance, workability and modifiability." (WK 2011hhh)

(b) The Straussian Paradigm

For Anselm Strauss, GT identified "three basic elements every grounded theory approach should include," as shown below: (WK 2011hhh; H. Legewie 2004)

- "Theoretical Sensitive Coding"—which refers to "generating theoretical strong concepts from the data to explain the phenomenon researched."
- "Theoretical Sampling"—which refers to "deciding whom to interview or what to observe next according to the state of theory generation, and that implies starting data analysis with the first interview, and writing down memos and hypotheses early."
- Comparison—which refers to "the need to compare between phenomena and contexts to make the theory strong."

However, for Strauss, GT does not do things that some other methodologies do, namely: (WK 2011hhh)

- that GT does not focus only on qualitative data, because "grounded theory reaches farther: it combines a specific style of research (or a paradigm) with pragmatic theory of action and with some methodological guidelines."
- that GT is not inductive nor deductive: "The research principle behind grounded theory is neither inductive nor deductive, but combines both in a way of abductive reasoning (coming from the works of Charles Sanders Peirce). This leads to a research practice where data sampling, data analysis and theory development are not seen as distinct and disjunct, but as different steps to be repeated until one can describe and explain the phenomenon that is to be researched. This stopping point is reached when new data does not change the emerging theory anymore."

In contrast, "Grounded theory according to Glaser emphasizes induction or emergence, and the individual researcher's creativity within a clear frame of stages"—"while Strauss is more interested in validation criteria and a systematic approach." (WK 2011hhh)

Problems iwth Grounded Theory

But GT has its own problems. Consider, for illustration, a few criticisms below (as summarized in *Table 3.8*).

Firstly, one main criticism of GT is that it is not clear what is meant by "theory" in generatig a GT theory.

For instance, for the critics, what is the epistemological status of a "theory" in GT, since "GT does not aim for the 'truth'" nor is concerned much with quantitative analysis, as Glaser claimed?

Secondly, another main criticism of GT is that it is also unclear what is meant by "ground" in a GT theory.

For instance, for the critics, "why is an idea of 'grounding' one's findings important in qualitative inquiry"? And "what are they 'grounded' in?" (WK 2011hhh)

Thirdly, still another main criticism of GT is that it is not possible "to free oneself of preconceptions" in the way as claimed by its proponents. (WK 2011hhh)

In other words, for the critics, "it is impossible to free oneself of preconceptions in the collection and analysis of data in the way that Glaser and Strauss say is necessary." (WK 2011hhh)

Fourthly, one additional main criticism of GT is that its usefulness is limited.

For instance, as U. Kelle (2005) warned us, "[m]any researchers may concur with that approach [GT], especially since qualitative research always had a relation to micro-sociological action theory, but others who want to employ a macro-sociological and system theory perspective may feel that the use of the coding paradigm [in GT] would lead them astray." (WK 2011hhh)

And fifthly, another additional main criticism of GT is that it is not quite coherent as a methodology.

For instance, for the critics, those in GT have incompatible aims, since "[t]hey also point to the formulaic nature of grounded theory and the lack of congruence of this with open and creative interpretation—which ought to be the hallmark of qualitative inquiry. They suggest that the one element of grounded theory worth keeping is constant comparative method." (WK 2011hhh)

But how can a theory have "formulaic" strength and also "open and creative interpretation" as its "hallmark," since the two aims are not quite compatible?

In fact, these criticisms against GT can also be used against other qualitative methods, because, historically, "grounded theory was developed in a period when other qualitative methods were often considered unscientific." (WK 2011hhh)

Beyond Grounded Theory

These criticisms of GT (in its history of the two competing paradigms) do not imply a total rejection of its usefulness but are to show us the two opposing sides of the debate, such that the possibility and desirability of qualitative research (from the perspective of culture with qualitative research and a tale of two paradigms in GT as a case study here) are not to the extent that the respective defenders would like us to believe.

Indeed, the analysis of qualitative research and a tale of two paradigms in GT constittues a didactic exercise to reveal the relevance of the ontological principles in existential dialectics, and good examples include the formalness-informalness principle, the absoluteness-relativeness principle, the predictability-unpredictability principle, the explicability-inexplicability principle, the fiction-reality principle, the preciseness-vagueness principle, the simpleness-complicatedness principle, the openness-hiddenness principle, the denseness-emptiness principle, the change-constancy principle, the order-chaos principle, the slowness-quickness principle, the expansion-contraction principle, the theory-praxis principle, the convention-novelty principle, the evolution-transformation principle, the symmetry-asymmetry principle, the softness-hardness principle, the seriousness-playfulness principle, the regression-progression principle, the same-difference principle, and the stability-reaction principle.

For instance, in relation to the formalness-informalness principle, if there is formalness (e.g., the formal logical requirement of consistency in a system of ideas, as in grounded theory), there is informalness (e.g., the non-formal existence of inconsistency in grounded theory, as shown in the criticism about incoherence in GT by the critics aforecited). And the reverse direction also holds true.

In relation to the absoluteness-relativeness principle, if there is absoluteness (e.g., the absolute concern of GT with generating new hypotheses, not aiming for the "truth"), there is relativeness (e.g., what is approriate for GT in regard to its concern with generating new hypotheses, not aiming for the "truth," is not necessarily so for those in quantitative research like social statistics in regard to its aim to confirm pre-determined hypotheses for truth-seeking). And the reverse direction also holds true.

In relation to the predictability-unpredictability principle, if there is predictability (e.g., the predictable tendency of those in GT to generate new hypotheses, instead of confirming them), there is unpredictability (e.g., the more difficult task to predict exactly the extent to which a particular new hypothesis generated by those in GT will ever be confirmed in a particular era of future history). And the reverse direction also holds true.

In relation to the explicability-inexplicability principle, if there is explicability (e.g., the explanation by those in GT that it is grounded on a methodology based on "codes," "concepts," "categories," and "theory"), there is inexplicability (e.g., the lack of sufficient explanation by those in GT of "why…an idea of 'grounding' one's findings" is "important in qualitative inquiry" and "what…they [are really] 'grounded' in," for the critics). And the reverse direction also holds true.

In relation to the fiction-reality principle, if there is fiction (e.g., the fictional aspect of GT when it uses all kinds of data, "even fiction" in literature), there is reality (e.g., the realistic aspect of GT when it also uses "surveys or statistical analyses" from the real world). And the reverse direction also holds true.

In relation to the preciseness-vagueness principle, if there is preciseness (e.g., the precise identification of four main steps in the methodology of GT, namely, "codes," "concepts," "categories," and "theory"), there is vagueness (e.g., the vagueness in the identification of the four steps in the methodology of GT, since it is not clear why there must be only four steps, not five, six, and so on). And the reverse direction also holds true.

In relation to the simpleness-complicatedness principle, if there is simpleness (e.g., the relatively simple analysis of GT by its defenders), there is complicatedness (e.g., the relatively more complicated analysis of GT, by challenging the claims and assumptions, as shown in the problems pointed out by the critics as cited). And the reverse direction also holds true.

In relation to the openness-hiddenness principle, if there is openness (e.g., the open exploration of the possibility of freeing "oneself of preconceptions in the collection and analysis of data," according to GT), there is hiddenness (e.g., the hidden bias in GT, since, for the critics, "it is impossible to free oneself of preconceptions in the collection and analysis of data in the way that Glaser and Strauss say is necessary"). And the reverse direction also holds true.

In relation to the denseness-emptiness principle, if there is denseness (e.g., the relatively denser concentration of concerns with how to generate new hypothesis in GT), there is emptiness (e.g., the relatively less dense, or more empty, concentration of concerns with how to seek the truth in GT). And the reverse direction also holds true.

In relation to the change-constancy principle, if there is change (e.g., the ever new ways to do qualitative research, as shown in GT), there is constancy (e.g., the ever constant existence of different problems with the different new ways to do qualitative research, as shown in the problems with GT pointed out by the critics aforedescribed). And the reverse direction also holds true.

In relation to the order-chaos principle, if there is order (e.g., the more or less orderly aspect of GT when it is structured on the basis of the four steps in its methodology), there is chaos (e.g,. the more or less chaotic aspect of GT when it follows "the Glaserian strategy" of "all is data," which means "whatever comes the researcher's way while studying a substantive area" can be used for data collection). And the reverse direction also holds true.

In relation to the slowness-quickness principle, if there is slowness (e.g., the relatively slower speed of doing codes in GT when "selective coding" is not used), there is quickness (e.g., the relatively quicker speed of doing codes in GT when "selective coding" is used, like "selectively sampl[ing]...new data with the core [variable] in mind," which "delimits the study" and therefore "makes it move fast"). (WK 2011hhh) And the reverse direction also holds true.

In relation to the expansion-contraction principle, if there is expansion (e.g., the relatively more developed ability of those in GT to deal with coding for the generation of new ideas), there is contraction (e.g., the relatively less developed ability of those in GT to deal with regression analysis for the confirmation of predetermined hypotheses). And the reverse direction also holds true.

In relation to the theory-praxis principle, if there is theory (e.g., the theoretical construction of GT), there is praxis (e.g., the practical application of GT to the domain of social organizations, as shown by the creation of "the Grounded Theory Institute in 1999 as a non-profit web-based organization"). (WK 2011hhh) And the reverse direction also holds true.

In relation to the convention-novelty principle, if there is convention (e.g., the conventional wisdom of doing qualitative research like "traditional ethnography, narratology, and storytelling"), there is novelty (e.g., the alternative novel challenge to the conventional wisdom about doing qualitative research like "traditional ethnography, narratology, and storytelling," by the new idea of doing codes in accordance to GT). (WK 2011hhh) And the reverse direction also holds true.

In relation to the evolution-transformation principle, if there is evolution (e.g., the natural evolution of humans to observe and analyze info in the state of nature in a primitive way), there is transformation (e.g., the technical transformation of human ability to observe and analyze info by the invention of grounded theory). And the reverse direction also holds true.

In relation to the symmetry-asymmetry principle, if there is symmetry (e.g., the existence of different paradigms of GT), there is asymmetry (e.g., the popularity of the Straussian paradigm among his followers—but the popularity of the Glaserian paradigm among his followers). And the reverse direction also holds true.

In relation to the softness-hardness principle, if there is softness (e.g., the defense of GT by its practitioners as cited), there is hardness (e.g., the critique of GT by the critics as cited). And the reverse direction also holds true.

In relation to the seriousness-playfulness principle, if there is seriousness (e.g., the serious business of those in GT to collect data for idea generation), there is playfulness (e.g., the playful part of those in GT when they play with all kinds of data, "even fiction" in literature). And the reverse direction also holds true.

In relation to the regression-progression principle, if there is regression (e.g., the regression made by GT, as shown in the problems pointed out by the critics aforecited), there is progression (e.g., the progress made by GT, as shown in its applications by the followers aforecited). And the reverse direction also holds true.

In relation to the same-difference principle, if there is similarity in outcome (e.g., the contribution to the molding and control of beliefs, values, and behaviors, regardless of whether this be done by focusing on GT, or by focusing on social statistics), there is difference in outcome (e.g., the contribution to the molding and control of beliefs, values, and behaviors by focusing on GT for a relatively more qualitative worldview—but the contribution to the molding and control of beliefs, values, and behaviors by focusing on social statistics for a relatively more quantitative worldview). And the reverse direction also holds true.

And in relation to the stability-reaction principle, if there is stability (e.g., the relatively more stable condition of coding in GT when it finally completes the last step of the process, namely, "theory" for the generation of new hypotheses), there is reaction (e.g., the coding of GT, even with the new theory so generated in mind, can then lead to further changes, since the hallmark of GT is its "open and creative interpretation," which is subject to changes whenever new data are available, so that its "comparative" analysis is "constant" and does not stop in interacting with the environment). And the reverse direction also holds true.

The Subjectivity of Qualitative Research

This comprehensive analysis of the possibility and desirability of qualitative research, from the perspectives of the mind, nature, society, and culture, serves to show us the different ways in which qualitative research is both possible and desirable, but not to the extent that the spokespersons from each side would like us to believe.

Qualitative research has its broadness, just as it has its subjectivity too, as the other side of the same coin.

But this is only one part of a larger story, since there is the other side of the story, which concerns quantitative research (as already discussed in Chapter Two).

These dual analyses should not be underestimated, because those on the side of quantitative research often dismisses qualitative research merely as the other side of data analysis without really appreciating it from the vantage point of qualitative research too. And the reverse is also true.

In any event, now that we have reached the end of this long intellectual journey, what then can be concluded about data analysis?

This is the question, for which we now turn to the concluding chapter, namely, Chapter Four, to answer.

Table 3.1. Qualitative Research and Its Limitation

- **Qualitative Research and the Mind**
  —Ex: Qualitative research, and the role of phenomenology
  —Ex: Qualitative research, interpretation, and psychology

- **Qualitative Research and Nature**
  —Ex: Qualitative research, and the promise of applied evolution
  —Ex: Qualitative research, and the modeling of GRNs

- **Qualitative Research and Society**
  —Ex: Qualitative research, and the critique of critical theory
  —Ex: Qualitative research, phronesis, and science wars

- **Qualitative Research and Culture**
  —Ex: Qualitative research, and the lies in ethnography
  —Ex: Qualitative research, and a tale of two paradigms in GT

Notes: The examples in the categories are solely illustrative (not exhaustive), and the comparison is relative (not absolute), nor are they necessarily mutually exclusive. And some can be easily re-classified elsewhere. As generalities, they allow exceptions.
Source: A summary of Ch.3 of *FPHDA*

Table 3.2. Qualitative Research, and the Role of Phenomenology

- **Main Concerns of Phenomenology**
  - —Ex: systematic reflection
  - —Ex: anti-reductionism
  - —Ex: bracketing

- **Main Components of Phenomenology**
  - —Ex: intentionality
  - —Ex: intuition
  - —Ex: evidence
  - —Ex: noesis and noema
  - —Ex: empathy and intersubjectivity
  - —Ex: lifeworld

- **Criticisms of Phenomenology**
  - —Firstly, one major criticism of Husserl's phenomenology concerns what Martin Heidegger called its "subjectivist tendencies." (WK 2011v)
  - —Secondly, another major criticism of Husserl's phenomenology concerns its holy grail of "transcendental ego" (as opposed to "empirical ego").
  - —And thirdly, still another major criticism of Husserl's phenomenology concerns its neglect of the study of "unconsciousness."

Notes: The examples in the categories are solely illustrative (not exhaustive), and the comparison is relative (not absolute), nor are they necessarily mutually exclusive. And some can be easily re-classified elsewhere. As generalities, they allow exceptions.

Source: From *Sec. 3.2.1* of *FPHDA*. See book for more details and citations.

Table 3.3. Qualitative Research, and the Hermeneutic Circle

- **Versions of the Hermeneutic Circle**
  —Ex: Friedrich Schleiermacher
  —Ex: Wilhelm Dilthey
  —Ex: Martin Heidegger
  —Ex: Hans-Georg Gadamer

- **Problems with the Hermeneutic Circle**
  —Firstly, one main criticism against the hermeneutic circle is that the "circle" in hermenutics remains unclear.
  —Secondly, another main criticism against the hermeneutic circle is that it may not make much sense.
  —Thirdly, still another main criticism against the hermeneutic circle is that it is too rigid.
  —And fourthly, one additional main criticism against the hermeneutic circle is that it still does not explain anything.

Notes: The examples in the categories are solely illustrative (not exhaustive), and the comparison is relative (not absolute), nor are they necessarily mutually exclusive. And some can be easily re-classified elsewhere. As generalities, they allow exceptions.
Source: From *Sec. 3.2.2* of *FPHDA*. See book for more details and citations.

Table 3.4. Qualitative Research, and the Promise of Applied Evolution

---

• **The Methodology of Genetic Algorithm**
—Ex: initialization
—Ex: selection
—Ex: reproduction
—Ex: termination

• **Criticisms of Applied Evolution**
—Firstly, one major criticism of genetic algorithm is that "the building block hypothesis (BBH)" is theoretically unjustified and empirically inaccurate.
—Secondly, another major criticism of genetic algorithm is that it can be computatiionally prohibitive for complex problems.
—Thirdly, still another major criticism of genetic algorithm is that it often reaches "local optima" (not "global optimum") solutions instead.
—Fourthly, one additional major criticism of genetic algorithm is that it is less useful for those problems with "a single right/wrong measure."
—And fifthly, another additional major criticism of genetic algorithm is that alternative optimization algorithms can be better genetic algorithm for some problems.

Notes: The examples in the categories are solely illustrative (not exhaustive), and the comparison is relative (not absolute), nor are they necessarily mutually exclusive. And some can be easily re-classified elsewhere. As generalities, they allow exceptions.
Source: From *Sec. 3.3.1* of *FPHDA*. See book for more details and citations.

Table 3.5. Qualitative Research, and the Critique of Critical Theory

- **Two Versions of Critical Theory**
  —Ex: literary critical theory
  —Ex: social critical theory

- **Main Aspects of Critical Theory**
  —Ex: the total view of society
  —Ex: the normative interest in social emancipation

- **Commonalities between the Two Versions of Critical Theory**
  —Ex: language and communication
  —Ex: social construction

- **Problems with Critical Theory**
  —Firstly, one major critique of critical theory in relation to the version in literary studies is that "literary critical theory" has no "normative" dimension." (WK 2011aa)
  —Secondly, another major critique of critical theory in relation to the version in the social sciences is that, while its emancipatory interest can be worthwhile, there is no consensual basis for everyone to agree upon, especially in regard to what "social transformation" is to be carried out in the human future.
  —Thirdly, still another major critique of critical theory in relation to the version in the social sciences is that its idea about the distinctiveness of the social sciences is too rigid.
  —Fourthly, one additional major critique of critical theory in relation to the two versions is that it is not broad enough in scope.
  —And fifthly, another additional major critique of critical theory in relation to the two versions is that it is too obsessed with the "construction" of reality and thus fails to consider a way to transcend it without commiting the sin of "scientism."

Notes: The examples in the categories are solely illustrative (not exhaustive), and the comparison is relative (not absolute), nor are they necessarily mutually exclusive. And some can be easily re-classified elsewhere. As generalities, they allow exceptions.
Source: From *Sec. 3.4.1* of *FPHDA*. See book for more details and citations.

Table 3.6. Qualitative Research, Phronesis, and Science Wars

- **The Case of Praxis Intervention**
  - —Ex: reasons for praxis intervention
  - —Ex: reflexive praxis in praxis intervention
  - —Ex: systematic intervention in praxis intervention
  - —Ex: three phases in praxis intervention
  - —Ex: different applications of praxis intervention

- **Problems with Praxis Intervention**
  - —Firstly, one major critique of praxis intervention is that it is not as bias-free as claimed.
  - —Secondly, another major critique of praxis intervention is that it is not as effective as claimed.
  - —Thirdly, still another major critique of praxis intervention is that it is not necessarily applicable in all situations.
  - —Fourthly, one additional major critique of praxis intervention is that it assumes some debatable values for the outcomes of praxis for all.
  - —And fifthly, another additional major critique of praxis intervention is that it commits the same mistake, though in the opposite direction, that its enemies in the science wars have.

Notes: The examples in the categories are solely illustrative (not exhaustive), and the comparison is relative (not absolute), nor are they necessarily mutually exclusive. And some can be easily re-classified elsewhere. As generalities, they allow exceptions.
Source: From *Sec. 3.4.2* of *FPHDA*. See book for more details and citations.

Table 3.7. Qualitative Research, and the Lies in Ethnography

• **Ethnographic Techniques**
  —Ex: participant observation
  —Ex: interviewing
  —Ex: questionnaires
  —Ex: informants

• **Applications of Ethnographic Technques**
  —Ex: anthropology
  —Ex: sociology
  —Ex: communication studies

• **Problems with Ethnographic Techniques**
  —Firstly, one main criticism of ethnogrphic techniques is that there is no consensus on how to evaluate them.
  —And secondly, another main criticism of ethnogrphic techniques is that ethnographers violate professional ethical standards more often than they want to openly admit. For instance, Gary Alan Fine Fine in "Ten Lies of Ethnography" (1993) argued that "ethnographic researchers…are typically not as ethical as they claim or assume to be," in the context of three categories of lies: (WK 2011ggg)
    • Lies on "classic virtues"
    • Lies on "technical skills"
    • Lies on the "ethnographic self"

Notes: The examples in the categories are solely illustrative (not exhaustive), and the comparison is relative (not absolute), nor are they necessarily mutually exclusive. And some can be easily re-classified elsewhere. As generalities, they allow exceptions.
Source: From *Sec. 3.5.1* of *FPHDA*. See book for more details and citations.

Table 3.8. Qualitative Research, and a Tale of Two Paradigms in GT

- **Four Stages of Grounded Theory**
  - —Ex: codes
  - —Ex: concepts
  - —Ex: categories
  - —Ex: theory

- **The Split between Two Paradigms in Grounded Theory**
  - —Ex: the Glaserian paradigm
  - —Ex: the Straussian paradigm

- **Problems with Grounded Theory**
  - —Firstly, one main criticism of GT is that it is not clear what is meant by "theory" in generatig a GT theory.
  - —Secondly, another main criticism of GT is that it is also unclear what is meant by "ground" in a GT theory.
  - —Thirdly, still another main criticism of GT is that it is not possible "to free oneself of preconceptions" in the way as claimed by its proponents. (WK 2011hhh)
  - —Fourthly, one additional main criticism of GT is that its usefulness is limited.
  - —And fifthly, another additional main criticism of GT is that it is not quite coherent as a methodology.

Notes: The examples in the categories are solely illustrative (not exhaustive), and the comparison is relative (not absolute), nor are they necessarily mutually exclusive. And some can be easily re-classified elsewhere. As generalities, they allow exceptions.
Source: From *Sec. 3.5.2* of *FPHDA*. See book for more details and citations.

• PART FOUR •

Conclusion

CHAPTER 4
CONCLUSION—THE FUTURE OF DATA ANALYSIS

> One's own world is one's world. What
> a person thinks is what he becomes.
> —Maitri Upanishad (QN 2011)

Beyond Quantitative and Qualitative Research

The analyses of quantitative research (in Chapter Two) and qualitative research (in Chapter Three) are to show us the extent of their possibility and desirability in data analysis.

This is important, in light of the opposing views about data analysis (as already introduced in *Sec. 1.1* and the rest of the book).

The challenge to these diverse views about data analysis does not mean, however, that data analysis is useless, or that those fields of study (related to data analysis) like methodology, epistemology, philosophy of science, psychology, linguistics, mathematics, logic, sociology, anthropology, and so on should be ignored too. Of course, neither of these extreme views is reasonable.

Instead, this book provides an alternative (better) way to understand the future of data analysis, especially in the dialectic context of quantitative and qualitative research—while learning from different approaches in the literature but without favoring any one of them (nor integrating them, since they are not necessarily compatible with each other).

In other words, this book offers a new theory (that is, *the complex theory of data analysis*) to go beyond the existing approaches in an original way.

My *complex theory of data analysis* has four major distinct features.

Firstly, my theory makes good use of all theoretical approaches to data analysis in relation to quantitative and qualitative research, be they about the mental argument, the natural argument, the societal argument, or the cultural argument—especially from the most comprehensive combined perspectives of the mind, nature, society, and culture (as already dealt with in Chapter Two and Chapter Three).

Secondly, just like many other theories of mine in my previous books, my theory here does not heavily favor any specific theory over others in the literature, nor trying to integrate them (as they are not necessarily compatible with each other).

Thirdly, my theory treats the issue of data analysis in the distinctive dialectic context of no quantitative research without qualitative research, and vice versa.

And fourthly, it contains twenty-five major theses, namely, (a) the first thesis: the formalness-informalness principle, (b) the second thesis: the absoluteness-relativeness principle, (c) the third thesis: the partiality-totality principle, (d) the fourth thesis: the predictability-unpredictability principle, (e) the fifth thesis: the explicability-inexplicability principle, (f) the sixth thesis: the fiction-reality principle, (g) the seventh thesis: the finiteness-transfiniteness principle, (h) the eighth thesis: the preciseness-vagueness principle, (i) the ninth thesis: the simpleness-complicatedness principle, (j) the tenth thesis: the openness-hiddenness principle, (k) the eleventh thesis: the denseness-emptiness principle, (l) the twelfth thesis: the change-constancy principle, (m) the thirteenth thesis: the order-chaos principle, (n) the fourteenth thesis: the slowness-quickness principle, (o) the fifteenth thesis: the expansion-contraction principle, (p) the sixteenth thesis: the theory-praxis principle, (q) the seventeenth thesis: the convention-novelty principle, (r) the eighteenth thesis: the evolution-transformation principle, (s) the nineteenth thesis: the symmetry-asymmetry principle, (t) the twentieth thesis: the softness-hardness principle, (u) the twenty-first thesis: the seriousness-playfulness principle, (v) the twenty-second thesis: the regression-progression principle, (w) the twenty-third thesis: the sameness-difference principle, (x) the twenty-fourth thesis: the stability-reaction principle, and (y) the twenty-fifth thesis: the post-human complexity—to be addressed in Chapter Four.

Because of the last thesis on "the post-human complexity," the word "complex" in the title of my theory is suggestive, since it is to stress the emergence of complexity in reality, both in the human and post-human future.

However, it is important to clarify at the outset that, lest any misunderstanding easily occurs, the word "complexity" here does not endorse complexity theory (which I already criticized and rejected in *FC*) but simply stresses the ever more complex world that future humans and their successors, post-humans, will face, in a way that current humans have never known.

With this clarification in mind—the seminal project here, if successful, will fundamentally change the way that we think about quantitative and qualitative research in data analysis, from the combined perspectives of the mind, nature, society, and culture, with enormous implications for the human future and what I originally called its "post-human" fate.

1st Thesis: The Formalness-Informalness Principle

The first thesis refers to the formalness-informalness principle (on the formal requirements of logical systems) in the category of "method" in existential dialectics, which was first proposed in *FPHML* (and later, *FPHH*, *FPHGAM*, and other books of mine), in that the formal requirements of a logical system (e.g., consistency, soundness, and completeness) have both usefulness and non-usefulness, to the extent that, if there are formal systems requiring them, there are alternative ones which do not.

The formalness-informalness principle reveals something interesting in existential dialectics, in that it goes beyond both classical and non-classical logics (while learning something from both).

Thus, it does *not* exclude classical logics but simply goes beyond both classical and non-classical logics, while learning something from each. There is no formalness without informalness—and vice versa.

For illustration, existential dialectics can make use of both classical logics under certain conditions (e.g., especially, though not exclusively, when they are clear-cut, etc.) and non-classical logics under alternative conditions (especially, though not exclusively, when they are "unknown," "irrelevant," "ambiguous," "possible," with "different degrees of truth," empirically inconsistent in a desirable way, etc.). (WK 2008u)

For instance, on the one hand, there is formalness (e.g., the formal logical requirement of completeness and soundness for a system of ideas, as in action science, in *Sec. 2.2.1*; the formal logical requirement of soundness for a system of ideas, as in the assumption of the freedom from cognitive biases in quantitative research, in *Sec. 2.2.2*; the formal logical requirement of completeness in a system of ideas, as a "paradigm," in *Sec. 2.3.1*; the formal logical requirement of soundness in a system of ideas, as scientific modeling, in *Sec. 2.3.2*; the formal logical requirement of soundness in a system of ideas, as in MPT, in *Sec. 2.4.1*; the formal logical requirement of completeness and soundness in a system of ideas, as in social statistics, in *Sec. 2.4.2*; the formal logical requirement of completeness in a system of ideas, as in manipulation theories, in *Sec. 2.5.1*; the formal logical requirement of completeness in a system of ideas, as in the theroy of frequentist probability, in *Sec. 2.5.2*; the formal logical requirement of completeness and soundness in a system of ideas, as in Husserl's phenomenology, in *Sec. 3.2.1*; the formal logical requirement of completeness in a system of ideas, as in the hermeneutic circle, in *Sec. 3.2.2*; the formal logical requirement of completeness and soundness in a system of ideas, as in genetic algorithm, in *Sec. 3.3.1*; the formal logical requirement of completeness and soundness in a system of ideas, as in the modeling of GRNs, in *Sec. 3.3.2*; the formal logical requirement of consistency in a system of ideas, as in the history of critical theory, in *Sec. 3.4.1*; the formal logical requirement of completeness in a system of ideas, as in praxis intervention, in *Sec. 3.4.2*; the formal logical requirement of consistency in a system of ideas, as in ethnography, in *Sec. 3.5.1*; and the formal logical requirement of consistency in a system of ideas, as in grounded theory, in *Sec. 3.5.2*).

On the other hand, there is also informalness (e.g., the nonformal existence of incompleteness and unsoudness in action science, as shown in the tension among its conflicting goals pointed out by the critics, in *Sec. 2.2.1*; the nonformal existence of unsoudness in quantitative research, as shown in the existence of numerous cognitive biases pointed out by the critics, in *Sec. 2.2.2*; the non-formal existence of incompleteness in a "paradigm," since a paradigm cannot justify itself to be "better" than an old one, because "the paradigms are incommensurable" and "no meaningful comparison between them is possible without fundamental modification of the concepts that are an intrinsic part of the paradigms being compared," in *Sec. 2.3.1*; the non-formal existence of unsoundness in scientific modeling, as shown in the problem of subjectivity in scientifiic modeling especially when necessary information is not available, in *Sec. 2.3.2*; the non-formal existence of unsoundness in MPT, as shown in the questionable assumptions aforecited, in *Sec. 2.4.1*; the non-formal existence of incompleteness and unsoundness in social statistics, as shown in the critique against social statistics in regard to different problems, such that "correlation does not imply causation" and so on, in *Sec. 2.4.2*; the non-formal existence of incompleteness in manipulation theories, as shown in the criticism that "these accounts are circular," in *Sec. 2.5.1*; the non-formal existence of incompleteness in the theroy of frequentist probability, as shown in the problem of circularity aforediscussed, in *Sec. 2.5.2*; the non-formal existence of incompletness and unsoundness in Husserl's phenomenology, as shown in the rejection of "transcendental ego" by its critics as non-existent, in *Sec. 3.2.1*; the non-formal existence of incompleteness in the hermeneutic circle, as shown in the criticism against Gadamer's version, since, for the critics, the focus on the rigid prejudice of tradition is question-begging or circular, since it can be used against the hermeneutic circle itself, in *Sec. 3.2.2*; the non-formal existence of incompleteness and unsoundness in genetic algorithm, as shown in the criticism by the critics that "the building block hypothesis has been sharply criticized on the grounds that it lacks theoretical justification, and experimental results have been published that draw the veracity of this hypothesis into question," in *Sec. 3.3.1*; the non-formal existence of unsoundness and incompleteness in the modeling of GRNs, as shown in the criticism that it is often too abstract, with little empirical relevance, and that its validity can only be inferred from knowledge outside the model, if available, in *Sec. 3.3.2*; the non-formal existence of inconsistency in the history of critical theory, as shown in "a certain ambivalence concerning the ultimate source or foundation of social domination, an ambivalence which gave rise to the 'pessimism' of the new critical theory over the possibility of human emancipation and freedom" optimistically held by those in the older days like the Kantians and the Marxists, in *Sec. 3.4.1*; the non-formal existence of incompleteness in praxis intervention, as shown in the criticism of inherent bias in praxis intervention, because of its Marxist framing in regard to false consciousness, in *Sec. 3.4.2*; the non-formal existence of inconsistency in ethnography, as shown by the ethical violations in ethnographic research pointed out by Fine, in *Sec. 3.5.1*; and the non-formal existence of inconsistency in grounded theory, as shown in the criticism about incoherence in GT by the critics aforecited, in *Sec. 3.5.2*).

And the reverse direction also holds true.

However, it should be stressed (as this is something that I regularly did in my previous books, *almost verbatim*) that there are different shades of gray (or different degrees of truth) in the two opposites, and the classification is not necessarily mutually exclusive either. In fact, even when some combinations of the two occur, they only end up sharing the same dialectic relationship, but in a different degree.

Also, there are some other relationships (as a kind of family resemblance) which have something in common with the principle, but they are not exactly the same but only more or less comparable, which varies from case to case.

Good examples of family resemblance in relation to the principle are logicalness-nonlogicalness, rationality-nonrationality, etc.

2nd Thesis: The Absoluteness-Relativeness Principle

The second thesis refers to the absoluteness-relativeness principle (on the multiplicity of things) in the category of "method" in existential dialectics, which was first proposed in *FPHK*, in that there is the multiplicity of things in reality, be they about entities, qualities (or properties), and relationships. If there is something absolute, there is likewise something relative. And there is no absoluteness without relativeness—and vice versa.

Both absoluteness and relativeness here are also relevant to different modalities often cited in the literature on ontology, such as possibility (e.g., something "can" happen) and its opposite (e.g., impossibility), probability (e.g., something "will" happen) and its opposite (e.g., improbability), and necessity (e.g., something "should" happen) and its opposite (e.g., contingency).

For instance, on the one hand, there is absoluteness (e.g., the absolute view about propositional knowing as the legitimate form of knowing in contemporary mainstream science, in *Sec. 2.2.1*; the absolute view of those with "egocentric bias" to "claim more responsibility for themselves for the results of a joint action than an outside observer would," in *Sec. 2.2.2*; the absolute view within each paradigm in relation to its own worldview, in *Sec. 2.3.1*; the absolute view of a scientific modeling which is "deterministic," in *Sec. 2.3.2*; the absolute view of MPT which "does not account for the personal, environmental, strategic, or social dimensions of investment decisions," in *Sec. 2.4.1*; the absolute "assignment rigidity" in cluster analysis, such that "an object can be assigned in one cluster only," in *Sec. 2.4.2*; the absolute view of "deterministic causation," in that, "if *A* causes *B*, then A must *always* be followed by *B*," in *Sec. 2.5.1*; the absolute view of the logical theorists, in that "[l]ogical probabilities are conceived to be objective logical relations between propositions…, and hence not to depend in any way upon belief," in *Sec. 2.5.2*; the absolute view of Gadamer in regard to the "prejudice" of an interpreter in a given tradition, in *Sec. 3.2.1*; the absolute view of Gadamer in regard to the "prejudice" of an interpreter in a given tradition, in *Sec. 3.2.2*; the absolute view in genetic algorithm to mimic "the process of natural evolution…to generate useful solutions to optimization and search problems," in *Sec. 3.3.1*; the absolute focus on discrete steps in boolean networks, in *Sec. 3.3.2*; the absolute concern in social critical theory by Habermas with "emancipation," in *Sec. 3.4.1*; the absolute view of those in praxis intervention to eliminate different biases of the participants through reflexive praxis, in *Sec. 3.4.2*; the absolute concern with ethnography in cultural anthropology, in *Sec. 3.5.1*; and the absolute concern of GT with generating new hypotheses, not aiming for the "truth," in *Sec. 3.5.2*).

On the other hand, there is relativeness (e.g., what is legitimate for the advocates of contemporary mainstream science in regard to propositional knowing is not necessarily sufficient for the advocates of practical knowing, as shown in John Heron and Peter Reason's cooperative inquiry, in *Sec. 2.2.1*; what is right for those with "egocentric bias" in regard to claiming more responsibility for themselves is not necessarily so for an outside observer, in *Sec. 2.2.2*; what is the right worldview within an older paradigm is not necessarily so for a newer paradigm which is to replace it later, in *Sec. 2.3.1*; what is right for a scientific modeling which is "deterministic" is not necessarily so for another which is "probabilistic" instead, in *Sec. 2.3.2*; what is right for MPT in regard to its failure to take into account "the personal, environmental, strategic, or social dimensions of investment decisions" is not necessarily so for its critics, who suggest their inclusion, in *Sec. 2.4.1*; what is right for cluster analysis in regard to "assignment rigidity" is not necessarily so for its critics, who call for more flexibility, in *Sec. 2.4.2*; what is true for "deterministic causation" is not necessarily so for "probabilistic causation," which suggests instead that "*A* probabilistically causes *B* if *A*'s occurrence increases the probability of *B*," in *Sec. 2.5.1*; what is right for the logical theorists is not necessarily so for the epistemic theorists like Ramsey who were "skeptical about the existence of such objective logical relations and argued that…evidential…probability is 'the logic of partial belief," in *Sec. 2.5.2*; what is right for Husserl in regard to his focus on "consciousness" is not necessarily so for Heidegger who "countered that consciousness is peripheral to the primacy of one's existence," or "the mode of being of Dasein," in *Sec. 3.2.1*; what is right for Gadamer in regard to the "prejudice" of an interpreter in a given tradition is not necessarily so for Shklar, who argued that, although "interpreters have multiple and sometimes conflicting cultural attachments, yet this do not prevent intercultural and/or interdisciplinary dialogue," so as to go beyond "prejudice," in *Sec. 3.2.2*; what is right for those in genetic algorithm in regard to its particular methodology is not necessarily so for the critics who offer alternative optimization algorithms like "simulated annealing, Gaussian adaptation, hill climbing, and swarm intelligence," in *Sec. 3.3.1*; what is right for "boolean networks" in regard to its simple non-continuous representation of the range of activity levels in GRNs is not necessarily so for "continuous networks" which allow continuous representation, in *Sec. 3.3.2*; what is right for Habermas in regard to his concern with "emancipation" is not necessarily so for those in literary social theory with their focus on the analysis of texts, not radical social change, in *Sec. 3.4.1*; what is feasible for those in praxis intervention in regard to their attempt to eliminate different biases is not necessarily so for the critics, who argue instead that false consciousness can be deep-rooted and that some individuals are simply not reflexive enough to overcome it, in *Sec. 3.4.2*; what is the right focus for those in cultural anthropology is not necessarily so for those in sociology, who also use ethnographic technqiues but are not so obsessed, in *Sec. 3.5.1*; and what is approriate for GT in regard to its concern with generating new hypotheses, not aiming for the "truth," is not necessarily so for those in quantitative research like social statistics in regard to its aim to confirm pre-determined hypotheses for truth-seeking, in *Sec. 3.5.2*).

And the reverse direction also holds true.

However, it should be stressed (as this is something that I regularly did in my previous books, *almost verbatim*) that there are different shades of gray (or different degrees of truth) in the two opposites, and the classification is not necessarily mutually exclusive either. In fact, even when some combinations of the two occur, they only end up sharing the same dialectic relationship, but in a different degree.

Also, there are some other relationships (as a kind of family resemblance) which have something in common with the principle, but they are not exactly the same but only more or less comparable, which varies from case to case.

Good examples of family resemblance in relation to the principle are uniformity-diversity, internalness-externalness, immanence-transcendence, independence-dependence, isolatedness-interconnectiveness, and so forth.

3rd Thesis: The Partiality-Totality Principle

The third thesis refers to the partiality-totality principle (on the relationships between whole and parts) in the category of "method" in existential dialectics, which was first proposed in *FC* and later *FPHGEOL*, based on earlier works like *FHC* and *FPHC*), in that any analysis of a phenomenon in the world requires both the examination of the individual parts and the whole of them (with the whole as being not the sum of its parts), without privileging the whole as more important than the parts, or vice-versa (that is, the parts as more important than the whole).

In fact, this principle targets two kinds of misleading methodologies as used over the ages, that is, again in the absence of better words, what I called (in *FHC*, *FPHC* and *FC*) "reductionism" and its opposite, "reverse-reductionism," which can take four major forms in relation to concept, theory, methodology, and ontology (as summarized in *Sec. 1.8*).

Thus, reductionism and reverse-reductionism can be (i) conceptual, (ii) theoretical, (iii) methodological, and (iv) ontological. And the whole is not the sum of its parts, and there is no partiality without totality—and vice versa.

Works on emergence (as in *FC*), gestalt psychology (as in *FPHGEOL*), and the hermeneutic circle (as in *FPHDA*), just to name a few of them, have contributed to the understanding of whole and parts.

For instance, on the one hand, there is partiality (e.g., the partial view within each paradigm in regard to its own worldview, in *Sec. 2.3.1*; the partial view of a specific element in "a system," in *Sec. 2.3.2*; the partial view of MPT to only "maximize risk-adjusted returns," in *Sec. 2.4.1*; the partial view of the parts in the hermeneutic circle, in *Sec. 3.2.2*; those cases of "local optima" in genetic algorithm, in *Sec. 3.3.1*; the partial view of each discipline in the social sciences, like economics, political science, etc., in *Sec. 3.4.1*; and the partial view of those in non-reflextive praxis because of their false consciousness, in *Sec. 3.4.2*).

On the other hand, there is also totality (e.g., the more holistic view of different paradigms over time, in the history of "paradigm shifts," such that the whole is not the sum of its parts, that is, the implications of the history of "paradigm shifts," which are still subject to a heated debate among historians of science, are not reducible to the sum of the individual paradigms, in *Sec. 2.3.1*; the more holistic view of the entire "collection of different elements" in a system, such that the whole is not the sum of its parts, that is, the whole collection with their multifaceted relationships "can produce results not obtainable by the elements alone" or the mere summation of them, in *Sec. 2.3.2*; the more holistic view of "the personal, environmental, strategic, or social dimensions of investment decisions," such that the whole is not the sum of its parts, that is, the possible emergence of "all the standard market failures such as those arising from information asymmetry, externalities, and public goods," which could not be well understood if only considering to "maximize risk-adjusted returns" alone within MPT, in *Sec. 2.4.1*; the holistic view of the whole in the hermeneutic circle, such that the whole cannot be reduced to the sum of its parts alone, in *Sec. 3.2.2*; those cases of "global optimum" in genetic algorithm, such that the whole is not the sum of its parts, that is, a "global optimum" is not simply the sum of "local optima," in *Sec. 3.3.1*; the more holistic view of the entirety of society, as in social critical theory, such that the whole is not the sum of its parts, that is, the interelations among different disciplines are not simply the sum of all disciplines, in *Sec. 3.4.1*; and the more holistic view of those in reflexive praxis in regard to the larger social structure of class domination and the need to overcome false consciousness, such that the whole is not the sum of its parts, that is, the reflexive view of the need to end false consciousness through praxis is not merely the sum of all individual forms of false consciousness, in *Sec. 3.4.2*).

And the reverse direction also holds true.

However, it should be stressed (as this is something that I regularly did in my previous books, *almost verbatim*) that there are different shades of gray (or different degrees of truth) in the two opposites, and the classification is not necessarily mutually exclusive either. In fact, even when some combinations of the two occur, they only end up sharing the same dialectic relationship, but in a different degree.

Also, there are some other relationships (as a kind of family resemblance) which have something in common with the principle, but they are not exactly the same but only more or less comparable, which varies from case to case.

Good examples of family resemblance in relation to the principle are individualisticness-holisticness, partness-wholeness, analysis-synthesis, and the like.

4th Thesis: The Predictability-Unpredictability Principle

The fourth thesis is called the predictability-unpredictability principle (on the occurrence of events) in the category of "method" in existential dialectics (which was first proposed in *FC* and, later, other books of mine), in that both predictability and unpredictability have a major role to play in the occurrence of things, so that neither determinism nor indeterminism wins the centuries-old fight. There is no predictability without unpredictability—and vice versa.

There are events which are predictable, just as there are those which are not. Or what is regarded as unpredictable at one point in time may turn out to be predictable later, and, conversely, what is deemed as predictable may turn out to not be so predictable. Even in predictability, outcomes are subject to uncertainty, the degree of which varies from case to case.

Works on complexity theory (as in *FC*), quantum mechanics (as in *FPHST* and *FPHG*), frequency probability, subjective (or Bayesian) probability, and "special causes" (as in *FPHGAM*), just to name a few of them, have contributed to the understanding of uncertainity.

For instance, on the one hand, there is predictability (e.g., the predictable tendency of those in action research to stress "problem solving actions" in their research, in *Sec. 2.2.1*; the predictable tendency of someone with "optimism bias" to be "over-optimistic about the outcome of planned actions," in *Sec. 2.2.2*; the predictable tendency of a paradigm to resolve some unresolved problems that an old paradigm cannot, in *Sec. 2.3.1*; it is easier to do "interpolation," in *Sec. 2.3.2*; the predictable tendency of those in MPT to measure "risk, return, and correlation" on the basis of "expected values, which means that they are mathematical statements about the future," in *Sec. 2.4.1*; the predictable tendency of non-probability sampling to produce surveys which "can only be said to be representative of the people that have actually completed" them, in *Sec. 2.4.2*; the predictable tendency, if viewed within manipulation theories, of those who "are interested in knowing the causes of crime" to "find ways of reducing it," in *Sec. 2.5.1*; the predictable tendency of those in frequentist interpretation to claim that, in the case of tossing a fair coin, "the probability of getting a heads is ½," in *Sec. 2.5.2*; the predictable tendency of those in Husserl's phenomenology to focus on "consciousness," in *Sec. 3.2.1*; the predictable tendency of those in Schleiermacher's hermeneutics to focus on the grammatical interpretation and psychological interpretation of a text by an author, in *Sec. 3.2.2*; the predictable tendency of genetic algorithm to converge"[i]n many problems…towards local optima or even arbitrary points rather than the global optimum of the problem," in *Sec. 3.3.1*; the predictable tendency of a model of GRNs to represent a network in terms of "nodes," with "input being proteins such as transcription factors, and outputs being the level of gene expression," in *Sec. 3.3.2*; the predictable tendency of social critical theory to undergo changes over time, as shown in the questioning of Marxist optimism by Horkheimer and Adorno, in *Sec. 3.4.1*; the predictable tendency of those in praxis intervention "to struggle against structurally ingrained discrimination," in *Sec. 3.4.2*; the predictable tendency of those in communication studies to use more and more ethnographic techniques nowadays, in *Sec. 3.5.1*; and the predictable tendency of those in GT to generate new hypotheses, instead of confirming them, in *Sec. 3.5.2*).

On the other hand, there is unpredictability (e.g., the more difficult task to predict exactly when, or the extent to which, a particular application of action research will be successful in resolving the inherent tension among its conflicting goals, as pointed out by the critics, in *Sec. 2.2.1*; the more difficult task to predict exactly the extent to which someone with "optimism bias" will be "over-optimistic about the outcome of planned actions" on a particular occasion, in *Sec. 2.2.2*; the more difficult task to predict exactly how long a new paradigm will last, before it is to be replaced by another one later, in *Sec. 2.3.1*; it is more difficult to do "extrapolation," in *Sec. 2.3.2*; the more difficult task to predict exactly the extent to which these mathematical statements are accurate, since "very often such expected values fail to take account of new circumstances which did not exist when the historical data were generated," in *Sec. 2.4.1*; the more difficult task to predict exactly the extent to which non-probability sampling can at least represent the target population, since "surveys that are not based on probability sampling have no way of measuring their bias or sampling error," in *Sec. 2.4.2*; the more difficult task to predict exactly the extent to which crimes will be reduced on a particular occasion, by those interested in knowing their causes, in *Sec. 2.5.1*; the more difficult task to predict exactly if a particular tossing of a coin in practice will turn out to be a head on a particular occasion, if viewed within the frequentist interpretation, in *Sec. 2.5.2*; the more difficult task to predict exactly the extent to which those in Husserl's phenomenology will reach "pure consciousness" in their version of "transcendental ego," in *Sec. 3.2.1*; the more difficult task to predict exactly the extent to which a particular interpreter will succeed in interpreting correctly "the peculiar combinations that characterize the work as a whole" by an author in relation to psychological interpretation, since there can be multiple conflicting interpretations, in *Sec. 3.2.2*; the more difficult task to predict exactly the extent to which "global optimum" will be achieved, because, according to the No Free Lunch theorem, "there is no general solution to this problem," in *Sec. 3.3.1*; the more difficult task to predict exactly the extent to which each "node" will function in increasing levels of complexity, because the modeling of GRNs is still in its infnacy and "[t]he gene networks are only beginning to be understood, and it is a next step for biology to attempt to deduce the functions for each gene 'node,' to help understand the behavior of the system in increasing levels of complexity, from gene to signaling pathway, cell or tissue level," in *Sec. 3.3.2*; the more difficult task to predict exactly the extent to which the pessimism of Horkheimer and Adorno will last in the future history of social critical theory, in *Sec. 3.4.1*; the more difficult task to predict exactly the extent to which those in praxis intervention "to struggle against structurally ingrained discrimination" will suceed on a particular occasion, because, as Alasdair MacIntyre argued, there is the problem of "the unpredictability of human beings," so even "one unpredictable human being can have a world-changing impact," in *Sec. 3.4.2*; the more difficult task to predict exactly the extent to which the spread of ethnogrpahic techniques will be in communication studies in a particular era of future history, in *Sec. 3.5.1*; and the more difficult task to predict exactly the extent to which a particular new hypothesis generated by those in GT will ever be confirmed in a particular era of future history, in *Sec. 3.5.2*).

And the reverse direction also holds true.

Again, it should be stressed, however, that there are different shades of gray (or different degrees of truth) in the two opposites, and the classification is not necessarily mutually exclusive either. In fact, even when some combinations of the two occur, they only end up sharing the same dialectic relationship, but in a different degree.

Also, there are some other relationships (as a kind of family resemblance) which have something in common with the principle, but they are not exactly the same but only more or less comparable, which varies from case to case.

Good examples of family resemblance in relation to the principle are sureness-arbitrariness, sureness-doubtfulness, likeliness-riskiness, deterministicness-indeterministicness, deterministicness-randomness, certainty-uncertainty, and so forth.

5th Thesis: The Explicability-Inexplicability Principle

The fifth thesis is called the explicability-inexplicability principle (on the underlying mechanisms of things) in the category of "method" in existential dialectics (which was proposed in *FPHU* and other books of mine), in that both explicability and inexplicability are part of the understanding of things. There is no explicability without inexplicability—and vice versa.

This principle tells us the dual nature of the research dilemma, in that, if reality can be explained in some ways, it also has its other ways which are not quite explainable, at a given point in time.

For instance, on the one hand, there is explicability (e.g., the explanation by the advocates of action research that it is desirable to move "beyond reflective knowledge" for "an active moment-to-moment theorizing, data collecting, and inquiry occurring in the midst of emergent structure," which contributes to social change, in *Sec. 2.2.1*; the explanation by those in the study of cognitive biases to propose "reference class forecasting" as a way "to eliminate or reduce the impact of cognitive biases on decision making," in *Sec. 2.2.2*; the explanation by Kuhn that its work on paradigm shifts does not suggest relativism, since "he reiterates five criteria...that...help determine...theory choice," in *Sec. 2.3.1*; the explanation of the relationship between two different variables in a scientific modeling on the basis of "correlation," in *Sec. 2.3.2*; the explanation by those in MPT that diversification is desirable because it "eliminates non-systematic risk," in *Sec. 2.4.1*; the explanation by the advocates of cluster analysis that it is desirable, because it provides "a quick review of data" and "a simple profile of individuals," in *Sec. 2.4.2*; the explanation, if viewed within "deterministic causation," of "*A* causes *B*" to mean that "if *A* causes *B*, then A must *always* be followed by *B*," in *Sec. 2.5.1*; the explanation by those in propensity interpretation that "propensities, or chances, are not relative frequencies, but purported causes of the observed stable relative frequencies," in *Sec. 2.5.2*; the explanation by Husserl that "bracketing" is desirable because it can eventually help us reach "pure consciousness" in "transcendental ego," in *Sec. 3.2.1*; the explanation by Gadamer that interpretation is valuable because "[w]e can reach the truth only by understanding or even mastering our experience," even though "we can never step outside of our tradition," so "all we can do is try to understand it," in *Sec. 3.2.2*; the explanation by the advocates of genetic algorithms that they are desirable because "these algorithms frequently succeed at generating solutions of high fitness when applied to practical problems," in *Sec. 3.3.1*; the explanation of the process of a GRN on the basis that "[s]ometimes, a 'self-sustaining feedback loop' ensures that a cell maintains its identity and passes it on," in *Sec. 3.3.2*; the explanation by those in social critical theory that "emancipation" is desirable because it aims to overthrow oppressive social domination, in *Sec. 3.4.1*; the explanation, by those in praxis intervention, that reflexiveness is desirable, because it leads to "creativity instead of sameness, autonomy instead of subordination, sociality instead of massification, rationality instead of blind reaction and intentionality rather than compliance," in *Sec. 3.4.2*; the explanation by those in ethnographic research that honesty is an important professional ethical virtue so that research outcomes are not "skewed," in *Sec. 3.5.1*; and the explanation by those in GT that it is grounded on a methodology based on "codes," "concepts," "categories," and "theory," in *Sec. 3.5.2*).

On the other hand, there is inexplicability (e.g., the lack of sufficient explanation by the advocates of action research of why it is necessarily desirable to focus more on action theorizing than on reflective knowledge, since not every researcher is driven by the desire for social change, in *Sec.*

2.2.1; the lack of sufficient explanation by those in the study of cognitive biases of why "reference class forecasting" is necessarily desirable, since "cognitive biases" can be "presumably adaptive," for example, when "they lead to more effective actions in given contexts or enable faster decisions when faster decisions are of greater value," in *Sec. 2.2.2*; the lack of sufficient explanation by Kuhn, since, for his critics, "the criteria still are not 'objective' in the usual sense of the word because individual scientists reach different conclusions with the same criteria due to valuing one criterion over another or even adding additional criteria for selfish or other subjective reasons," in *Sec. 2.3.1*; the more difficult task to explain "causality" between two different variables in a scientific modeling, as shown in the problem of inability to extrapolate in scientific modeling, insofar as the causal relation is not well understood, in *Sec. 2.3.2*; the more difficult task for those in MPT to explain why diversification is necessarily desirable, since it occurs "at the cost of increasing the systematic risk," because "[t]his artificially increased demand pushes up the price of assets…and, as a result,…the risk of the portfolio increases," in *Sec. 2.4.1*; the lack of sufficient explanation by the advocates of why cluster analysis is necessarily desirable, since it also produces "detrimental effects" as cited by the critics, in *Sec. 2.4.2*; the lack of sufficient explanation, if viewed within "deterministic causation," of why war "causes" deaths, why smoking "causes" cancer," etc., in *Sec. 2.5.1*; the lack of sufficient explanation by those in propensity interpretation of "exactly what propensity means," since "none of the well-recognised accounts of propensity comes close to meeting this challenge," in *Sec. 2.5.2*; the lack of sufficient explanation by Husserl of why "bracketing" is necessarily desirable because, for Heidegger, "consciousness is peripheral to the primacy of one's existence," that is, "the mode of being of Dasein," which "cannot be reduced" to consciousness and is more important, in *Sec. 3.2.1*; the lack of sufficient explanation by Gadamer of why "we can never step outside of our tradition," since, for the critics like Habermas, "the focus on tradition seemed to undermine possibilities for social criticism and transformation," in *Sec. 3.2.2*; the lack of sufficient explanation by the advocates of genetic algorithms of why genetic algorithms can generate these solutions as claimed, because, for the critics, "their behavior is difficult to understand," since, "[i]n particular it is difficult to understand why these algorithms frequently succeed at generating solutions of high fitness when applied to practical problems" as claimed, in *Sec. 3.3.1*; the lack of sufficient explanation about "the mechanism of epigenetics by which chromatin modification may provide cellular memory by blocking or allowing transcription" in a GRN, which is still "less understood," in

Sec. 3.3.2; the lack of sufficient explanation by those in social critical theory of why "emancipation" is necessarily desirable, since the critics can then ask, "who" is to say what is "good" or "right" for society to be transformed?" in *Sec. 3.4.1*; the lack of sufficient explanation by those in praxis intervention of why these ends like "creativity," "autonomy," "sociality," "rationality," "intentionality," etc. are necessarily desirable, since, for the critics, some individuals may instead prefer "sameness," "subordination," "compliance," etc., in *Sec. 3.4.2*; the lack of sufficient explanation by those in ethnographic research of why in actually they often lie, in spite of the official idealization of honesty as a professonal ethical standard, as pointed out by Fine, in *Sec. 3.5.1*; and the lack of sufficient explanation by those in GT of "why…an idea of 'grounding' one's findings" is "important in qualitative inquiry" and "what…they [are really] 'grounded' in," for the critics, in *Sec. 3.5.2*).

And the reverse direction also holds true.

Once more, it should be stressed, however, that there are different shades of gray (or different degrees of truth) in the two opposites, and the classification is not necessarily mutually exclusive either. In fact, even when some combinations of the two occur, they only end up sharing the same dialectic relationship, but in a different degree.

Also, there are some other relationships (as a kind of family resemblance) which have something in common with the principle, but they are not exactly the same but only more or less comparable, which varies from case to case.

Good examples of family resemblance in relation to the principle are underlyingness-regularness, causation-regularity, causation-correlation, explanation-description, explanation-narration, explanation-interpretation, function-nonfunction, and so forth.

6th Thesis: The Fiction-Reality Principle

The sixth thesis is called the fiction-reality principle (on the different faces of reality) in the category of "method" in existential dialectics (as first proposed in *FPHLIT*), in that both fiction and reality co-exist. There is no reality without fiction—and vice versa.

This principle tells us the different categories of reality, in that, if there is reality, there is also some fiction in it—just as, if there is fiction, there is also some reality in it.

For instance, on the one hand, there is fiction (e.g., the subjective experience of "living contradictions" in 1st-person action research, as in living theory, in *Sec. 2.2.1*; the fictional aspects of cognitive biases when they "can lead to perceptual distortion, inaccurate judgment, or illogical interpretation," in *Sec. 2.2.2*; the fictional aspect of a paradigm, as shown in "paradigm paralysis," which refers to "the inability or refusal to see beyond the current models of thinking," that is, the inability to see what is wrong with one's worldview, in *Sec. 2.3.1*; the fictional aspect of a scientific modeling when it relies on incorrect subjective information, in *Sec. 2.3.2*; the fictional aspect of MPT, as shown in its questionable assumptions which are not really true, in *Sec. 2.4.1*; the fictional aspect of factor analysis when the researcher also uses "subjective attributes" which may not accurately reflect reality, in *Sec. 2.4.2*; the fictional aspect of Aristotle's types of causality, as shown in the rejection of "formal" and "final" causes by the critics like Niccolò Machiavelli and Francis Bacon, in *Sec. 2.5.1*; the fictional aspect of subjectivist interpretation, since it gives "the notion of probability a subjective status by regarding it as a measure of the 'degree of belief' of the individual assessing the uncertainty of a particular situation," in *Sec. 2.5.2*; the fictional aspect of Husserl's phenomenology, as shown in the rejection of "transcendental ego" by its critics as non-existent, in *Sec. 3.2.1*; the fictional aspect of Dilthey's hermeneutic circle, when "incomprehension" produces "wrong" understanding, in *Sec. 3.2.2*; the fictional aspect of genetic algorithm, as shown in the criticism by the critics that "it lacks theoretical justification, and experimental results have been published that draw the veracity of this hypothesis into question," in *Sec. 3.3.1*; the fictional aspect of the modeling of GRNs, when it is simplistic and does not accurately represent reality, as pointed out by the critics, in *Sec. 3.3.2*; the fictional aspect of critical theory, insofar as it deals with reality as "constructed" in the analysis of texts, in *Sec. 3.4.1*; false consciousness in non-reflexive praxis, in *Sec. 3.4.2*; the fictional aspect of ethnographic research, when "researchers take near-fictions and turn them into claims of fact," in *Sec. 3.5.1*; and the fictional aspect of GT when it uses all kinds of data, "even fiction" in literature, in *Sec. 3.5.2*).

On the other hand, there is reality (e.g., the better understanding of reality after going through "action reflection cycles" and learning from the evaluations, in *Sec. 2.2.1*; the realistic aspects of cognitive biases whey "they lead to more effective actions in given contexts or enable faster decisions when faster decisions are of greater value," in *Sec. 2.2.2*; the realistic aspect of a paradigm, in that it can resolve some unresolved problems that an old paradigm cannot, in *Sec. 2.3.1*; the realistic aspect of a scientific modeling when it uses correct information, in *Sec. 2.3.2*; the realistic aspect of MPT, as shown in some successful applications in finance and non-finance aforecited, in *Sec. 2.4.1*; the realistic aspect of factor analysis when the researcher uses "objective attributes" which may more likely reflect reality, in *Sec. 2.4.2*; the realistic aspect of Aristotle's types of causality, as shown in the acceptance of "efficient" causation by some modern epistemic realists, in *Sec. 2.5.1*; the realistic aspect of subjectivist interpretation, when its assignment of a probability to a proposition is "based on the evidence presented," in *Sec. 2.5.2*; the realistic aspect of Husserl's phenomenology, as shown in its "general effort to subject experience to fundamental, critical scrutiny" as pointed out by Maurice Natanson, in *Sec. 3.2.1*; the realistic aspect of Dilthey's hermeneutic circle, when "comprehension" produces "correct" understanding, in *Sec. 3.2.2*; the realistic aspect of genetic algorithm, when it "succeed[s] at generating solutions of high fitness when applied to practical problems," in *Sec. 3.3.1*; the realistic aspect of the modeling of GRNs, when it can help us understand better more simple cases, as shown by Boolean networks, which "have been used due to their simplicity and ability to handle noisy data," in *Sec. 3.3.2*; the realistic aspect of critical theory, insofar as it deals with the real existence of oppressive "social dominaton," in *Sec. 3.4.1*; reflexion to wake up from "epistemic sleep" in reflexive praxis, in *Sec. 3.4.2*; the realistic aspect of ethnographic research, when researchers try their best to report as accurately as possible, in *Sec. 3.5.1*; and the realistic aspect of GT when it also uses "surveys or statistical analyses" from the real world, in *Sec. 3.5.2*).

And the reverse direction also holds true.

Once more, it should be stressed, however, that there are different shades of gray (or different degrees of truth) in the two opposites, and the classification is not necessarily mutually exclusive either. In fact, even when some combinations of the two occur, they only end up sharing the same dialectic relationship, but in a different degree.

Also, there are some other relationships (as a kind of family resemblance) which have something in common with the principle, but they are not exactly the same but only more or less comparable, which varies from case to case.

Good examples of family resemblance in relation to the principle are fiction-nonfiction, fiction-fact, imagination-fact, imagination-nonfiction, subjectivity-objectivity, and so forth.

7th Thesis: The Finiteness-Transfiniteness Principle

The seventh thesis is called the finiteness-transfiniteness principle (on the nature of numbers) in the category of "structure" in existential dialectics (as first proposed in *FPHG*), in that, if there are finite things, there are likewise transfinite ones—and vice versa.

To avoid confusion, my usage of the word "transfinite" here differs radically from the one used by Cantor (and other mathematicians) for "relative" infinity—and is more limited, in light of the problems confronting any attempt to understand the idea of infinity, be it by intuition, imagination, and conception (as detailedly analyzed in *Sec. 2.2.3* of *FPHG*).

Instead, by "transfinity," I allow numbers which can be many times larger—or smaller, for that matter—than the finite things that we encounter in daily life, but they do not have to be related to the idea of infinity at all (which may not exist).

Of course, there may be some *borderline* cases, in which it is not clear whether the number in question is transfinite (in my usage) or simply a mathematical convenience. A good example of a borderline case is the Planck unit of length for "the smallest space possibly measured in nature," which is "less than billionths of trillionths of trillionths of an inch" (or something like 1.6×10^{-35} meters). (P. Baofu 2006a; N. McAleer 1987: 219; WKV 2008; D. Corbett 2008a)

That said—my usage of "transfinity" can also resolve (or better, dissolve) an age-old problem in philosophy known as "Zeno's paradoxes" (as already explained in *Sec. 2.2.3* of *FPHG*).

In any event, in relation to the finiteness-transfiniteness principle, on the other hand, there is finiteness (e.g., the finite number of separate points in some "discrete" scientific systems, in *Sec. 2.3.2*; the finite amount of "credit" that an investor can have in practice, in *Sec. 2.4.1*; "only a finite number of repetitions" is performed in the process of determining frequency probability, in *Sec. 2.5.2*; and the finite number of "discrete steps" in the modeling of a Boolan network, in *Sec. 3.3.2*).

On the other hand, there is transfinitty (e.g., the transfinite number of different points in some "continuous" scientific systems, in *Sec. 2.3.2*; the transfinite amount of "credit" that an investor can have in principle, within MPT, in *Sec. 2.4.1*; according to the law of large numbers, "the empirical frequency converges to the limit 1/2 as the number of trials goes to infinity," in *Sec. 2.5.2*; and the transfinite number of "the continuous range of activity levels" in the modeling of "continuous networks," in *Sec. 3.3.2*).

Once more, it should be stressed, however, that there are different shades of gray (or different degrees of truth) in the two opposites, and the classification is not necessarily mutually exclusive either. In fact, even when some combinations of the two occur, they only end up sharing the same dialectic relationship, but in a different degree.

Also, there are some other relationships (as a kind of family resemblance) which have something in common with the principle, but they are not exactly the same but only more or less comparable, which varies from case to case.

Good examples of family resemblance in relation to the principle are boundedness-quasiunboundedness, smallness-largeness, microness-macroness, boundedness-unboundedness, and the like.

8th Thesis: The Preciseness-Vagueness Principle

The eighth thesis is called the preciseness-vagueness principle (on the refinement of things) in the category of "structure" in existential dialectics (which was first out worked out in *FIA*), in that both preciseness and vagueness are important, not that one is better than the other, but that both are used, in different degrees of preference, in accordance to the contextual application from the perspectives of nature, the mind, culture, and society. There is no preciseness without vagueness—and vice versa.

For instance, on the one hand, there is preciseness (e.g., the precise identification of "four different types of knowledge" by Heron and Reason in the context of cooperative inquiry, in *Sec. 2.2.1*; the precise identification of the examples of different categories of cognitive biases aforecited, in *Sec. 2.2.2*; the precise identification of the "five criteria…that…help determine…theory choice" by Kuhn, in *Sec. 2.3.1*; the precise identification of six factors to evaluate a scientific model aforecited, in *Sec. 2.3.2*; the precise identification of three categories of criticisms against MPT as cited, in *Sec. 2.4.1*; the precise identification of three advantages and three disadvantages of factor analysis as cited, in *Sec. 2.4.2*; the precise identification of four types of causality by Aristotle, in *Sec. 2.5.1*; the precise classification of four types of "evidential" probabilities in the literature aforecited, in *Sec. 2.5.2*; the precise identification of five major compoents in Husserl's phenomenology for discussion in the literature aforecited, in *Sec. 3.2.1*; the precise identification by Dilthey of three levels of interpretation from the vantage point of an individual, in *Sec. 3.2.2*; the precise identification of four steps in the methodology of genetic algorithm aforecited, in *Sec. 3.3.1*; the precise identification of four steps in the modeling of Boolean networks aforedescribed, in *Sec. 3.3.2*; the precise identification of four ways in which reality is "constructed" in critical theory, in *Sec. 3.4.1*; the precise classification of three phases in praxis intervention in the literature aforecited, in *Sec. 3.4.2*; the precise identification of "five criteria that ethnographers might find helpful" by Richardson, in *Sec. 3.5.1*; and the precise identification of four main steps in the methodology of GT, namely, "codes," "concepts," "categories," and "theory," in *Sec. 3.5.2*).

On the other hand, there is vagueness (e.g., the vagueness in the identificaiton, since it is not clear why there must be only "four different types of knowledge," not five, six, seven, and so on, in *Sec. 2.2.1*; the vagueness in the precise identification of the examples of different categories of cognitive biases aforecited, since it is not clear why there must be these specific examples, but not others, in *Sec. 2.2.2*; the vagueness in the identification of the five criteria, since it is not clear why there must be only five criteria or whether or not they should be valued equally, as, for the critics, "individual scientists reach different conclusions with the same criteria due to valuing one criterion over another or even adding additional criteria for selfish or other subjective reasons," in *Sec. 2.3.1*; the vagueness in the identification of the six factors, since it is not clear why there must be only six, not seven, eight, and so on, in *Sec. 2.3.2*; the vagueness in the identification of the three categories of criticisms against MPT as cited, since it is not clear why there must be only three, not two, four, five, and so on, in *Sec. 2.4.1*; the vagueness in the identification of the three advantages and three disadvantages of factor analysis as cited, since it is not clear why there must be only three, not four, five, and so on, in *Sec. 2.4.2*; the vagueness in the identification of the four types of causality by Aristotle, since it is not clear why there must be only four, not one, two, three, five, and so on, in *Sec. 2.5.1*; the vagueness in the classification of the four types of "evidential" probabilities in the literature aforecited, since it is not clear why there must be four in the classification, not three, five, six, and so on, in *Sec. 2.5.2*; the vagueness in the identification of the five major compoents in Husserl's phenomenology for discussion in the literature aforecited, since it is not clear why there must be only five, not four, six, seven, and so on, in *Sec. 3.2.1*; the vagueness in the identification by Dilthey of the three levels of interpretation from the vantage point of an individual, since it is not clear why there must be only three, not two, four, five, and so on, in *Sec. 3.2.2*; the vagueness in the identification of the four steps in the methodology of genetic algorithm aforecited, since it is not clear why there must be only four, not three, five, six, and so on, as there is an alternative classification which has only three steps, for instance, in *Sec. 3.3.1*; the vagueness in the identification of the four steps in the modeling of Boolean networks aforedescribed, since it is not clear why there must be only four, not three, five, six, and so on, in *Sec. 3.3.2*; the vagueness in the identification of the four ways in which reality is "constructed" in critical theory, since it is not clear why there must be only four, not five, six, and so on, in *Sec. 3.4.1*; the vagueness in the classification of the three phases in praxis intervention in the literature afore-cited, since it is not clear why there must be only three phases, not four, five, and so on, in *Sec. 3.4.2*; the vagueness in the identification of the "five criteria that ethnographers might find helpful" by Richardson, since it is not clear why there must be only five, not six, seven, and so on, in *Sec. 3.5.1*; and the vagueness in the identification of the four steps in the methodology of GT, since it is not clear why there must be only four steps, not five, six, and so on, in *Sec. 3.5.2*).

And the reverse direction also holds true.

Once more, it should be stressed, however, that there are different shades of gray (or different degrees of truth) in the two opposites, and the classification is not necessarily mutually exclusive either. In fact, even when some combinations of the two occur, they only end up sharing the same dialectic relationship, but in a different degree.

Also, there are some other relationships (as a kind of family resemblance) which have something in common with the principle, but they are not exactly the same but only more or less comparable, which varies from case to case.

Good examples of family resemblance in relation to the principle are clarity-ambiguity, directness-indirectness, quantitativeness-qualitativeness, specificity-obscurity, thickness-thinness, describability-nondescribability, concretness-abstractness, translatability-untranslatability, specificity-generality, clearcutness-transitionalness, and the like.

9th Thesis: The Simpleness-Complicatedness Principle

The ninth thesis refers to the simpleness-complicatedness principle (on the interconnection among things) in the category of "structure" in existential dialectics (which was first out worked out in *FIA*), in that both simpleness and complicatedness are vital, without favoring one over the other, and each is utilized, depending on the basis of the perspectives of nature, the mind, culture, and society. There is no simpleness without complicatedness—and vice versa.

For instance, on the one hand, there is simpleness (e.g., the relatively simple portrayal of action research by its advocates, in *Sec. 2.2.1*; the relatively simple analysis of cognitive biases by the scholars who study them, in *Sec. 2.2.2*; the relatively simple analysis of "paradigm" by Kuhn, in *Sec. 2.3.1*; for Thomas Kuhn, "a paradigm shift offers radical simplification" of scientific explanations, in *Sec. 2.3.2*; the relatively simple analysis of investment decisions by MPT to only "maximize risk-adjusted returns," in *Sec. 2.4.1*; the relatively simple task for probability sampling to construct "a suitable sample frame" in a small population like "sampling the employees of a company by using payroll list," in *Sec. 2.4.2*; the relatively simple analysis of the different theories of causality by their respective advocates, in *Sec. 2.5.1*; the relatively simple analysis of probability in the classical interpretation, in *Sec. 2.5.2*; the relatively simple experience of the world in everyday life, according to Husserl's phenomenology, in *Sec. 3.2.1*; the relatively simple analysis of the hermeneutic circle by its advocates, in *Sec. 3.2.2*; the relatively simple problems in genetic algorithm when they are computationally feasible, in *Sec. 3.3.1*; "[t]he approaches used to model gene regulatory networks have been constrained to be interpretable and, as a result, are generally simplified versions of the network," in *Sec. 3.3.2*; the relatively simple optimism about liberation in orthodox Marxism, in *Sec. 3.4.1*; the relatively simple analysis of praxis intervention by its advocates afore-described, in *Sec. 3.4.2*; the relatively simple view of idealized ethical claims by the practitioners in ethnography, in *Sec. 3.5.1*; and the relatively simple analysis of GT by its defenders, in *Sec. 3.5.2*).

On the other hand, there is complicatedness (e.g., the relatively more complicated analysis of action resarch by questionings its claims and assumptions, as shown in the criticisms pointed out by the critics, in *Sec. 2.2.1*; the relatively more complicated analysis of cognitive biases by questioning the claims and assumptions in the study, as shown in the problems pointed out by the critics as cited, in *Sec. 2.2.2*; the relatively more complicated analysis of "paradigm" by challenging the claims and assumptions, as shown in the criticisms pointed out by the critics as cited, in *Sec. 2.3.1*; "Kuhn argues that as science progresses, explanations tend to become more complex, before a paradigm shift offers radical simplification," in *Sec. 2.3.2*; the relatively more complicated analysis of investment decisions by the critics of MPT to also consider "the personal, environmental, strategic, or social dimensions of investment decisions," in *Sec. 2.4.1*; the relatively more complicated task for probability sampling to construct "a suitable sample frame" in "large, disorganized populations," in *Sec. 2.4.2*; the relatively more complicated analysis of the different theories of causality, by challenging the claims and assumptions, as shown in the problems pointed out by the critics as cited, in *Sec. 2.5.1*; the relatively more complicated analysis of probability nowadays, as shown in the diverse probability interpretations and the problems in each, as pointed out by the critics as cited, in *Sec. 2.5.2*; the relatively more complicated analysis of one's experience of the world by the use of "bracketing" in Husserl's phenomenology, such that it subjects "experience to fundamental, critical scrutiny: to take nothing for granted, as pointed out by Maurice Natanson, in *Sec. 3.2.1*; the relatively more complicated analysis of the hermeneutic circle, by challenging the claims and assumptions, as shown in the different problems pointed out by the critics as cited, in *Sec. 3.2.2*; the relatively complex problems in genetic algorithm when they are computationally prohibitive, in *Sec. 3.3.1*; "[u]sing a model that is not constrained to be interpretable, a more accurate model can be produced," but it becomes more complicated, in *Sec. 3.3.2*; the relatively more complicated analysis of liberation, by challenging the claims and assumptions, as shown in the questioning of optimism pointed out by Horkheimer and Adorno afore-discussed, in *Sec. 3.4.1*; the relatively more complicated analysis of praxis intervention, by challenging the claims and assumptions, as shown in the problems pointed out by the critics afore-discussed, in *Sec. 3.4.2*; the relatively more complicated view of idealized ethical claims by the practitioners in ethnography, by challenging the claims and assumptions, as shown in the lies pointed out by the critics like Fine, in *Sec. 3.5.1*; and the relatively more complicated analysis of GT, by challenging the claims and assumptions, as shown in the problems pointed out by the critics as cited, in *Sec. 3.5.2*).

And the reverse direction also holds true.

Again, it should be stressed, however, that there are different shades of gray (or different degrees of truth) in the two opposites, and the classification is not necessarily mutually exclusive either. In fact, even when some combinations of the two occur, they only end up sharing the same dialectic relationship, but in a different degree.

Also, there are some other relationships (as a kind of family resemblance) which have something in common with the principle, but they are not exactly the same but only more or less comparable, which varies from case to case.

Good examples of family resemblance in relation to the principle are inflexibility-flexibility, standardization-specialization, imperfectness-perfectness, superficiality-depth, shallowness-deepness, economicalness-elaboratedness, plainness-circumspection, onesidedness-multisidedness, and the like.

10th Thesis: The Openness-Hiddenness Principle

The tenth thesis refers to the openness-hiddenenss principle (on the detection of things) in the category of "structure" in existential dialectics (which was already worked out in my previous books, especially in *FPHU*), in that reality has its hidden face, just as it is open to outside view in some other ways. There is no openness without hiddenness—and vice versa.

For instance, on the one hand, there is openness (e.g., the open exploration of the need to combine "problem solving actions" with "data-driven collaborative analysis or research" in action research, in *Sec. 2.2.1*; the open exploration of the relationship between personality and behaviors, in *Sec. 2.2.2*; the open exploration of "new phenomena or new relationships among phenomena" by a new paradigm, in *Sec. 2.3.1*; the open exploration of the possibility of doing scientific modeling without doing experiments, in *Sec. 2.3.2*; the open exploration of investment decisions on the basis of diversification by MPT, in *Sec. 2.4.1*; the open exploration of correlation among multiple attributes in factor analysis, in *Sec. 2.4.2*; the open exploration of eight different ways to evaluate causality by Hume, in *Sec. 2.5.1*; the open exploration of probability and belief in subjectivist interpretation, in *Sec. 2.5.2*; the open exploration of the possibility of "pure consciousness" in Husserl's phenomenology, in *Sec. 3.2.1*; the open exploration of a new way of understanding in the hermeneutic circle, for Gadamer, in *Sec. 3.2.2*; the open exploration of using genetic algorithm to generate "solutions of high fitness when applied to practical problems," in *Sec. 3.3.1*; the open exploration of the nature of "artificial neural networks" by the modeling of GRNs, in *Sec. 3.3.2*; the open exploration of "social domination" and its liberation by those in orthodox Marxism, in *Sec. 3.4.1*; the open exploration of the values in praxis intervention, like "creativity," "autonomy," "sociality," "rationality," "intentionality," etc., as pointed out by Markoviç, in *Sec. 3.4.2*; the open exploration of different professional ethical standards to follow for ethnographic research, in *Sec. 3.5.1*; and the open exploration of the possibility of freeing "oneself of preconceptions in the collection and analysis of data," according to GT, in *Sec. 3.5.2*).

On the other hand, there is hiddenness (e.g., the hidden bias in action research, because of its favoring action theorizing over reflective knowledge, in *Sec. 2.2.1*; the hidden bias of those with "fundamental attribution error" to "over-emphasize personality-based explanations for behaviors observed in others while under-emphasizing the role and power of situational influences on the same behavior," in *Sec. 2.2.2*; the hidden bias in a new paradigm, because of its tendency to be eventually consolidated inside "the box," in *Sec. 2.3.1*; the hidden bias in scientific modeling, because of its "simplified reflections of reality," in *Sec. 2.3.2*; the hidden bias in MPT, as shown in the problems pointed out by the critics aforecited, in *Sec. 2.4.1*; the hidden nature of some "multiple attributes" which "can be highly correlated for no apparent reason," in *Sec. 2.4.2*; the hidden bias in Hume's eight ways to evaluate causality, since, for the critic Max Born, "relativity and quantum mechanics have forced physicists to abandon these assumptions as exact statements of what happens at the most fundamental levels," in *Sec. 2.5.1*; the hidden bias of subjectivity in subjectivist interpretation, since "for a given problem, multiple thought experiments could apply, and choosing one is a matter of judgement: different people may assign different prior probabilities," in *Sec. 2.5.2*; the hidden bias in the search for "pure consciousness" in Husserl's phenomenology, because of its hidden Kantian influence in regard to the holy grail of "transcendental ego," which, for the critics, does not exist, in *Sec. 3.2.1*; the hidden bias in the hermeneutic circle, because of the "prejudice" of an interpreter in a given tradition, for Gadamer, in *Sec. 3.2.2*; the hidden nature of genetic algorithm, since, for the critics, "their behavior is difficult to understand," since, "[i]n particular it is difficult to understand why these algorithms frequently succeed at generating solutions of high fitness when applied to practical problems" as claimed, in *Sec. 3.3.1*; the hidden bias in the modeling of GRNs, since "artificial neural networks omit using a hidden layer so that they can be interpreted, losing the ability to model higher order correlations in the data," in *Sec. 3.3.2*; the hidden bias in orthodox Marxism, as shown in the questioning of optimism about liberation by Horkheimer and Adorno, in *Sec. 3.4.1*; the hidden bias in the advocation of these values by Markoviç since, for the critics, some individuals may just prefer "sameness," "subordination," "compliance," etc., in *Sec. 3.4.2*; the hidden bias in holding these professional ethical stndards, as in actually researchers do not follow them, more often than they would like to admit in public, according to Fine, in *Sec. 3.5.1*; and the hidden bias in GT, since, for the critics, "it is impossible to free oneself of preconceptions in the collection and analysis of data in the way that Glaser and Strauss say is necessary," in *Sec. 3.5.2*).

And the reverse direction also holds true.

Again, it should be stressed, however, that there are different shades of gray (or different degrees of truth) in the two opposites, and the classification is not necessarily mutually exclusive either. In fact, even when some combinations of the two occur, they only end up sharing the same dialectic relationship, but in a different degree.

Also, there are some other relationships (as a kind of family resemblance) which have something in common with the principle, but they are not exactly the same but only more or less comparable, which varies from case to case.

Good examples of family resemblance in relation to the principle are overtness-covertness, publicness-privateness, openness-closedness, transparency-secrecy, openness-biasedness, participation-obstruction, exposition-illusion, and so on.

11th Thesis: The Denseness-Emptiness Principle

The eleventh thesis concerns the denseness-emptiness principle (on the distribution of entities in space) in the category of "structure" in existential dialectics, which is first proposed in *FPHUP*, in that both density and void are needed, in relation to the mind, nature, culture, and society, albeit in different ways. There is no denseness without emptiness—and vice versa.

Lest any misunderstanding occurs, the term "void" is used here only as an approximation of emptiness (depending on the degree of the lack of density), since, in physics, it is well known that "empty" space is not really empty all the way, because it can be full of energy (e.g., random quantum fluctuations at the sub-atomic level, and, for that matter, dark energy in the universe) and matter (e.g., different versions of sub-atomic particles, and, for that matter, dark matter in the universe). (F. Wilczek 2008)

For instance, on the one hand, there is denseness (e.g., the relatively denser concern with action in "action research," in *Sec. 2.2.1*; the relatively denser concentration of concerns, by those with "system justification bias," with defending and bolstering "the status quo," in *Sec. 2.2.2*; the relatively denser concentration of "grant funding" in "experiments" which do not deviate too much from "the standard model of physics," in *Sec. 2.3.1*; the relatively denser concentration of "subjective" information in Bayesian statistics, in *Sec. 2.3.2*; the relatively denser concentration of diversified assets if invested according to MPT, in *Sec. 2.4.1*; the relatively denser concentration of samples in non-probability sampling which have no external validity, in *Sec. 2.4.2*; the relatively denser concentration of concerns with counterfactual conditionals, by those in counterfactual theories, in *Sec. 2.5.1*; the relatively denser concentration of concerns with the "degree of belief" in subjectivist interpretation, in *Sec. 2.5.2*; the relatively denser concentration of concerns with "bracketing" by those in Husserl's phenomenology, in *Sec. 3.2.1*; the relatively denser concentration of concerns with "prejudice" in Gadamer's hermeneutics, in *Sec. 3.2.2*; the relatively denser concentration of "local optima" in many problems toward which GAs tend to converge, in *Sec. 3.3.1*; the relatively denser concentration of "single step multiple delayed reactions" in "stochastic gene networks," in *Sec. 3.3.2*; the relatively denser concentration of concerns with the "construction" of reality in critical theory, in *Sec. 3.4.1*; the relatively denser concentration of values like "creativity," "autonomy," "sociality," "rationality," "intentionality," etc. in those for praxis intervention, in *Sec. 3.4.2*; the relatively denser concentration of concerns with ethnographic technqiues in cultural anthropology, in *Sec. 3.5.1*; and the relatively denser concentration of concerns with how to generate new hypothesis in GT, in *Sec. 3.5.2*).

On the other hand, there is emptiness (e.g., the relatively less dense, or more empty, concern with action in "contemporary science," as pointed out by the advocates of action research, in *Sec. 2.2.1*; the relatively less dense, or more empty, concentration of concerns, by those with "negativity bias," with defending and bolstering the status quo, since these individuals tends "to pay more attention and give more weight to negative than positive experiences or other kinds of information," in *Sec. 2.2.2*; the relatively less dense, or more empty, concentration of "grant funding" in "experiments" which deviate greatly from "the standard model of physics," in *Sec. 2.3.1*; the relatively less dense, or more empty, concentration of "subjective" information in frequency-based statistics, in *Sec. 2.3.2*; the relatively less dense, or more empty, concentration of diversified assets if invested according to the critics of MPT, in *Sec. 2.4.1*; the relatively less dense, or more empty, concentration of samples in probability sampling which have no external validity, in *Sec. 2.4.2*; the relatively less dense, or more empty, concentration of concerns with counterfactual conditionals, by those in theories of deterministic causation, in *Sec. 2.5.1*; the relatively less dense, or more empty, concentration of concerns with the "degree of belief" in frequentist interpretation, in *Sec. 2.5.2*; the relatively less dense, or more empty, concentration of concerns with "bracketing" by those in Plato's rationalism, in *Sec. 3.2.1*; the relatively less dense, or more empty, concentration of concerns with "prejudice" in Schleiermacher's hermeneutics, in *Sec. 3.2.2*; the relatively less dense, or more empty, concentration of "local optima" in some other problems to which GAs tend to converge, depending on "the shape of the fitness landscape," such that "certain problems may provide an easy ascent towards a global optimum" instead, in *Sec. 3.3.1*; the relatively less dense, or more empty, concentration of "single step multiple delayed reactions" in Boolean networks, in *Sec. 3.3.2*; the relatively less dense, or more empty, concentration of concerns with the "construction" of reality in positivism, in *Sec. 3.4.1*; the relatively less dense, or more empty, concentration of values like "sameness," "subordination," "compliance," etc. in those for praxis intervention, in *Sec. 3.4.2*; the relatively less dense, or more empty, concentration of concerns with formal modeling in cultural anthropology, in *Sec. 3.5.1*; and the relatively less dense, or more empty, concentration of concerns with how to seek the truth in GT, in *Sec. 3.5.2*).

And the reverse direction also holds true.

Again, it should be stressed, however, that there are different shades of gray (or different degrees of truth) in the two opposites, and the classification is not necessarily mutually exclusive either. In fact, even when some combinations of the two occur, they only end up sharing the same dialectic relationship, but in a different degree.

Also, there are some other relationships (as a kind of family resemblance) which have something in common with the principle, but they are not exactly the same but only more or less comparable, which varies from case to case.

Good examples of family resemblance in relation to the principle are fullness-voidness, nearness-farness, concentration-dispersion, and the like.

12th Thesis: The Change-Constancy Principle

The twelfth thesis concerns the change-constancy principle (on the alteration of things) in the category of "process" in existential dialectics, which is first named in *BCPC* (and were worked out in my books like *FHC*, *FCD*, and *FPHC)*, in that change occurs over time, although constancy is also allowed.

Old players fade away, and new ones emerges, with ever new causes and ever new forms—and asymmetry undergoes changes over time too, so does symmetry. There is no change without constancy—and vice versa.

For instance, on the one hand, there is change (e.g., the ever new ways to do research, as shown in the emergence of action research, in *Sec. 2.2.1*; the ever new ways that a person can be cognitively biased, as shown in the numerous examples as cited, in *Sec. 2.2.2*; the ever new paradigms to emerge in science over time, in *Sec. 2.3.1*; the ever new ways to do science over time, as in the use of "scientific modeling," in *Sec. 2.3.2*; the ever new ways to invest assets, as shown in the rise of MPT, in *Sec. 2.4.1*; the ever new ways to do social statistics, as shown in the three versions aforediscussed, in *Sec. 2.4.2*; the ever new ways to understand causality, as shown in the different theories aforediscussed, in *Sec. 2.5.1*; the ever new ways to study probability, as shown in the diversity of probability interpretations, in *Sec. 2.5.2*; the ever new ways to understand human experience with reality, as shown in Husserl's phenomenology, in *Sec. 3.2.1*; the ever new ways to understand the nature of interpretation, as shown in the different scholars on the hermeneutic circle, in *Sec. 3.2.2*; the ever new ways to achieve optimization, as shown in the use of GAs, in *Sec. 3.3.1*; the ever new ways to do modeling, as shown in the three models of GRNs, in *Sec. 3.3.2*; the ever new ways to understand society, as shown in the rise of critical theory, in *Sec. 3.4.1*; the ever new ways to do praxis, as shown in the new method of praxis intervention, in *Sec. 3.4.2*; the ever new ways to do qualitative research, as shown in the four ethnographic techniques afore-described, in *Sec. 3.5.1*; and the ever new ways to do qualitative research, as shown in GT, in *Sec. 3.5.2*).

On the other hand, there is constancy (e.g., the ever constant existence of different problems with different ways of doing research, as shown in the problems with action research pointed out by the critics, in *Sec. 2.2.1*; the ever constant existence of different problems with the examples, as pointed out by the critics as cited, in *Sec. 2.2.2*; the ever constant existence of "paradigm shifts" in science over time, in *Sec. 2.3.1*; the ever constant existence of different problems in doing science over time, as shown in the problems of scientific modeling pointed out by the critics, in *Sec. 2.3.2*; the ever constant existence of different problems with investment decisions, as shown in the problems with MPT pointed out by the critics, in *Sec. 2.4.1*; the ever constant existence of different problems with social statistics, as shown in the problems of the three versions pointed out by the critics, in *Sec. 2.4.2*; the ever constant existence of different problems with the different ways to understand causality, as shown in the problems with the different theories of causality aforediscussed, in *Sec. 2.5.1*; the ever constant existence of different problems with different probability interpretations pointed out by the critics, in *Sec. 2.5.2*; the ever constant existence of different problems with the new ways to understand human experience with reality, as shown in the problems of Husserl's phenomenology pointed out by the critics, in *Sec. 3.2.1*; the ever constant existence of different problems with the different new ways to understand the nature of interpretation, as shown in the different problems with the different versions of the hermeneutic circle pointed out by the critics, in *Sec. 3.2.2*; the ever constant existence of different problems in the different ways to achieve optimization, as shown in the problems of GA's pointed out by the critics aforecited, in *Sec. 3.3.1*; the ever constant existence of different problems with doing modeling, as shown in the problems with the modeling of GRNs pointed out by the critics afore-described, in *Sec. 3.3.2*; the ever constant existence of different problems with the new ways to understand society, as shown in the problems of critical theory pointed out by the critics, in *Sec. 3.4.1*; the ever constant existence of different problems with the new ways to do praxis, as shown in the problems with praxis intervention pointed out by the critics, in *Sec. 3.4.2*; the ever constant existence of different problems with the different ways to do qualitative research, as shown in the problems of ethnographic techniques pointed out by the critics, in *Sec. 3.5.1*; and the ever constant existence of different problems with the different new ways to do qualitative research, as shown in the problems with GT pointed out by the critics aforedescribed, in *Sec. 3.5.2*).

And the reverse direction also holds true.

Again, it should be stressed, however, that there are different shades of gray (or different degrees of truth) in the two opposites, and the classification is not necessarily mutually exclusive either. In fact, even when some combinations of the two occur, they only end up sharing the same dialectic relationship, but in a different degree.

Also, there are some other relationships (as a kind of family resemblance) which have something in common with the principle, but they are not exactly the same but only more or less comparable, which varies from case to case.

Good examples of family resemblance in relation to the principle are dynamicness-staticness, instability-stability, and the like.

13th Thesis: The Order-Chaos Principle

The thirteenth thesis is called the the order-chaos principle (on the pattern of things) in the category of "process" in existential dialectics (as first worked out in *FC*), in that both order and chaos are vital in the process of change in the world.

The preference for order is biased, since it does not give sufficient attention to the vital role of chaos in the transformation of the world (without somehow reducing it for the understanding of order). There is no order without chaos—and vice versa.

The scientific search for order in the world is often a hidden bias in its ontological obsession with order, since chaos is often treated as the "bad" guy, with order as the "good" guy (for the end goal of science).

Neither order nor chaos is the final end of the world, and one is not to be treated as the means for the other in the transformation of things. Both are fundamental in their recurrent dialectical interactions with each other over time, without reducing one for the other.

For instance, on the one hand, there is order (e.g., in cybernetics, the use of "protoprogramme for reducing the chaotic mass to some form of order," as "[a] paradigm there would be a sort of prohibition to proceed with any action that would increase the total entropy of the system," in *Sec. 2.3.1*; the more or less orderly reality to be described by a scientific modeling which is "linear," in *Sec. 2.3.2*; the more or less orderly markets for investment decisions if viewed within MPT, in *Sec. 2.4.1*; the more or less orderly macro-world governed by "regular sequences and correlations of natural phenomena," as pointed out by David Hume, in *Sec. 2.5.1*; according to G. Syswerda [1989] in his critique of "the building block hypothesis" in general algorithm, "one and two-point crossover are much less disruptive" or relatively more orderly, in *Sec. 3.3.1*; the more or less orderly reality as represented by "Boolean networks," in *Sec. 3.3.2*; the more or less orderly state when institutionalized by the oppressive class for social domination, accordance to the Marxists, in *Sec. 3.4.1*; the more or less orderly nature of non-reflexive praxis, because of its orientation for conformity, in *Sec. 3.4.2*; and the more or less orderly aspect of GT when it is structured on the basis of the four steps in its methodology, in *Sec. 3.5.2*).

On the other hand, there is chaos (e.g., the result in chaotic mass, when there is no "paradigm," that is, when there is no "prohibition to proceed with any action that would increase the total entropy of the system," in *Sec. 2.3.1*; the more or less chaotic reality to be described by a scientific modeling which is "nonlinear," in *Sec. 2.3.2*; the more or less chaotic world for investment decisions, if viewed by the critics of MPT, who believe that "the markets are actually dependent upon a much bigger and more complicated chaotic system—the world," in *Sec. 2.4.1*; the more or less chaotic micro-world of quantum mechanics, as pointed out by Max Born, in *Sec. 2.5.1*; according to Syswerda, "uniform crossover is highly disruptive of short schemata" or relatively more chaotic, in *Sec. 3.3.1*; the more or less chaotic reality as represented by "continuous networks" for "gene expression in a…non-linear, way" or, alternatively, by "stochastic gene networks" in a non-deterministic way, in *Sec. 3.3.2*; the more or less chaotic state when brought about by social revolutionaries, at the beginning stage, for social liberation, in accordance to the Marxists, in *Sec. 3.4.1*; the more or less chaotic nature of reflexive praxis, because of its orientation for disruption or unconformity, in *Sec. 3.4.2*; and the more or less chaotic aspect of GT when it follows "the Glaserian strategy" of "all is data," which means "whatever comes the researcher's way while studying a substantive area" can be used for data collection, in *Sec. 3.5.2*).

And the reverse direction also holds true.

Once more, it should be stressed, however, that there are different shades of gray (or different degrees of truth) in the two opposites, and the classification is not necessarily mutually exclusive either. In fact, even when some combinations of the two occur, they only end up sharing the same dialectic relationship, but in a different degree.

Also, there are some other relationships (as a kind of family resemblance) which have something in common with the principle, but they are not exactly the same but only more or less comparable, which varies from case to case.

Good examples of family resemblance in relation to the principle are lawfulness-disorder, order-disorder, linearity-nonlinearity, linearity-irreversalness, pattern-chaos, etc.

14th Thesis: The Slowness-Quickness Principle

The fourteenth thesis concerns the slowness-quickness principle (on the speed of change) in the category of "process" in existential dialectics (which was first worked out in *FIA*), in that both slowness and quickness co-exist, with their own internal tension, to the extent that each fights for its own relevance with the other, in accordance to the perspectives of nature, the mind, culture, and society, without one being the victor and the other being the vanquished in the long haul. There is no slowness without quickness—and vice versa.

For instance, on the one hand, there is slowness (e.g., the relatively slower speed of learning, if a double loop learning process is used, since what is to learn is more complex, by trying " not only to achieve the intended consequences, but also to openly inquire about conflict and to possibly transform the governing variables," in *Sec. 2.2.1*; the relatively slower readiness of someone with "confirmation bias" to "search for or interpret information in a way" that contradicts his "preconceptions," in *Sec. 2.2.2*; the relatively slower readiness of a scientist to question the existing paradigm in the stage of "normal science," where the paradigm dominates science, in *Sec. 2.3.1*; the relatively slower speed of a scientific model to solve computatonal problems if it adds "complexity" into the system to "improve...realism," in *Sec. 2.3.2*; the relatively slower readiness of those who reject MPT to diversify their assets, in *Sec. 2.4.1*; the relatively slower review of data if unstructured or disorganized, in *Sec. 2.4.2*; the relatively slower speed for somone to have cancer if he never smokes nor suffers from second-handed smoking and has lived a reasonable healthy lifestyle, in *Sec. 2.5.1*; the relatively slower readiness of those in logical interpretation to argue that "[l]ogical probabilities are conceived to be objective logical relations between propositions..., and hence not to depend in any way upon belief," in *Sec. 2.5.2*; the relatively slower readiness of those in Plato's rationalism to subject "experience to fundamental, critical scrutiny: to take nothing for granted," in *Sec. 3.2.1*; the relatively slower readiness of those in the hermeneutic circle for psychological study to engage in quantitative analysis for statistical hypthothesis testing, in *Sec. 3.2.2*; the relatively slower speed of.GAs to solve problems, when they are complex and computationally prohibitive, in *Sec. 3.3.1*; the relatively slower readiness of those in the modeling of Boolean networks to model "gene expression" as "a stochastic process," in *Sec. 3.3.2*; the relatively slower speed of the market, if modified by "centralized planning and socialized ownership," to ever lead towards a successful social revolution as originally conceived by Marx against state capitalism, to the point that it could not happen at all in practice, as pointed out by Horkheimer and Adorno, based on what they witnessed in the 1st half of the 20th century, in *Sec. 3.4.1*; the relatively slower readiness of a "reflexive" individual to conform to settled mentalities, in *Sec. 3.4.2*; the relatively slower readiness of an ethnographic researcher to pubicly admit that he lies in different ways and different degrees, so as to defend both personal and occupationl reputation, according to Fine, in *Sec. 3.5.1*; and the relatively slower speed of doing codes in GT when "selective coding" is not used, in *Sec. 3.5.2*).

On the other hand, there is quickness (e.g., the relatively quicker speed of learning, if a single loop learning process is used, since what is to learn is less complex, by trying to achieve only "the intended consequences and to suppress conflict about the governing variables," in *Sec. 2.2.1*; the relatively quicker readiness of someone with "confirmation bias" to "search for or interpret information in a way" that confirms his "preconceptions," in *Sec. 2.2.2*; the relatively quicker readiness of a scientist to question the existing paradigm and to look for a new paradigm in the stage of "revolutionary science," where the existing paradigm is in crisis, in *Sec. 2.3.1*; the relatively quicker speed of a scientific model to solve computational problems if it only requires "simplicity" in the system for expediency, in *Sec. 2.3.2*; the relatively quicker readiness of those who accept MPT to diversify their assets, in *Sec. 2.4.1*; the relatively quicker review of data if organized in clusters as in cluster analysis, in *Sec. 2.4.2*; the relatively quicker speed for somone to have cancer if he has smoked a lot for decades, in *Sec. 2.5.1*; the relatively faster readiness of those in epistemic interpretation to argue that "probability is 'the logic of partial belief,'" in *Sec. 2.5.2*; the relatively quicker readiness of those in Husserl's phenomenology to subject "experience to fundamental, critical scrutiny: to take nothing for granted," in *Sec. 3.2.1*; the relatively quicker readiness of those in the hermeneutic circle for psychological study to engage in qualitative analysis for meaning and interpretation, in *Sec. 3.2.2*; the relatively faster speed of.GAs to solve problems, when they are simple and computationally feasible, in *Sec. 3.3.1*; the relatively quicker readiness of those in the modeling of "stochastic gene networks" to model "gene expression" as "a stochastic process," in *Sec. 3.3.2*; the relatively quicker speed of the market, when "replaced by centralized planning and socialized ownership," to lead instead towards "fascism and totalitarianism," as this happened in the 1st half of the 20th century, as pointed out by Horkheimer and Adorno, in *Sec. 3.4.1*; the relatively quicker readiness of a "routine" individual to conform to settled mentalities, in *Sec. 3.4.2*; the relatively quicker readiness of an ethnographic researcher to lie secretly in his ethnographic work in different ways and different degrees, according to Fine, in *Sec. 3.5.1*; and the relatively quicker speed of doing codes in GT when "selective coding" is used, like "selectively sampl[ing]...new data with the core [variable] in mind," which "delimits the study" and therefore "makes it move fast," in *Sec. 3.5.2*).

And the reverse direction also holds true.

Again, it should be stressed, however, that there are different shades of gray (or different degrees of truth) in the two opposites, and the classification is not necessarily mutually exclusive either. In fact, even when some combinations of the two occur, they only end up sharing the same dialectic relationship, but in a different degree.

Also, there are some other relationships (as a kind of family resemblance) which have something in common with the principle, but they are not exactly the same but only more or less comparable, which varies from case to case.

Good examples of family resemblance in relation to the principle are inconvenience-convenience, passiveness-activenessness, gradualness-abruptness, deceleration-acceleration, and the like.

15th Thesis: The Expansion-Contraction Principle

The fifteenth thesis is called the expansion-contraction principle in the category of "process" in existential dialectics, in that entities in the world can both expand in some ways and contract in other ones, as part of their nature. There is no expansion without contraction—and vice versa.

This principle, although not so explicitly called, was already used in my previous works on different topics (e.g., the theory of floating consciousness in *FCD* and *FPHC*, the union of the unions in *BWT*, and the cyclical progression of system fragmentation and integration in *FCD*).

For instance, on the one hand, there is expansion (e.g., the relatively more developed ability of those in action research to cope with the problems relating to "action" in practice, in *Sec. 2.2.1*; the relatively more developed ability of someone with "overconfidence effect" to answer questions, in *Sec. 2.2.2*; the relatively more developed ability of a scientist who accepts a new paradigm in normal science to solve unresolved problems that an older paradigm could not, in *Sec. 2.3.1*; the relatively more developed ability of a scientist to deal with realism if he uses "complexity" in a system, in *Sec. 2.3.2*; the relatively more developed ability of those in MPT to deal with investment decisions solely on how to "maximize risk-adjusted returns," in *Sec. 2.4.1*; the relatively more developed ability of those in factor analysis to discover "unobserved latent variables," in *Sec. 2.4.2*; the relatively more developed ability of someone like Aristotle to interprete the world according to the four types of causation, in *Sec. 2.5.1*; the relatively more developed ability of those in logical interpretation to "determine a unique logical probability for any statement, relative to any body of evidence," in *Sec. 2.5.2*; the relatively more developed ability of those in Husserl's phenomenology to deal with "subjectivist tendencies," in *Sec. 3.2.1*; the relatively more developed ability of those in Gadamer's version of the hermeneutic circle to deal with the "prejudice" of an individual in a given tradition, in *Sec. 3.2.2*; the relatively more developed ability of those in GAs to deal with optimization problems from the standpoint of evolutionary theory, in *Sec. 3.3.1*; the relatively more developed ability of those in the modeling of Boolean networks to deal with simple networks on the basis of the four-step methodology, in *Sec. 3.3.2*; the relatively more developed ability of those in critical theory to deal with issues concerning "interpretation," "construction of reality," etc., in *Sec. 3.4.1*; the relatively more developed ability of those in reflexive praxis to deal with the challenge of creativity and autonomy, in *Sec. 3.4.2*; the relatively more developed ability of a cultural anthropologist to do ethnography, in *Sec. 3.5.1*; and the relatively more developed ability of those in GT to deal with coding for the generation of new ideas, in *Sec. 3.5.2*).

On the other hand, there is contraction (e.g., the relatively less developed ability of those in action research to cope with the problems relating to "reflective knowledge," in *Sec. 2.2.1*; the relatively less developed ability of someone with "overconfidence effect" to realize that his answers to questions are not really that great, in *Sec. 2.2.2*; the relatively less developed ability of a scientist to solve new anomalous or unresolved problems that the existing paradigm cannot, until another new paradigm in revolutionary science emerges later, in *Sec. 2.3.1*; the relatively less developed ability of a scientist to deal with expediency if he uses "complexity" in a system, in *Sec. 2.3.2*; the relatively less developed ability of of those in MPT to deal with problems related to "personal, environmental, strategic, or social dimensions of investment decisions," in *Sec. 2.4.1*; the relatively less developed ability of those in factor analysis to work on action research, in *Sec. 2.4.2*; the relatively less developed ability of someone like Aristotle to interprete the world according to the quantum mechanical view of causation by Max Born, in *Sec. 2.5.1*; the relatively less developed ability of those in logical interpretation to assign "degrees of rational belief" to any statement concerning probability, in *Sec. 2.5.2*; the relatively less developed ability of those in Husserl's phenomenology to deal with "the mode of being of Dasein," in *Sec. 3.2.1*; the relatively less developed ability of those in Gadamer's version of the hermeneutic circle to deal with "the detailed experience of everyday existence by an individual" as advocated by those in Heidegger's version, in *Sec. 3.2.2*; the relatively less developed ability of those in GAs to deal with optimization problems from the standpoint of alternative optimizaiton methods like swarm intelligence, in *Sec. 3.3.1*; the relatively less developed ability of those in the modeling of Boolean networks to deal with the "continuous representation" of gene expression as modeled by those in "continuous networks," which allow the latter to capture "several properties of gene regulatory networks not present in the Boolean model," in *Sec. 3.3.2*; the relatively less developed ability of those in critical theory to deal with issues concerning "truth-seeking" as in positivism, in *Sec. 3.4.1*; the relatively less developed ability of those in reflexive praxis to deal with the challenge of conformity and dependency, in *Sec. 3.4.2*; the relatively less developed ability of a cultural anthropologist to do computer simulation, in *Sec. 3.5.1*; and the relatively less developed ability of those in GT to deal with regression analysis for the confirmation of predetermined hypotheses, in *Sec. 3.5.2*).

And the reverse direction also holds true.

Again, it should be stressed, however, that there are different shades of gray (or different degrees of truth) in the two opposites, and the classification is not necessarily mutually exclusive either. In fact, even when some combinations of the two occur, they only end up sharing the same dialectic relationship, but in a different degree.

Also, there are some other relationships (as a kind of family resemblance) which have something in common with the principle, but they are not exactly the same but only more or less comparable, which varies from case to case.

Good examples of family resemblance in relation to the principle are conquest-autarky, rise-fall (or up-down), spread-shrink, extendingness-shorteningness, and so forth.

16th Thesis: The Theory-Praxis Principle

The sixteenth thesis is the theory-praxis principle (on the duality of knowledge) in the category of "agency" in existential dialectics (which was first worked out in *FPHE*), in that, if there is theoretical construction, there is likewise its practical application, both technical and normative. There is no theory without praxis—and vice versa.

For instance, on the one hand, there is theory (e.g., the theoretical construction of "participatory action research" by Paulo Freire, in *Sec. 2.2.1*; the theoreitcal construction of cognitive biases in psychology, in *Sec. 2.2.2*; the theoretical construction of "paradigm" study by Kuhn and others, in *Sec. 2.3.1*; the theoretical construction of the basics of scientific modeling, in *Sec. 2.3.2*; the theoretical construction of MPT, in *Sec. 2.4.1*; the theoretical construction of cluster analysis, in *Sec. 2.4.2*; the theoretical construction of causality in philosophy, in *Sec. 2.5.1*; the theoretical construction of probability in mathematics, in *Sec. 2.5.2*; the theoretical construction of Heidgger's version of phenomenology in his critique of Husserl's, in *Sec. 3.2.1*; the theoretical construction of hermeneutics in philosophy, in *Sec. 3.2.2*; the theoretical construction of natural evolution in biology, in *Sec. 3.3.1*; the theoretical construction of GRNs in biology, in *Sec. 3.3.2*; the theoretical construction of communiative competence in the fields like linguistics and communication studies, in *Sec. 3.4.1*; the theoretical construction of praxis intervention, in *Sec. 3.4.2*; the theoretical construction of ethnographic methods, in *Sec. 3.5.1*; and the theoretical construction of GT, in *Sec. 3.5.2*).

On the other hand, there is praxis (e.g., the practical application of "participatory action research" by Orlando Fals-Borda to the field of adult education in Latin America, in *Sec. 2.2.1*; the practical application of the theoreitcal construction of cognitive biases in psychology to the field of jurisprudence, when "a fair jury trial" now requires that "the jury ignore irrelevant features of the case, weigh the relevant features appropriately, consider different possibilities open-mindedly and resist fallacies," in *Sec. 2.2.2*; the practical application of Kuhn's idea to the field of the social sciences, as Kuhn was honored "as a foundational force behind the post-Mertonian Sociology of Scientific Knowledge," in *Sec. 2.3.1*; the practical application of the basics of scientific modeling to the field of "Modelling and Simulation" for the understanding of "Defence capability development process," in *Sec. 2.3.2*; the practical application of MPT to the field of micro-sociology or social psychology, in *Sec. 2.4.1*; the practical application of cluster analysis to the field of educational research as shown in "the project 'School Like Mine,'" in *Sec. 2.4.2*; the practical application of the theoretical construction of causality in philosophy to the field of psychology, as shown in manipulation theories, in *Sec. 2.5.1*; the practical application of probability in mathemtics to the fields of both natural and social sciences, for hypothesis testing, for example, in *Sec. 2.5.2*; the practical application of Heidgger's version of phenomenology to the field of psychoanalysis, because Heidegger criticized Husserl for failing to take into account other domains of reality that "one is not conscious of," such as unconsciousness, which is "both personal and mysterious," in *Sec. 3.2.1*; the practical application of hermeneutics in philosophy to the field of humanisitc psychology in the 20th century, in *Sec. 3.2.2*; the practical application of natural evolution in biology to the field of computation, wth the use of GAs for applied evolution, in *Sec. 3.3.1*; the practical application of GRNs in biology to the fields of mathematics and computing science for modeling, in *Sec. 3.3.2*; the practical application of communiative competence in the fields like linguistics and communication studies to the field of critical theory, in *Sec. 3.4.1*; the practical application of praxis intervention to the fields of education, medical services, and social work afore-described, in *Sec. 3.4.2*; the practical application of ethnographic methods to the field of communication studies, for example, in *Sec. 3.5.1*; and the practical application of GT to the domain of social organizations, as shown by the creation of "the Grounded Theory Institute in 1999 as a non-profit web-based organization," in *Sec. 3.5.2*).

And the reverse direction also holds true.

Again, it should be stressed, however, that there are different shades of gray (or different degrees of truth) in the two opposites, and the classification is not necessarily mutually exclusive either. In fact, even when some combinations of the two occur, they only end up sharing the same dialectic relationship, but in a different degree.

Also, there are some other relationships (as a kind of family resemblance) which have something in common with the principle, but they are not exactly the same but only more or less comparable, which varies from case to case.

Good examples of family resemblance in relation to the principle are discovery-application, knowledge-action, invention-innovation, theory-intentionality, discourse-action, knowledge-interest, knowledge-motivation, universality-particularity, and so forth.

17th Thesis: The Convention-Novelty Principle

The seventeenth thesis is the convention-novelty principle (on the nature of creative thinking) in the category of "agency" in existential dialectics (which was first worked out in *FPHCT*), in that, if there is conventional wisdom, there is likewise novel challenge, to the extent that both convergent and divergent thinking are part of life. There is no convention without novelty—and vice versa.

As srummarized from *FPHCT* (almost verbatim here), there are (a) "creative techniques" and (b) "creative traits," which, when satisfied—in relation to the larger context of the mind, nature, society, and culture—can be used to enhance creative works.

In addition, creative thinking has its own possibilities and limits (in relation to invention), just as it has its own promises and pitfalls (in relation to innovation)— as already analyzed in *FPHCT*).

In the end, creative thinking has its own desirability and dark sides (as also already analyzed in *FPHCT*).

With these clarifications in mind—there are good empirical examples for the convention-novelty princple.

For instance, on the one hand, there is convention (e.g., the conventional wisdom of doing research by focusing on "reflective knowledge," in *Sec. 2.2.1*; the conventional view that quantitative research can be conducted in a way free of cognitive biases, in *Sec. 2.2.2*; the conventional wisdom, before Kuhn's work, about the quantitiative modeling of science as helping us understand reality "better" over time, in *Sec. 2.3.1*; the conventional wisdom of doing science by experiments, in Sec. *2.3.2*; the conventional way of analyzing finance without the use of portfolio analytics, in *Sec. 2.4.1*; the conventional wisdom of doing statistics by probability sampling, in *Sec. 2.4.2*; the conventional view about the four types of causality in ancient Greece, in *Sec. 2.5.1*; the conventional notion of probability like the classical interpretation by Laplace, in *Sec. 2.5.2*; the conventional view of phenomenology dominated by Husserl's version, in *Sec. 3.2.1*; the conventional version of the hermeneutic circle as first proposed by Schleiermacher to give it a psychological touch, in *Sec. 3.2.2*; the conventional wisdom of "natural evolution" as understood in evolution theory, in *Sec. 3.3.1*; the conventional way of modeling GRNs on the basis of "discret" steps as in Boolean networks, in *Sec. 3.3.2*; the conventional view in the 1960s and 1970s about "language, symbolism, text, and meaning...as the theoretical foundation for the humanities," in *Sec. 3.4.1*; the conventional routines in "non-reflexive" praxis for conformity, in *Sec. 3.4.2*; the conventional wisdom about ethical claims in the ethnographic profession, in *Sec. 3.5.1*; and the conventional wisdom of doing qualitative research like "traditional ethnography, narratology, and storytelling," in *Sec. 3.5.2*).

On the other hand, there is novelty (e.g., the alternative novel challenge to the conventional wisdom about "reflective knowledge" by the new idea of "moving beyond reflective knowledge...to an active moment-to-moment theorizing," in *Sec. 2.2.1*; the alternative novel challenge to the convnetional wisdom about quantitative research, by the new view in modern psychology that people have all kinds of cognitive biases, in *Sec. 2.2.2*; the alternative novel challenge to the conventional wisdom about the progress in the quantitative modeling of science, by Kuhn's new idea that "a new paradigm which replaces an old paradigm is not necessarily better, because the criteria of judgment are controlled by the paradigm itself, and by the conceptual framework which defines the paradigm and gives it its explanatory value," in *Sec. 2.3.1*; the alternative novel challenge to the conventional wisdom about doing experiments in science, by the new idea of doing science by scientific modeling instead, in *Sec. 2.3.2*; the alternative novel challenge to the conventional way of analyzing finance, by the new idea of portfolio analytics which makes good use of mathematical modeling, together with the use of "computer technology, operational research, and statistics," in *Sec. 2.4.1*; the alternative novel challenge to the conventional wisdom about doing statistics by probability sampling, by the new idea of factor analysis for "unobserved latent variables," in *Sec. 2.4.2*; the alternative novel challenge to the conventional view about the four types of causality in ancient Greece, by the new idea of Hume about "regular sequences and correlations of natural phenomena," as "the natural sciences describe[e] *how* things happen rather than asking *why* they happen, in *Sec. 2.5.1*; the alternative novel challenge to the classical interpretation by Laplace, by the different new interpretations later on, as shown in the diverse probability interpretations aforediscussed, in *Sec. 2.5.2*; the alternative novel challenge to the conventional view of phenomenology dominated by Husserl's version, by the new idea of phenomenology proposed by Heideigger, with "its increased relevance to psychoanalysis" in relation to unconsciousness, in *Sec. 3.2.1*; the alternative novel challenge to the conventional version of the hermeneutic circle as first proposed by Schleiermacher, by the new idea of Habermas to give it a practical turn for social change, in *Sec. 3.2.2*; the alternative novel challenge to the conventional wisdom about "natural evolution," by the new idea of "artificial evolution" as used in applied evolution like GAs, in *Sec. 3.3.1*; the alternative novel challenge to the conventional way of modeling GRNs on the basis of "discret" steps as in Boolean networks, by the new way of modeling on the basis of "continuous networks," in *Sec. 3.3.2*; the alternative novel challenge to the conventional view in the 1960s and 1970s about "language, symbolism, text, and meaning…as the theoretical foundation for the humanities," by the new idea of Habermas to use the theoretical studies of communication to combine the two versions of critical theory, in *Sec. 3.4.1*; the alternative novel challenge to the conventional routines in non-reflexive praxis, by the new form known as "reflexive" praxis for creativity and autonomy, in *Sec. 3.4.2*; the alternative novel challenge to the conventional wisdom about ethical claims in the ethnographic profession, by Fine's new view that ethnographic researchers lie more often than they would like to admit in public, in *Sec. 3.5.1*; and the alternative novel challenge to the conventional wisdom about doing qualitative research like "traditional ethnography, narratology, and storytelling," by the new idea of doing codes in accordance to GT, in *Sec. 3.5.2*).

And the reverse direction also holds true.

Again, it should be stressed, however, that there are different shades of gray (or different degrees of truth) in the two opposites, and the classification is not necessarily mutually exclusive either. In fact, even when some combinations of the two occur, they only end up sharing the same dialectic relationship, but in a different degree.

Also, there are some other relationships (as a kind of family resemblance) which have something in common with the principle, but they are not exactly the same but only more or less comparable, which varies from case to case.

Good examples of family resemblance in relation to the principle are convergence-divergence (or convergent thinking vs. divergent thinking), normalness-nonnormalness, conformity-nonconformity, and so on.

18th Thesis: The Evolution-Transformation Principle

The eighteenth thesis refers to the evolution-transformation principle (on the multiple kinds of agency). in the category of "agency" in existential dialectics (which was first worked out in *FAE* and then in other books of mine).

This principle—and the symmetry-asymmetry principle, for instance—are both about agency. More precisely, the word "agency," in a formal definition, refers to "a person or thing through which power is exerted or an end is achieved." (MWD 2007) It therefore does not have to necessarily involve an intelligent lifeform.

Because of this dual meaning in agency, the evolution-transformation principle is more concerned with *the kind of agency*, that is, both about the *evolution* in the state of nature (e.g., an object of natural beauty) and the *transformation* in the world of intelligent lifeforms (e.g., a work of art). There is no evolution without transformation—and vice versa.

In classical Darwinian evolutionary theory (as more detailedly analyzed in *BNN*), evolution is "blind." But in the human world, change often takes place because of the conscious intervention of humans in transforming society and culture, just to cite two instances.

And the transformative part of the principle precisely refers to the other dimension in the dual meaning of agency, in giving technology (as invented by intelligent lifeforms like humans and, soon, post-humans) a major role to play in the change of the world, which is something that I extensively analyzed in *FHC* in the context of the technophilic lifeworld, espeically though not exclusively since modern times. The same theme is also elaborated in many other books of mine.

For instance, on the one hand, there is evolution (e.g., the natural evolution of humans to learn from each other in small groups, in *Sec. 2.2.1*; the natural evolution of humans to engage in all kinds of cognitive biases in the state of nature, in *Sec. 2.2.2*; the natural evolution of humans to study nature in the state of nature, albeit in a primitive way, in *Sec. 2.3.1*; the natural evolution of humans to observe, act, and understand reality in the state of nature in a primitive way, in *Sec. 2.3.2*; the natural evolution of humans to take risks in the state of nature, in *Sec. 2.4.1*; the natural evolution of humans to observe and measure things in the state of nature in a primitive way, in *Sec. 2.4.2*; the natural evolution of humans to experience relationships among different "events…, objects, processes, properties, variables, facts, and states of affairs" in the state of nature, in

Sec. 2.5.1; the natural evolution of humans to count things in the state of nature in a primitive way, in *Sec. 2.5.2*; the natural evolution of humans to experience consciousness in the state of nature, in *Sec. 3.2.1*; the natural evolution of humans to understand things in the state of nature in a primitive way, in *Sec. 3.2.2*; the natural evolution of humans to understand reproduction by going through the process of reproduction in the state of nature, in *Sec. 3.3.1*; the natural evolution of intelligent life like humans and animals to have gene regulatory networks in the state of nature, in *Sec. 3.3.2*; the natural evolution of humans to communicate with each other in the state of nature in a primitive way, in *Sec. 3.4.1*; the natural evolution of humans to understand group dynamics by living in groups in the state of nature, in *Sec. 3.4.2*; the natural evolution of humans to observe and participate in the state of nature, in *Sec. 3.5.1*; and the natural evolution of humans to observe and analyze info in the state of nature in a primitive way, in *Sec. 3.5.2*).

On the other hand, there is transformation (e.g., the technical transformation of humans to learn by the invention of "international development agencies and university programs, as well as countless local community organizations around the world" for Participatory Action Research, in *Sec. 2.2.1*; the technical transformation of humans to undestand cognitive biases with the invention of modern psychology to provide a systematic study of them, together with some scientific methods like "reference class forecasting" to deal with them, in *Sec. 2.2.2*; the technical transformation of human ability to study nature by the invention of systemtic fields of knowledge, like quantitative research, philosophy of science, and sociology of knowledge, in *Sec. 2.3.1*; the technical transformation of human ability to understand nature by the invention of scientific modeling, in *Sec. 2.3.2*; the technical transformation of human ability to take risks by the invention of portfolio analytics for the diversification of investment decisions, in *Sec. 2.4.1*; the technical transformation of human ability to measure things by the invention of social statistics, in *Sec. 2.4.2*; the technical transformation of human ability to understand causality by the invention of different theories of causality, in *Sec. 2.5.1*; the technical transformation of human ability to measure things by the invention of probability theory and its diverse interpretations for practical applications, in *Sec. 2.5.2*; the technical transformation of human ability to understand consciousness by the invention of phenomenology like the one by Husserl, in *Sec. 3.2.1*; the technical transformation of human ability to understand things by the invention of the hermeneutic circle for humanistic psychology, in *Sec. 3.2.2*; the technical transformation of human understanding of reproduction by the invention of evolutionary theory, in *Sec. 3.3.1*; the technical transformation of intelligent life like humans to understand gene regulatory networks by the invention of the modeling of GRNs, in *Sec. 3.3.2*; the technical transformation of human ability to communicate by the invention of communication studies like the one used by Habermas, in *Sec. 3.4.1*; the technical transformation of human ability to understand group dynamics by the invention of praxis intervention, in *Sec. 3.4.2*; the technical transformation of human ability to observe and participate by the invention of ethnographic techniques, in *Sec. 3.5.1*; and the technical transformation of human ability to observe and analyze info by the invention of grounded theory, in *Sec. 3.5.2*).

And the reverse direction also holds true.

Again, it should be stressed, however, that there are different shades of gray (or different degrees of truth) in the two opposites, and the classification is not necessarily mutually exclusive either. In fact, even when some combinations of the two occur, they only end up sharing the same dialectic relationship, but in a different degree.

Also, there are some other relationships (as a kind of family resemblance) which have something in common with the principle, but they are not exactly the same but only more or less comparable, which varies from case to case.

Good examples of family resemblance in relation to the principle are inorganicness-volition, naturalness-technologicalness, naturalness-nonnaturalness, nonwillingness-willingness, inorganicness-motivation, evolution-creation, naturalness-artificialness, and so on.

19th Thesis: The Symmetry-Asymmetry Principle

The nineteenth thesis is labeled as the symmetry-asymmetry principle (on the relationships among existents) in the category of "agency" in existential dialectics (which was also already worked out in my previous books), in that there is no asymmetry without symmetry—and vice versa.

For instance, the Same can be symmetric and asymmetric towards the Others. But in case of asymmetry (as analyzed in *BDPD*), oppression and self-oppression can occur. So, when the Same is asymmetric towards the Others, the Same can also be relatively asymmetric towards itself in self-oppression, just as the Others can be likewise towards themselves. The subsequent oppressiveness is dualistic, as much by the Same against the Others and itself, as by the Others against the Same and themselves. Both oppression and self-oppression can be achieved by way of downgrading differences between the Same and the Others and of accentuating them.

This is true, even though not all forms of asymmetry have to be about oppression and self-oppression.

In addition, from Chapter Three of *FPHG*, symmetry is not perfect, to be understood in an approximate sense under many life circumstances. With this clarification in mind, hereafter are some empirical examples.

For instance, on the one hand, there is symmetry (e.g., the co-existence of different kinds of knowing, in *Sec. 2.2.1*; the co-existence of different cogntive biases, in *Sec. 2.2.2*; the existence of different views about the scientifiic way of thinking, in *Sec. 2.3.1*; the existence of different ways to do science, in *Sec. 2.3.2*; the existence of different views about risk, be it focused on losses or returns, in *Sec. 2.4.1*; the existence of different versions of social statistics, in *Sec. 2.4.2*; the existence of different views about causality, in *Sec. 2.5.1*; the existence of different ways to understand probability, in *Sec. 2.5.2*; the existence of different schools of philosophy, in *Sec. 3.2.1*; the existence of different versions of the hermeneutic circle, in *Sec. 3.2.2*; the existence of different ways to undrestand the change of things in nature over time, in *Sec. 3.3.1*; the existence of different models of representing reality in GRNs, in *Sec. 3.3.2*; the existence of different views about language and communication, in *Sec. 3.4.1*; the existence of different forms of praxis, in *Sec. 3.4.2*; the existence of different ways to do research, in *Sec. 3.5.1*; and the existence of different paradigms of GT, in *Sec. 3.5.2*).

On the other hand, there is asymmetry (e.g., the popular pursuit of propositional knowing in contemporary science—but the popular pursuit of action knowing in action science, in *Sec. 2.2.1*; some cognitive biases "are presumably adaptive"—but some others are not, as they are "presumably result from a lack of appropriate mental mechanisms, or from the misapplication of a mechanism that is adaptive under different circumstances," in *Sec. 2.2.2*; the popularity of the term "Scientific Revolution" in the singular "in the late Renaissance"—but the popularity of "the term 'scientific revolutions' in the plural, taking place at widely different periods of time and in different disciplines" nowadays, because of the work of Kuhn and others, in *Sec. 2.3.1*; the popularity of doing science by experiments in the older days—but the popularity of doing science by modeling nowadays, especially with the advance in computers, in *Sec. 2.3.2*; for the critics of MPT, "in reality, investors are only concerned about losses, and do not care about the dispersion or tightness of above-average returns," so "our intuitive concept of risk is fundamentally asymmetric in nature," in *Sec. 2.4.1*; the strength of factor analysis is to discover "unobserved latent variables"—but the strength of cluster analysis is to provide "a quick review of data," in *Sec. 2.4.2*; the popularity of Aristotle's idea of the four types of causality in ancient Greece—but the popularity of Hume's idea of regularity in the modern West, in *Sec. 2.5.1*; propensity theorists focus more on propensities—but subjectivist theorists focus more on the degrees of belief, in *Sec. 2.5.2*; the popularity of Plato's rationalism in antiquity—but the popularity of Husserl's phenomenology in the 20th century, in *Sec. 3.2.1*; the popularity of Schleiermacher's version of the hermeneutic circle in the older days of the 19th century—but the popularity of Habermas's version of the hermeneutic circle in our time, in *Sec. 3.2.2*; the popularity of creation theory in pre-modern times—but the popularity of evolutionary theory in modern times, in *Sec. 3.3.1*; Boolean networks model GRNs on the basis of a "discrete" representation—but continuous networks model GRNs on the basis of a "continuous" representation, in *Sec. 3.3.2*; the popularity of the view in the 1960s and 1970s about "language, symbolism, text, and meaning...as the theoretical foundation for the humanities"—but the popularity of the view in the 1970s and 1980s about redefining critical theory "as a theory of communication," in *Sec. 3.4.1*; non-reflexive praxis reflects social domination for the status quo—but reflexive praxis reflexts social rebellion against the status quo, in *Sec. 3.4.2*; the popularity of ethnography among cultural anthropologists—but the popularity of regression analysis among microeconomists, in *Sec. 3.5.1*; and the popularity of the Straussian paradigm among his followers—but the popularity of the Glaserian paradigm among his followers, in *Sec. 3.5.2*).

And the reverse direction also holds true.

Again, for the last time, it should be stressed, however, that there are different shades of gray (or different degrees of truth) in the two opposites, and the classification is not necessarily mutually exclusive either. In fact, even when some combinations of the two occur, they only end up sharing the same dialectic relationship, but in a different degree.

Also, there are some other relationships (as a kind of family resemblance) which have something in common with the principle, but they are not exactly the same but only more or less comparable, which varies from case to case.

Good examples of family resemblance in relation to the principle are balance-extremity, reflexiveness-unreflexiveness, equality-inequality, harmoniousness-unharrmoniousness, center-periphery, compatibility-incompatibility, and something like that.

20th Thesis: The Softness-Hardness Principle

The twentieth thesis refers to the softness-hardness principle (on the force of change) in the category of "agency" in existential dialectics (which was first worked out in *ALD*), in that any change by an agent, be it organic (like humans) or non-organic (like natural objects), can occur in a forceful (aggressive) or gentle (pacific) way, which can come in all shapes and sizes, of course. There is no softness without hardness—and vice versa.

For instance, on the one hand, there is softness (e.g., the praise of action research by its advocates as a new form of doing research to challenge "traditional social science," in *Sec. 2.2.1*; the defense of some cognitive biases as "presumably adaptive," in *Sec. 2.2.2*; the praise of Kuhn's work, as shown in honoring him "as a foundational force behind the post-Mertonian Sociology of Scientific Knowledge," in *Sec. 2.3.1*; the appreciation of scientific modeling, as shown in its use of "Modelling and Simulation" for the understanding of "Defence capability development process," in *Sec. 2.3.2*; the support of portfolio analytics, as shown in its applications in other fields like micro-sociology, in *Sec. 2.4.1*; the appreciation of cluster analysis, as shown in identifying its ability to provide "a quick review of data," in *Sec. 2.4.2*; the appreciation of Aristotle's idea of the four types of casuality, as shown in the acceptance of "efficient" causality by some modern epistemic realists, in *Sec. 2.5.1*; the defense of logical interpretation by Carnap, in *Sec. 2.5.2*; the appreciation of Husserl's phenomenology by its followers, in *Sec. 3.2.1*; the defense of the hermeneutic circle by its supporters like Luis Schokel, in *Sec. 3.2.2*; the defense of GAs, as shown by those who use to solve optimization problems, in *Sec. 3.3.1*; the appreciation of the modeling of GRNs, as shown by the use of the three models aforedescribed, in *Sec. 3.3.2*; the appreciation of Marxism, as shown in its incorporation into social critical theory, in *Sec. 3.4.1*; the defense of praxis intervention by its advocates, in *Sec. 3.4.2*; the use of ethnography by those in cultural anthropology, in *Sec. 3.5.1*; and the defense of GT by its practitioners as cited, in *Sec. 3.5.2*).

On the other hand, there is hardness (e.g., the critique of action research by its critics, as shown in the criticisms aforediscussed, in *Sec. 2.2.1*; the critique of some other cognitive biases as "presumably…a lack of appropriate mental mechanisms," in *Sec. 2.2.2*; the critique of Kuhn's work, as shown by the criticism of "relativism" pointed out by the critics as cited, in *Sec. 2.3.1*; the critique of scientific modeling, as shown in the problems pointed out by the critics as cited, in *Sec. 2.3.2*; the critique of portfolio analytics, as shown in the problems pointed out by the critics as cited, in *Sec. 2.4.1*; the critique of cluster analysis, as shown in the problems pointed out by the critics as cited, in *Sec. 2.4.2*; the critique of Aristotle's idea of the four types of casuality, as shown in the modern rejection of "formal" and "final" causality by Machiavelli and Bacon, in *Sec. 2.5.1*; the critique of logical interpretation by Ramsey, in *Sec. 2.5.2*; the critique of Husserl's phenomenology by its critics, in *Sec. 3.2.1*; the critique of the hermeneutic circle by its critics like Judith Shklar aforecited, in *Sec. 3.2.2*; the critique of GAs, as shown in the problems pointed out by the critics as cited, especially by those who favor alternative methods, in *Sec. 3.3.1*; the critique of the modeling of GRNs, as shown in the criticisms aforecited, in *Sec. 3.3.2*; the critique of Marxism, as shown in the problems pointed out by Horkheimer and Adorno as cited, in *Sec. 3.4.1*; the critique of praxis intervention by its critics, in *Sec. 3.4.2*; the critique of ethnography by the critics like Fine, in *Sec. 3.5.1*; and the critique of GT by the critics as cited, in *Sec. 3.5.2*).

And the reverse direction also holds true.

Again, it should be stressed, however, that there are different shades of gray (or different degrees of truth) in the two opposites, and the classification is not necessarily mutually exclusive either. In fact, even when some combinations of the two occur, they only end up sharing the same dialectic relationship, but in a different degree.

Also, there are some other relationships (as a kind of family resemblance) which have something in common with the principle, but they are not exactly the same but only more or less comparable, which varies from case to case.

Good examples of family resemblance in relation to the principle are peacefulness-violence, cooperation-competition, pacificity-aggression, dovishness-hawkishness, reward-punishment, peace-war, friendliness-unfriendliness, kindness-hostility, love-hatred, acceptance-discrimination, and the like.

21st Thesis: The Seriousness-Playfulness Principle

The twenty-first thesis refers to the seriousness-playfulness principle (on the extent of seriousness) in the category of "agency" in existential dialectics (which is first worked out in *FPHH*), in that, if there is seriousness, there is also playfulness, especially for any agent with some kind of intelligence life, be it about humans, animals, or, later, post-humans. There is no seriousness without playfulness—and vice versa.

Lest any misunderstanding occurs, the word "playfulness" here should not be confused with other terms like "laughter," "smile," "joke," "tickleness," and "laugh-like vocalization" (in some animals), for instance.

Being playful can come in all shapes and sizes and therefore is not necessarily tied up with any of these terms, alhough it can result from any of them, or a combination of them, or something else altogether.

This principle, together with some other principles (especially though not exclusively, like the formalness-informalness principle), have important implications for what constitutes rationality, be it about "rationality of application" at the practical level or "rationality of knowledge" at the meta-theoretical level (as summarized in *Sec. 4.20* of *FPHH*).

For instance, on the one hand, there is seriousness (e.g., the serious business of action research for social change, as shown in the version by John Heron and Peter Reason, in *Sec. 2.2.1*; the seriousness of someone with "hindsight bias" to understand past events, in *Sec. 2.2.2*; the serious business of a scientist in modeling to "represent empirical objects, phenomena, and physical processes in a logical and objective way," in *Sec. 2.3.2*; the serious business of those in MPT to understand investment decisions, especially on how to "maximize risk-adjusted returns," in *Sec. 2.4.1*; the serious business of those in non-probability sampling to find "respondents to complete" an survey, in *Sec. 2.4.2*; the serious business of those in "counterfactual theories" to understand counterfactual scenarios, in *Sec. 2.5.1*; the serious business of those in subjectivist interpretation to understand the nature of probability, in *Sec. 2.5.2*; the serious business of phenomenologists to study "consciousness," in *Sec. 3.2.1*; the serious business of those in the hermeneutic circle to study the nature of meaning and interpretation, in *Sec. 3.2.2*; the serious business of those in GAs to achieve optimization, in *Sec. 3.3.1*; the serious business of those in GRNs, when they work hard to discover the complexity of gene expressions, in *Sec. 3.3.2*; the serious business of those in literary critical theory, when they work hard to understand "language, symbolism, text, and meaning," in *Sec. 3.4.1*; the serious business of those in praxis intervention to understand their mentalities better by waking up from their "epistemic sleep," in *Sec. 3.4.2*; the serious business of those in cultural ethnographic research to study a particular people in a given culture, in *Sec. 3.5.1*; and the serious business of those in GT to collect data for idea generation, in *Sec. 3.5.2*).

On the other hand, there is playfulness (e.g., the playful dimension of action research, as shown in the version by John Heron and Peter Reason for "presentational knowing," where the researcher can rehearse or "craft" in a creative or "artistic" way different possible "new practices," in *Sec. 2.2.1*; the playfulness of someone with "hindsight bias" to "filter…memory of past events through present knowledge, so that those events look more predictable than they actually were," in *Sec. 2.2.2*; the playful part of a scientist in modeling, when he plays with different factors like "simplicity," "aesthetic appeal," "user friendliness," etc. to evaluate a model, in *Sec. 2.3.2*; the playful part of those in MPT to mess with different questionable assumptions, in *Sec. 2.4.1*; the playful part of those in non-probability sampling when they choose a sample "composed of whatever persons can be most easily accessed to fill out the survey," in *Sec. 2.4.2*; the playful part of those in "counterfactual theories" to play with different "counterfactual alternatives," in *Sec. 2.5.1*; the playful part of those in subjectivist interpretation, when they play with "multiple thought experiments" and different "judgements" in assigning different "prior probabilities," in *Sec. 2.5.2*; the playful part of phenomenologists, when different scholars give it a different interpretation, like the ones by Max Scheler, Martin Heidegger, Alfred Schütz, Graham Harman, and so on, in *Sec. 3.2.1*; the playful part of those in the hermeneutic circle, since "interpretation must oscillate between particular and general, part and whole" in a circle, as pointed out by Heidegger, so an interpreter can play with different possible interpretations, in *Sec. 3.2.2*; the playful part of those in GAs, since there are many different fitness functions, selection techniques, termination conditions, and so on for them to choose from, or to play with, in the four steps of the methodology, and, according to the No Free Lunch theorem, "there is no general solution to this problem," in *Sec. 3.3.1*; the playful part of those in GRNs, when they play with different assumptions and constraints to obtain some desired results, as shown in the case of Boolean networks which "have been used due to their simplicity and ability to handle noisy data but lose data information by having a binary representation of the genes," in *Sec. 3.3.2*; the playful part of those in literary critical theory, when they play with numerous possible diferent interpretations about the meanings of texts, in *Sec. 3.4.1*; the playful part of those in praxis intervention, when they play around with their settled mentalities to find out the hidden assumptions, in *Sec. 3.4.2*; the playful part of those in cultural ethnographic research, when they play with the data for the results that they like to report, according to Fine, in *Sec. 3.5.1*; and the playful part of those in GT when they play with all kinds of data, "even fiction" in literature, in *Sec. 3.5.2*).

And the reverse direction also holds true.

Again, it should be stressed, however, that there are different shades of gray (or different degrees of truth) in the two opposites, and the classification is not necessarily mutually exclusive either. In fact, even when some combinations of the two occur, they only end up sharing the same dialectic relationship, but in a different degree.

Also, there are some other relationships (as a kind of family resemblance) which have something in common with the principle, but they are not exactly the same but only more or less comparable, which varies from case to case.

Good examples of family resemblance in relation to the principle are seriousness-jokingness, seriousness-humorousness, seriousness-wittiness, seriousness-nonseriousness, seriousness-restingness, seriousness-relaxedness, and so forth.

22nd Thesis: The Regression-Progression Principle

The twenty-second thesis is called the regression-progression princple (on the direction of history) in the category of "outcome" in existential dialectics (which was also already worked out in my previous books), in that neither the cyclical nor the linear views are adequate for explaining many phenomena at all levels. There is no regression without progression—and vice versa.

History progresses to more advanced forms, but with a regressive touch. Examples include no freedom without unfreedom, no equality without inequality, and no civilization without barbarity. This is not an inevitable law, but merely a highly likely empirical trend.

For instance, on the one hand, there is regression (e.g., the regression made by action research, as shown in the problems pointed out by the critics, in *Sec. 2.2.1*; the regression made by the study of cognitive biases, as shown in the problems pointed out by the critics as cited, in *Sec. 2.2.2*; the regression made by Kuhn's work, as shown in the problems pointed out by the critics, in *Sec. 2.3.1*; the regression made by scientific modeling, as shown in the limits pointed out by the critics, in *Sec. 2.3.2*; the regression made by MPT, as shown in the problems pointed out by the critics, in *Sec. 2.4.1*; the regression made by social statistics, as shown in the disadvantages and problems pointed out by the critics aforecited, in *Sec. 2.4.2*; the regression made by different theories of causality, as shown in the problems pointed out by the critics, in *Sec. 2.5.1*; the regression made by probability interpretations, as shown in the different problems for each type of probability interpretation aforediscussed, in *Sec. 2.5.2*; the regression made by Husserl's phenomenology, as shown in the problems pointed out by the critics, in *Sec. 3.2.1*; the regression made by the hermeneutic circle, as shown in the problems pointed out by the critics aforecited, in *Sec. 3.2.2*; the regression made by GAs, as shown in the problems pointed out by the critics, in *Sec. 3.3.1*; the regression made by the modeling of GRNs, as shown in the problems pointed out by the critics, in *Sec. 3.3.2*; the regression made by critical theory, as shown in the problems pointed out by the critics, in *Sec. 3.4.1*; the regression made by praxis intervention, as shown in the problems pointed out by the critics afore-described, in *Sec. 3.4.2*; the regression made by ethnographic research, as shown in the problems pointed out by the critics aforecited, in *Sec. 3.5.1*; and the regression made by GT, as shown in the problems pointed out by the critics aforecited, in *Sec. 3.5.2*).

On the other hand, there is progress (e.g., the progress made by action research, as shown in the successes of different versions in the last six decades, in *Sec. 2.2.1*; the progress made by the study of cognitive biases, as shown in the enhancement of our understanding of different categories of cognitive biases, in *Sec. 2.2.2*; the progress made by Kuhn's work, as shown in "the enormous impact of Kuhn's work" in "the philosophy of science," in *Sec. 2.3.1*; the progress made by scientific modeling, as shown in the useful applications aforecited, in *Sec. 2.3.2*; the progress made by MPT, as shown in its applications to other fields aforecited, in *Sec. 2.4.1*; the progress made by social statistics, as shown in the advantages of the three versions aforecited, in *Sec. 2.4.2*; the progress made by different theories of causality, as shown in the applications by the practitioners aforecited, in *Sec. 2.5.1*; the progress made by probability interpretations, as shown in their applications for hypothesis testing, for example, in *Sec. 2.5.2*; the progress made by Husserl's phenomenology, as shown in its influence in the 20th century, in *Sec. 3.2.1*; the progress made by the hermeneutic circle, as shown in its influence on humanistic psychology in our time aforediscussed, in *Sec. 3.2.2*; the progress made by GAs, as shown in applications by the advocates to solve optimization problems, in *Sec. 3.3.1*; the progress made by the modeling of GRNs, as shown in their uses to help us better undersand gene expressions aforedescribed, in *Sec. 3.3.2*; the progress made by critical theory, as shown in the works aforecited which help us understand society in a critical way, "drawing from knowledge across the social sciences and humanities," in *Sec. 3.4.1*; the progress made by praxis intervention, as shown in the applications by their advocates aforecited, in *Sec. 3.4.2*; the progress made by ethnographic research, as shown in their applications to different fields afore-described, in *Sec. 3.5.1*; and the progress made by GT, as shown in its applications by the followers aforecited, in *Sec. 3.5.2*).

And the reverse direction also holds true.

Again, it should be stressed, however, that there are different shades of gray (or different degrees of truth) in the two opposites, and the classification is not necessarily mutually exclusive either. In fact, even when some combinations of the two occur, they only end up sharing the same dialectic relationship, but in a different degree.

Also, there are some other relationships (as a kind of family resemblance) which have something in common with the principle, but they are not exactly the same but only more or less comparable, which varies from case to case.

Good examples of family resemblance in relation to the principle are undesirability-desirability, risk-opportunity, badness-goodness, cost-benefit, and so on.

23rd Thesis: The Same-Difference Principle

The twenty-third thesis refers to the same-difference principle (on the metamorphosis of change) in the category of "outcome" in existential dialectics (which was worked out in *ALD* and other books of mine), in that an entity, as it evolves over time, can be both different from and similar to its opposing alternatives and does not have to be solely more different from them over time. There is no similarity without difference—and vice versa.

Opposites are not absolute in a black-or-white fashion; so, an entity can become relatively more similar to (or more different from) its opposite over time.

This is further constrained by another principle, that is, the symmetry-asymmetry principle about the relationships among existents under the category about agency in existential dialectics, in that if there is symmetry (equality) between two entities, there is likewise asymmetry (inequality) emerging in a different way.

For instance, on the one hand, there is similarity in outcome (e.g., the contribution to the molding and control of beliefs, values, and behaviors, regardless of whether this be done by way of action teaching, as in PAR, or by way of traditional teaching, in *Sec. 2.2.1*; the contribution to the molding and control of beliefs, values, and behaviors, regardless of whether this be done by the use of reference class forecasting or by the appeal to bounded rationality, in Sec. 2.2.2; the contribution to the molding and control of beliefs, values, and behaviors, regardless of whether this be done by the belief in a logic of discovery, as in modern science before Kuhn's work, or by the belief in a psychology of discovery, as in Kuhn's work, in *Sec. 2.3.1*; the contribution to the molding and control of beliefs, values, and behaviors, regardless of whether this be done by way of scientific modeling, or by way of scientific experiment, in *Sec. 2.3.2*; the contribution to the molding and control of beliefs, values, and behaviors, regardless of whether this be done by way of following MPT, or by way of following the alternative to MPT like behavioral economics, in *Sec. 2.4.1*; the contribution to the molding and control of beliefs, values, and behaviors, regardless of whether this be done by way of probability sampling, or by way of non-probability sampling, in *Sec. 2.4.2*; the contribution to the molding and control of beliefs, values, and behaviors, regardless of whether this be done by the belief in "deterministic causation" as in Christian religion or by the belief in "probabilistic causation" as in quantum mechanics, in *Sec. 2.5.1*; the contribution to the molding and control of beliefs, values, and behaviors, regardless of whether this be done by way of physical probability interpreations, or by way of evidential probability interpreations, in *Sec. 2.5.2*; the contribution to the molding and control of beliefs, values, and behaviors, regardless of whether this be done by way of Husserl's phenomenology, or by way of Plato's rationalism, in *Sec. 3.2.1*; the contribution to the molding and control of beliefs, values, and behaviors, regardless of whether this be done by way of Gadamer's hermeneutics, or by way of Habermas's hermeneutics, in *Sec. 3.2.2*; the contribution to the molding and control of beliefs, values, and behaviors, regardless of whether this be done by way of applying evolutionary theory, or by way of applying creation theory, in *Sec. 3.3.1*; the contribution to the molding and control of beliefs, values, and behaviors, regardless of whether this be done by using Boolean networks, or by using stochastic gene networks, in *Sec. 3.3.2*; the contribution to the molding and control of beliefs, values, and behaviors, regardless of whether this be done by way of social critical theory, or by way of literary critical theory, in *Sec. 3.4.1*; the contribution to the molding and control of beliefs, values, and behaviors, regardless of whether this be done by way of reflexive praxis, or by way of routine praxis, in *Sec. 3.4.2*; the contribution to the molding and control of beliefs, values, and behaviors, regardless of whether this be done by a devotion to ethnographic ethical standards, or by a devotion to phenomenological reductions, in *Sec. 3.5.1*; and the contribution to the molding and control of beliefs, values, and behaviors, regardless of whether this be done by focusing on GT, or by focusing on social statistics, in *Sec. 3.5.2*).

On the other hand, there is difference in outcome (e.g., the contribution to the molding and control of beliefs, values, and behaviors by way of action teaching for a relatively more egalitarian lifeworld—but the contribution to the molding and control of beliefs, values, and behaviors by way of traditional teaching for a relatively more hierarchical lifeworld, in *Sec. 2.2.1*; the contribution to the molding and control of beliefs, values, and behaviors by the use of reference class forecasting for a relatively more active treatment of human rationality—but the contribution to the molding and control of beliefs, values, and behaviors by the appeal to bounded rationality for a relatively more acceptance of the limits on human rationality, in *Sec. 2.2.2*; the contribution to the molding and control of beliefs, values, and behaviors by the belief in a logic of discovery for a relatively more progressive worldview—but the contribution to the molding and control of beliefs, values, and behaviors by the belief in a psychology of discovery for a relatively more relativist worldview, in *Sec. 2.3.1*; the contribution to the molding and control of beliefs, values, and behaviors by way of scientific modeling for a relatively more simplified worldview—but the contribution to the molding and control of beliefs, values, and behaviors by way of scientific experiment for a relatively more realistic worldview, in *Sec. 2.3.2*; the contribution to the molding and control of beliefs, values, and behaviors by way of following MPT for a relatively more simplified worldview—but the contribution to the molding and control of beliefs, values, and behaviors by way of following the alternative to MPT like behavioral economics for a relatively more realistic worldview, in *Sec. 2.4.1*; the contribution to the molding and control of beliefs, values, and behaviors by way of probability sampling for a relatively more scientific mindset—but the contribution to the molding and control of beliefs, values, and behaviors by way of non-probability sampling for a relatively more subjective mindset, in *Sec. 2.4.2*; the contribution to the molding and control of beliefs, values, and behaviors by the belief in "deterministic causation" for a relatively more certain worldview—but the contribution to the molding and control of beliefs, values, and behaviors by way of the belief in "probabilistic causation" for a relatively more indeterministic worldview, in *Sec. 2.5.1*; the contribution to the molding and control of beliefs, values, and behaviors by way of physical probability interpreations for a relatively more objective mindset—but the contribution to the molding and control of beliefs, values, and behaviors by way of evidential probability interpreations for a relatively more subjective mindset, in *Sec. 2.5.2*; the contribution to the molding and control of beliefs, values, and behaviors by way of Husserl's phenomenology for a relatively more subjectivist lifeworld—but the contribution to the molding and control of beliefs, values, and behaviors by way of Plato's rationalism for a relatively more rationalist lifeworld, in *Sec. 3.2.1*; the contribution to the molding and control of beliefs, values, and behaviors by way of Gadamer's hermeneutics for a relatively more conservative lifeworld—but the contribution to the molding and control of beliefs, values, and behaviors by way of Habermas's hermeneutics for a relatively more liberal lifeworld, in *Sec. 3.2.2*; the contribution to the molding and control of beliefs, values, and behaviors by way of applying evolutionary theory for a relatively more secular worldview—but the contribution to the molding and control of beliefs, values, and behaviors by way of applying creation theory for a relatively more sacred worldview, in *Sec. 3.3.1*; the contribution to the molding and control of beliefs, values, and behaviors by using Boolean networks for a relatively more deterministic worldview—but the contribution to the molding and control of beliefs, values, and behaviors by using stochastic gene networks for a relatively more probabilistic worldview, in *Sec. 3.3.2*; the contribution to the molding and control of beliefs, values, and behaviors by way of social critical theory for a relatively more society-transforming lifeworld—but the contribution to the molding and control of beliefs, values, and behaviors by way of literary critical theory for a relatively more text-analyzing lifeworld, in *Sec. 3.4.1*; the contribution to the molding and control of beliefs, values, and behaviors by way of reflexive praxis for a relatively more rebellious lifeworld—but the contribution to the molding and control of beliefs, values, and behaviors by way of routine praxis for a relatively more conformist lifeworld, in *Sec. 3.4.2*; the contribution to the molding and control of beliefs, values, and behaviors by a devotion to ethnographic ethical standards for a relatively more objective worldview—but the contribution to the molding and control of beliefs, values, and behaviors by a devotion to phenomenological reductions for a relatively more intuitive worldview, in *Sec. 3.5.1*; and the contribution to the molding and control of beliefs, values, and behaviors by focusing on GT for a relatively more qualitative worldview—but the contribution to the molding and control of beliefs, values, and behaviors by focusing on social statistics for a relatively more quantitative worldview, in *Sec. 3.5.2*).

And the reverse direction also holds true.

Again, it should be stressed, however, that there are different shades of gray (or different degrees of truth) in the two opposites, and the classification is not necessarily mutually exclusive either. In fact, even when some combinations of the two occur, they only end up sharing the same dialectic relationship, but in a different degree.

Also, there are some other relationships (as a kind of family resemblance) which have something in common with the principle, but they are not exactly the same but only more or less comparable, which varies from case to case.

Good examples of family resemblance in relation to the principle are homogeneity-heterogeneity, we-they, and so forth.

24th Thesis: The Stability-Reaction Principle

The twenty-fourth thesis is called the stability-reaction princple (on the feedback mechanism for further changes) in the category of "outcome" in existential dialectics (which was first worked out in *FPHCHEM*), in that once an outcome becomes stabilized over time (which varies from case to case), it will undergo further changes in its multifaceted interactions with the outside world (e.g., with other entities, energy, space, and so on).

There is no stability without reaction for further changes—and vice versa.

For instance, on the one hand, there is stability in outcome (e.g., the relatively more stable condition of an "output" when it is ready for "refreezing," in the general framework of action research by Lewin, in *Sec. 2.2.1*; the relatively more stable condition of cognitive biases when they are formed and internalized in an individiual's psychology, in *Sec. 2.2.2*; the relatively more stable condition of "normal science" when its problem-solving ability is enhanced by the acceptance of a central, dominant paradigm, in *Sec. 2.3.1*; the relatively more stable condition of a paradigm when, for Kuhn, it can offer "explanations" which "tend to become more complex" as science progresses, in *Sec. 2.3.2*; the relatively more stable condition of diversification when it "eliminates non-systematic risk" in the short run, in *Sec. 2.4.1*; the relatively more stable condition of clusters in cluster analysis when each "object" is "assigned in one cluster only," in *Sec. 2.4.2*; the relatively more stable condition of an individual's deep-rooted interest in "knowing the causes of crime," for instance, in *Sec. 2.5.1*; the relatively more stable condition of a posterior probability in subjectivist interpretation, that is, the degree of belief about a hypothesis when the outcome of an experiment is known, in *Sec. 2.5.2*; the relatively more stable condition of one's subjectivist tendencies for bracketing, if one follows Husserl's phenomenology, in *Sec. 3.2.1*; the relatively more stable condition of an individual's understanding, when it is within the boundary of "prejudice" in a given "tradition," according to Gadamer, in *Sec. 3.2.2*; the relatively more stable condition of a GA, when it reaches termination in the final step of its methodology, in accordance to some specified conditions like fitness function, number of generations, computational time, etc., in *Sec. 3.3.1*; the relatively more stable condition of a chemical reaction in a GRN, when it finishes the entire process after a time delay, according to "stochastic gene networks," in *Sec. 3.3.2*; the relatively more stable condition of society, when it is subtlely imposed by the dominating class through a system of social domination like the use of "mass culture industry," "socialized ownership," etc., in accordance to social critical theorists, in *Sec. 3.4.1*; the relatively more stable condition of an individual's mentality, when it becomes settled in routine praxis, in *Sec. 3.4.2*; the relatively more stable condition of ethnographic practices, when they are formalized by ethical stndards for professional reputation, in *Sec. 3.5.1*; and the relatively more stable condition of coding in GT when it finally completes the last step of the process, namely, "theory" for the generation of new hypotheses, in *Sec. 3.5.2*).

On the other hand, there is reaction in outcome (e.g., output can be subject to further changes, once it is "unfreezed" and "transformed" in the cycle of the action research process, so that "action research…sets in motion a long-range, cyclical, self-correcting mechanism, in *Sec. 2.2.1*; cognitive biases can be subject to further changes, once the individuals interact with others in the larger environment, such that some biases can be modified, some new ones are formed, or some others ones are eliminated with the application of "reference class forecasting", and so on, in *Sec. 2.2.2*; normal science can be subject to further changes, as it increasingly shifts to the stage of "revolutionary science," with the accumulation of "anomalous problems," in *Sec. 2.3.1*; a paradigm can lead to further changes for simplification, as "Kuhn argues that as science progresses,…a paradigm shift [towards a new paradigm] offers radical simplification," in *Sec. 2.3.2*; diversification can lead to further changes over time, when it has its own effect on the larger market by "increasing the systematic risk," because "diversification forces the portfolio manager to invest in assets without analyzing their fundamentals, solely for the benefit of eliminating the portfolio's non-systematic risk" and "[t]his artificially increased demand pushes up the price of assets that, when analyzed individually, would be of little fundamental value," so "the result is that the whole portfolio becomes more expensive and, as a result, the probability of a positive return decreases" and "the risk of the portfolio increases," in *Sec. 2.4.1*; the clustering can lead to further changes later, when the analysis produces "detrimental effects to teachers who work in low-decile schools, students who are educated in them, and parents who support them, by telling them the schools are classified as ineffective, when in fact many are doing well in some unique aspects that are not sufficiently illustrated by the clusters formed," and this in turn may put pressue on the researchers to change the cluster analysis so as to make the results less offensive to some groups, in *Sec. 2.4.2*; this individual's interest in "knowing the causes of crime" can lead to further changes, when he "might find ways of reducing it," in accordance to manipulation theories, in *Sec. 2.5.1*; the probability in subjectivist interpretation can be subject to further changes later on, when new experiments are carried out with different new evidences and outcomes, in *Sec. 2.5.2*; one's subjectivist tendencies for bracketing can lead to further changes later on, because, if going far enough in the process of interacting with others and the world, they can eventually lead one closer and closer to "pure consciousness," after myraid cycles of reduction, according to Husserl's phenomenology, in *Sec. 3.2.1*; an individual's understanding can lead to further changes over time, since, for Gadamer, "experience isn't fixed but rather changing and always indicating new perspectives" in the endless process of the hermeneutic circle, in *Sec. 3.2.2*; a GA can lead to further changes later on, since the process can be repeated, after certain evaluation, for the next process of the search heursitic, with different specified new conditions, in *Sec. 3.3.1*; a chemical reaction can lead to further reactions, because a gene expression can take "many reactions and could not be correctly modeled as an instantaneous reaction in a single step," as it interacts with different entities in a cell and others in the larger environment, according to "stochastic gene networks," in *Sec. 3.3.2*; this oppressive condition of society can lead to further changes over time, as those who are oppressed will work for their social emancipation, in accordance to social critical theorists, in *Sec. 3.4.1*; an individual's settled mentality can lead to further changes when it interacts with others and the environment, and then experiences different mentalities with different assumptions, in *Sec. 3.4.2*; these ethnographic practices can lead to further changes over time, as they receive evaluative feedbacks in practice, which then call for professional changes in regard to the techniques involved, as proposed by Fine, in *Sec. 3.5.1*; and the coding of GT, even with the new theory so generated in mind, can then lead to further changes, since the hallmark of GT is its "open and creative interpretation," which is subject to changes whenever new data are available, so that its "comparative" analysis is "constant" and does not stop in interacting with the environment, in *Sec. 3.5.2*).

And the reverse direction also holds true.

Again, it should be stressed, however, that there are different shades of gray (or different degrees of truth) in the two opposites, and the classification is not necessarily mutually exclusive either. In fact, even when some combinations of the two occur, they only end up sharing the same dialectic relationship, but in a different degree.

Also, there are some other relationships (as a kind of family resemblance) which have something in common with the principle, but they are not exactly the same but only more or less comparable, which varies from case to case.

Good examples of family resemblance in relation to the principle are stability-forwardness, normalcy-feedback, balance-forwardness, balance-reaction, consolidation-modification, normalcy-feedback, stability-breakup, and the like.

25th Thesis: The Post-Human Complexity

And the twenty-fifth thesis is about the role of "post-humans," which I originally proposed in my previous books, starting with the first book titled *The Future of Human Civilization* in 2000 and all others afterwards.

As already pointed out in *Sec. 1.7*, I need to emphasize, as this is something that I used to repeat (*almost verbatim*) from my previous books, two clarifications here about the term "post-human" as a neologism in my works.

Firstly, the word "post-human" here should *not* be confused with another term which looks similar but has a totally different meaning in the literature of postmodernism, namely, "post-humanism"—which constitutes a critique of "humanism" as tradionally understood (especially, though not exclusively, in relation to the idea of progress in science and reason in the Enlightenment project). (WK 2008)

My works reject the project of "postmodernism" and propose the future world of what I orignally called "after-postmodernity" in *FHC* and *FCD*, for instance.

And secondly, the word "post-human" here should also *not* be confused with a similar term which is used to champion the ideology of technology for the future co-existence between humans and cyborgs in "trans-humanism." (WK 2008a)

Instead, my term "post-human" in relation to "posthuman-ism" also rejects "transhumanism" (especially, though not exclusively, in relation to the promises of technology) and refers to something else altogether, that is, the future extinction of humans and its post-human successors in deep space and beyond unto multiverses.

My critique of "transhumanism" was more extensively elabroated in *Sec. 2.4.1* of *BEPE*—just as my critique of different futuristic scenarios of fictional artificial intelligence in science fiction was given in *Sec. 3.5.2* of *FPHCOM.*

With these two clarifications in mind (as summarized in *Table 1.31*)—the post-human complexity in data analysis can be addressed in relation to five main directions, based on my complex theory of data analysis as analysed in this book (together with my visions as already worked out in my previous books).

Firstly, the post-human complexity in data analysis will learn from my complex theory of data analysis in relation to the dialectic context of no quantitative research without qualitative research (and vice versa), in the context of data analysis.

Secondly, the post-human complexity in data analysis will learn from my complex theory of data analysis in relation to the four perspectives of the mind, nature, society, and culture.

Thirdly, the post-human complexity in data analysis will learn from my complex theory of data analysis in relation to the non-privilege of any specific theory over others in the literature and the non-integration of them all (as they are not necessarily compatible with each other).

Fourthly, the post-human complexity in data analysis will have to confront the new challenges as posed by the long-term civilizational development of intelligent life in the distant future, both here on this planet Earth and elsewhere in deep space until multiverses.

As this is something that I regularly pointed out in my previous books for background information (and summarize here, *almost verbatim*, with some updates), I already worked out, in my numerous books, what these new challenges will be and provide different original visions to meet them—especially, in regard to the future evolutions of the mind (e.g., "the supersession computing," "metamorphic humor," "the contrarian personality," "the hyper-martial body," "the hyper-sexual body," "floating consciousness," "hyper-spatial consciousness," "thinking machines," "unfolding unconsciousness," "thinking robots," "genetically altered superior beings," "cyborgs," and others), of nature (e.g., "creational chemistry," "multilateral sound," "resettlement geology," "post-cosmology," "the alteration of space-time," "the creation of new matter-energy," "selective geometry," and the like), of society (e.g., "detached gambling," "heterodox education," "multifaceted war and peace," "virtual organizations," different versions of "post-capitalism" and "post-democracy," the movement of "cyclical progression" at both structural and systemic levels, "ambivalent technology," and so forth), and of culture (e.g., "comparative-impartial literature," "post-human mind games," "comparative-substitutive religion," "post-ethics," "post-civilization," "transformative aesthetic experience," "contrastive mathematical-logic," and the whatnot) in history (e.g., the age of "after-postmodernity"), in the context of my approaches in relation to methodology (e.g., "sophisticated methodologial holism," "critical-dialectic formal science," etc,) and ontology (e.g., "existential dialectics," "contrastive rationality").

Of course, the examples (as listed above) are not exhaustive but illustrative, since my numerous books have worked out many other visions (as already roughly summarized in *Sec.1.7* and *Sec. 1.8*). Many, though not all, of my visions on the mind, nature, society, culture, history, methodology, and ontology (as cited above) are summarized in the tables as shown in Chapter One and Chapter Four (especially in *Table 4.43* about my original theories on numerous topics).

And finally, or fifthly, the post-human complexity in data analysis will learn from my complex theory of data analysis in regard to the need to go beyond quantitative and qualitative research (as further elaborated in the next section).

Towards the Post-Human Complexity

This "post-human" complexity in data analysis to transcend both quantitative and qualitative research can be understood in terms of six great future transformations of data analysis, as explained below (and summarized in *Table 1.2*).

Six Great Future Transformations of Data Analysis

The six great future transformations of data analysis can be explained hereafter, with the caveat, however, that the classification in terms of six is solely aesthetic and utilitarian (illustrative), as there is no objective basis that it must be classified in term of six (as this is something that I often stressed and repeated in my previous books, albeit in the context of different topics).

The First Great Future Transformation of Data Analysis

The first great future transformation of data analysis in the post-human era concerns what I originally call *new sub-fields*.

This search for "new sub-fields" gives future advanced humans and later their successors, namely, the post-humans, the ability to come up with different new studies of data analysis in a way that current humans cannot.

There are already some indications that the future of data analysis is moving toward this direction.

Consider, for illustration, the different "sub-fields" as developed in the history of data analysis, such as "data mining" (in *Sec. 1.3*), "statistical applications" (in *Sec. 1.3*), "predictive analytics" (in *Sec. 1.3*), "text analytics" (in *Sec. 1.3*), "action science" (in *Sec. 2.2.1*), "philosophy of science" (in *Sec. 2.3.1*), "portfolio analytics" (in *Sec. 2.4.1*), "social statistics" (in *Sec. 2.4.2*), "applied evolution" (in *Sec. 3.3.2*), "phronetic social science" (in *Sec. 3.4.2*), and "grounded theory" (in *Sec. 3.5.2*), for instance.

All these are just some illustrative (relatively primitive) examples in our time, of course, but the important point to remember here is that the search for different new sub-fields will further transform data analysis at the more advanced level in the post-human future that the human world has never known.

The Second Great Future Transformation of Data Analysis

The second great future transformation of data analysis in the post-human era concerns what I originally call *new subjectivities*.

This search for "new subjectivities" gives future advanced humans and later their successors, namely, the post-humans, the ability to make good use of subjectivity (like emotions, intuition, imagination, heuristics, experience, and so on) for data analysis in a way that current humans cannot.

There are already some indications that the future of data analysis is moving toward this direction.

Consider, for illustration, the different subjectivities as developed in the history of data analysis, such as "cognitive biases" (in *Sec. 2.2.2*), "manipulation theories" (in *Sec. 2.5.1*), "intuition" and "empathy" (in *Sec. 3.2.1*), "interpretation and psychology" (in *Sec. 3.2.2*), "self-reflextion" (in *Sec. 3.4.1*), and "reflexive praxis" (in *Sec. 3.4.2*), for instance.

All these are just some illustrative (relatively primitive) examples in our time, of course, but the important point to remember here is that the quest for different new subjectivities will further transform data analysis at the more advanced level in the post-human future that the human world has never known.

The Third Great Future Transformation of Data Analysis

The third great future transformation of data analysis in the post-human era concerns what I originally call *new actions*.

This search for "new actions" gives future advanced humans and later their successors, namely, the post-humans, the ability to engage in praxis (like personal experience, social change, and so on) in a way that current humans cannot.

There are already some indications that the future of data analysis is moving toward this direction.

Consider, for illustration, the different forms of praxis as developed in the history of data analysis, such as "action" (in *Sec. 2.2.1*), "emancipation" (in *Sec. 3.4.1*), "phronesis" (in *Sec. 3.4.2*), and "participation" (in *Sec. 3.5.1*), for instance.

All these are just some illustrative (relatively primitive) examples in our time, of course, but the important point to remember here is that the search for different new actions will further transform data analysis at the more advanced level in the post-human future that the human world has never known.

The Fourth Great Future Transformation of Data Analysis

The fourth great future transformation of data analysis in the post-human era concerns what I originally call *new narratives.*

This search for "new narratives" gives future advanced humans and later their successors, namely, the post-humans, the ability to come up with different techniques of interpretation in a way that current humans cannot.

There are already some indications that the future of data analysis is moving toward this direction.

Consider, for illustration, the different ways of interepretation as made in the history of data analysis, in regard to "assumptions " (in *Sec. 2.4.1*), "probability interpretations" (in *Sec. 2.5.2*), "bracketing" (in *Sec. 3.2.1*), "construction" (in *Sec. 3.4.1*), "self-criticism" (in *Sec. 3.4.2*), "ethnography" (in *Sec. 3.5.1*), and "paradigmatic split" (in *Sec. 3.5.2*), for instance.

All these are just some illustrative (relatively primitive) examples in our time, of course, but the important point to remember here is that the search for different new interpretations will further transform data analysis at the more advanced level in the post-human future that the human world has never known.

The Fifth Great Future Transformation of Data Analysis

The fifth great future transformation of data analysis in the post-human era concerns what I originally call *new technologies.*

This search for "new technologies" gives future advanced humans and later their successors, namely, the post-humans, the ability to make good use of advanced technologies (like statistics, experiments, modeling, and the like) for data analysis in a way that current humans cannot.

There are already some indications that the future of data analysis is moving toward this direction.

Consider, for illustration, the different technologies as already invented in the history of data analysis, such as "scientific modeling" (in *Sec. 2.3.1*), "statistical tools" (in *Sec. 2.4.2*), "simulation" for applied evolution (in *Sec. 3.3.1*), and the "modeling" of GRNs (in *Sec. 3.3.2*), for instance.

In my previous books, I also singled out some important future technological advances for data analysis, and good examples include new forms of computing (in *FPHCOM*), of IT (in *FCD*), of systems (in *FPHFS)*, of complexity (in *FC*), and so on.

All these are just some illustrative (relatively primitive) examples in our time, of course, but the important point to remember here is that the continuation of different new technologies will further transform data analysis at the more advanced level in the post-human future that the human world has never known.

The Sixth Great Future Transformation of Data Analysis

And the sixth great future transformation of data analysis in the post-human era concerns what I originally call *new minds.*

This search for "new minds" gives future advanced humans and later their successors, namely, the post-humans, the ability to deal with complex problems in a way that current humans cannot.

There are already some indications that the future of data analysis is moving toward this direction.

Consider, for illustration, the different ways to solve problems as evolved in the history of data analysis, such as the use of "scientific modeling" (in *Sec. 2.3.2*), of "social statistics" (in *Sec. 2.4.2*), of "applied evolution" (in *Sec. 3.3.1*), of "gene regulatory networks" (in *Sec. 3.3.2*), and of "coding" (in *Sec. 3.5.2*), for instance.

But these different ways of problem-solving must be understood in relation to the future developments of the mind in the long haul, which I had already elaborated in my previous books, and good examples include "floating consciousness" (as in *FCD* and *FPHC*), "hyper-spatial consciousness" (as in *FPHC* and *FPHG*), "unfolding unconsciousness" (as in *FPHU*), "the hyper-martial body" (as in *FPHMA*), "the hyper-sexual body" (as in *FPHS*), and others.

Lest any misunderstanding occurs, it is important to note that the future "post-human" supersession of the human mind has to be understood from the combined perspectives of the mind, nature, society, and culture, to the extent that artificial intelligence (with the use of computing technologies) will be only one of many factors to contribute to the long hall of future history of the mind in different directions.

Therefore, it is necessary to move beyond AI in order to understand the future evolution of the mind (intelligence life) in the universe, as this is something I specifically addressed in *FPHCOM.*

Also, other visions of mine in relation to nature, society, and culture are summarized in *Sec. 1.7* and in the tables of Chapter Four (especially *Table 4.43*, for instance).

All these are just some illustrative (relatively primitive) examples in our time, of course, but the important point to remember here is that the search for new minds will further transform data analysis at the more advanced level in the post-human future that the human world has never known.

Enormous Implications

These six great future transformations of data analysis have enormous implications for some deep questions which have attracted some of the best minds in human history.

For instance, are the natural sciences the model for the social sciences? Is quantitative analysis better than qualitative one? Is the point, in the end, to change the world in praxis, not to keep analyzing it? Are reflection and intuition important for understanding the world, so as to counter the rationalist bias for theoretical construction? Is there a way out of the hermeneutic circle? And…?

The science wars in our time (as already discussed in *Sec. 3.4.2*) and other debates like them (as already discssed in the rest of the book) constitute only the latest chapters to this long history of different views about data analysis.

Questions and disagreements like this will receive new answers in the future history of advanced human and later post-human data analysis, especially in the current context of going beyond both quantitative and qualitative research—such that their possibility and desirability are not to the extent that the respective defenders would like us to believe.

The various forms of data analysis—be they focused on praxis (like action science, phronetic social science, social critical theory, etc.), on subjectivity (like phenomenology, psychoanalysis, etc.), on field work (e.g., participant observation, interviewing, etc.), on modeling (e.g., computer simulation, etc.), on statistics (e.g., regression analysis, etc.), on experiment (e.g., lab work, etc.), on interpretation (e.g., hermenuetics, etc.), and the like—are just different directions to take in understanding the world.

Each has its own advantages and disadvantages and is preferred over others by different practitioners with their own different interests, but none of them is possible and desirable to the extent that its respective advocates would like us to believe.

In the long haul, the debate about data analysis will be shifted to different directions, as shown in the six great future transformations that I have envisioned and outlined above.

But it is important to remember that these six great future transformations of data analysis will bring neither utopia nor dystopia—as this is something that I have warned time and again in all of my previous books.

Ultimately, it is not up to us current humans to decide whether or not, or to what extent, the post-human future of data analysis will be for better and for worse, because it will be judged by the values and beliefs of future post-humans on Earth and in deep space unto multiverses—in a way that the entire history of the human world has never known. This is also something that I have warned time and again in my previous works.

Yet, current humans, like many before them, have wasted so much time in debating and fighting for their respective preferred forms of data analysis (as shown in the latest version like the science wars) that this future human (and post-human) vision of data analysis is lost (or ignored) in the midst of their sound and fury over nothing.

Table 4.1. Sophisticated Methodological Holism (Part I)

• "My methodological holism implies the partiality-totality principle in the ontology of existential dialectics (see the table on the partiality-totality principle for summary), which is against the varieties of (a) reductionism and (b) reverse-reductionism, in relation to (i) concept, (ii) theory, (iii) methodology, and (iv) ontology." (*FC*)

• "[M]y methodological holism here is not opposed to methodological individualism but includes it (and, for that matter, other methodologies too)…." (*FPHC*) For this reason (and others too, as summarized hereafter), my version of methodological holism is sophisticated—not vulgar as sometimes used by inapt scholars using the same term. (*FC*)

• "[M]y methodological holism does not democratically presume that all levels are equally valid, as all levels are not created equal. In other words, in relation to issue X, level A may be more relevant than level B, but in relation to Y, level B can be more relevant than level A instead." (*FPHC*) One excellent example of this vulgar democratic presumption is what I called in *BNN* "the compromise fallacy." (*FC*)

• My methodological holism does not presume that a lower level of analysis is more important than a higher level, solely because the former serves as the foundation for the latter—and vice versa, for that matter. One excellent example of this reductionistic presumption is what I called in *FPHST* "the foundation fallacy." (*FPHST, FC*)

(continued on next page)

Table 4.1. Sophisticated Methodological Holism (Part II)

• "[M]y methodological holism does not make any a-priori postulation that there must be a definite (and, for that matter, indefinite) number of levels" in any analysis. (*FPHC*) Nor does it dogmatically require that there must be a certain combination of levels of analysis in a given inquiry. (*FC*)

• "[M]ethodological holism, in my usage, does not assume that all levels…can necessarily be integrated, since methodological holism is not aimed to search for the holy grail of 'an integral theory…' (as is the case for Wilber). In other words, it allows that sometimes some levels may experience irreducible gaps between them, to be understood, at best, as empirical correlations, not as causal relations…." (*FPHC*)

• "[D]ifferent levels may overlap and even interact with each other in a given context (but sometimes may not), and the fact that I even proposed different ways of re-classifying the levels (whenever needed) in *FDC* reinforces this point….The dual danger here is either forcefully making different levels interact when they are just different (or, metaphorically speaking, apples and oranges) or inappropriately ignoring their interactions when some situations instead require them." (*FPHC*)

• "[T]o understand different levels from their own (unique) perspectives (as required by my methodological holism) is not the same as trying to reduce them to a preferred level in the process of learning from other levels. This second kind of multidisciplinary work is not genuine and does no justice to the unique complexities and merits inherent at each level." (*FPHC*)

(continued on next page)

Table 4.1. Sophisticated Methodological Holism (Part III)

• "[My] methodological holism walks a fine line between the artificial classification (separation) of levels and the simultaneous incorporation of them, if only for the sake of human scholarly endeavor. It should be reminded that nature does not impose upon itself the academic classification of the levels of analysis as humans have. The enterprise of classification is therefore anthropocentric." (*FPHC*)

• "[M]y methodological holism advocates neither epistemic subjectivism nor epistemic non-subjectivism (e.g., realism, idealism, and historicism), neither epistemic relativism (e.g., subjectivism, historicism) nor epistemic absolutism (e.g., realism, positivism), neither epistemic reductionism nor epistemic emergencism, and neither epistemic objectivism (e.g., realism, idealism) nor epistemic historicism….Neither does methodological holism, in my usage, accept the false meta-conceptual dichotomy between nominalism and realism….These false dichotomies…are to be transcended. In other words, methodological holism does not fully accept epistemic realism, positivism (a form of epistemic idealism), historicism, subjectivism, and reductionism in epistemology and philosophy of science but learns from the strengths and weaknesses of all of the opposing approaches without siding with any of them…." (*FPHC*)

• "Sophisticated methodological holism is subject to the constraints as imposed by the syntax of existential dialectics (e.g., the partiality-totality principle and the predictability-unpredictability principle). Even in predictability, outcomes are subject to uncertainty, the degree of which varies from case to case." (*FC*)

(continued on next page)

Table 4.1. Sophisticated Methodological Holism (Part IV)

• "Sophisticated methodological holism—when applied, especially though not exclusively, as illustrated in my numerous works—can enrich the understanding of reality in some distinctive ways. Here are three examples (as revealed in each of my books)." (*FPHCT*; *FPHL*)

—"Firstly, it provides a comprehensive analysis of a subject matter, from which much can be learned about reality," "both in relation to the perspectives of the mind, nature, society, and culture and also in relation to a new classification of the subject matter."

—"Secondly, it suggests some visions of the future in relation to the subject matter in question."

—"And thirdly, it proposes some insights on meta-theory (e.g., methodology and ontology) in general—with the clear understanding, however, of the dilemma of specific vs. general ontology (as shown in the table on the syntax of existential dialectics in the context of the dilemma of ontology). For this very reason, all of these ways are important, without reducing one into the analysis of another."

• "In the end, my meta-theory (both sophisticated methodological holism and existential dialectics) serves as a foundation to unify all domains of knowledge into *an unified theory of everything* (by way of some ontological principles and the comprehensive perspectives of the mind, nature, society, and culture). This is so, without commiting the sins of reductionism and reverse-reductionism (as often seen in many holistic approaches, with the fad of sysems approach—be it about systems theory, chaos theory, complexity theory, or else—as a most recent notorious example, which I debunked in *The Future of Complexity* and also in *The Future of Post-Human Formal Science*). My distinctive approach makes good use of different schools of thought without favoring any of them nor trying to integrate them (as they are not necessarily compatible with each other), so as to adjust for subjectivity, diversity, conflict, and complexity, for example. In this sense, the word "unified" does not have to mean integrative approach, in a narrow sense, nor systems approach, in a broad one, both of which I reject while learning from them." (*FPHFS*; *FPHS*)

Sources: A summary of *Sec.1.2* in *FPHC*—and also from *BNN*, *FPHST*, *ALD*, *FC*, *FPHCT*, and the rest of all other books of mine. See the books for more detail.

Table 4.2. On Reductionism and Reverse-Reductionism (Part I)

- **The Partiality-Totality Principle**
  —The partiality-totality principle in the ontology of existential dialectics targets against the varieties of reductionism and reverse-reductionism (as already worked out in my previous books).

- **Against the Varieties of Reductionism**
  —*Conceptual Reductionism*
  - Some illustrative instances involve myriad dualities like mind vs. body, self vs. world, democracy vs. non-democracy, and the like (as already addressed in *FHC*, *FPHC*, and *BDPD*, for instance).

  —*Theoretical Reductionism*
  - A fascinating case study concerns what I originally called "the foundation fallacy" in *FPHST*, in any attempt to naively understand space-time from the physical perspective as the foundation and, consequently, to dangerously dismiss other perspectives.
  - In *FAE*, I elaborated further these versions of reductionism in the literature on aesthetics (e.g., form vs. content, representation vs. expression, critics vs. artists, and externalism vs. internalism).
  - In *FIA*, I revealed other forms of reductionism in the literature on information architecture (e.g., the constructivist argument).
  - In *FPHU*, I showed the persistent legacy of reductionism, this time, in the literature on anomalous experience (e.g., the obsession with physics, chemistry, and biology for explaining anomalous experience).
  - In *FPHE*, I examined another case of reductionism in action, in the context of engineering (e.g., technical constraints vs. normative constraints).
  - In *FPHMM*, I elaborated one more version of reductionism, in relation to the three domains of communication (e.g., the competing views on sending, connecting, and receiving).
  - In *FPHCT*, I explored another version of reductionism, in relation to invention and innovation (e.g., the bio-psychological argument vs. the socio-cultural arguments).

(continued on next page)

Table 4.2. On Reductionism and Reverse-Reductionism (Part II)

- **Against the Varieties of Reductionism** ***(cont'd)***
  - *—Theoretical Reductionism (cont'd)*
    - In *FPHG*, I identify another version of reductionism, in relation to infinity, symmetry, and dimensionality (e.g., the Euclidean argument vs. the non-Euclidean arguments).
    - In *FPHUP*, I analyzed another version of reductionism, in relation to density and void (e.g., the engineering argument and the ecology argument).
    - In *FPHL*, I examined another version of reductionism, in relation to structure and context (e.g., the structuralist argument and the contextualist argument).
    - In *PFHO*, I explored another version of reductionism, in relation to communcation, decision-making, and leadership (e.g., the rational-system argument vs. the natural-system argument).
    - In *PFHMA*, I revealed another version of reductionism, in relation to the martial body and spirit (e.g., the spiritual argument vs. the materialist argument vs. the defensive argument).
    - In *PFHS*, I analyzed another version of reductionism, in relation to the sexual body and spirit (e.g., the naturalist argument vs. the constructivist argument).
    - In *PFHLAW*, I examined another version of reductionism, in relation to law (e.g., the necessity argument vs. the contengency argument).
    - In *FPHWP*, I showed another version of reductionism, in relation to war and peace (e.g., the aggressivist argument vs. the pacifist argument).
    - In *BEPE*, I analyzed another version of reductionism, in relation to morality and immorality (e.g., the objectivist argument vs. the non-objectivist argument vs. the skeptical argument).
    - In *BCOS*, I examined another version of reductionism, in relation to the contested beginnings and speculative ends of the universe (e.g., the scientific argument vs. the religious argument vs. the esoteric argument vs. the metaphysical argument).
    - In *FPHP*, I show another version of reductionism, in relation to normality and abnormality (e.g., the natural argument vs. the social argument vs. the cultural argument vs. the mental argument).

(continued on next page)

Table 4.2. On Reductionism and Reverse-Reductionism (Part III)

- **Against the Varieties of Reductionism** ***(cont'd)***
  —*Theoretical Reductionism (cont'd)*
  - In *FPHGEOL*, I scrutinized another version of reductionism, in relation to statics and dynamics (e.g., the catastrophe argument vs. the uniformity argument vs. the revision argument).
  - In *FPHCHESS*, I showed another version of reductionism, in relation to tactics and strategy (e.g., the natural argument vs. the social argument vs. the cultural argument vs. the mental argument).
  - In *FPHR*, I revealed another version of reductionism, in relation to secularness and sacredness (e.g., the critical argument vs. the skeptical argument vs. the theist argument).
  - In *FPHEDU*, I examined another version of reductionism, in relation to teaching and learning (e.g., the teacher-centered argument vs. the student-centered argument vs. the balanced argument).
  - In *FPHH*, I explored another version of reductionism, in relation to joking and laughing (e.g., the natural argument vs. the social argument vs. the cultural argument vs. the mental argument).
  - In *FPHA*, I proposed another version of reductionism, in relation to sound and silence (e.g., the natural argument vs. the social argument vs. the cultural argument vs. the mental argument).
  - In *FPHGAM*, I proposed another version of reductionism, in relation to risk and caution (e.g., the natural argument vs. the social argument vs. the cultural argument vs. the mental argument).
  - In *FPHCOM*, I worked out another version of reductionism, in relation to hardware and software (e.g., the natural argument vs. the social argument vs. the cultural argument vs. the mental argument).
  - In *FPHLIT*, I proposed another version of reductionism, in relation to fiction and non-fiction (e.g., the natural argument vs. the social argument vs. the cultural argument vs. the mental argument).
  - In *FPHCHEM*, I proposed another version of reductionism, in relation to substances and their changes (e.g., the natural argument vs. the social argument vs. the cultural argument vs. the mental argument).

(continued on next page)

Table 4.2. On Reductionism and Reverse-Reductionism (Part IV)

- **Against the Varieties of Reductionism** ***(cont'd)***
  —*Theoretical Reductionism (cont'd)*
    - And in *FPHDA*, I propose another version of reductionism, in relation to quantitative and qualitative research (e.g., the natural argument vs. the social argument vs. the cultural argument vs. the mental argument.
  —*Methodological Reductionism*
    - A good instance concerns the debate between different versions of qualitative and quantitative methods (as already analyzed in *FC* and also *FHC*). In *FPHML*, I examined similar reductionism, this time, in the literature on mathematical logic (e.g., the obsession with consistency, soundness, and completeness). And in *FPHFS*, I also explored the problems of reductionism in the context of formal science (e.g., the analytical argument).
  —*Ontological Reductionism*
    - An excellent example is the debate between emergentism and reductionism in complexity theory and also in psychology (as elaborated in *FPHC*, in the context of Being and Becoming).

- **Against the Varieties of Reverse-Reductionism**
  —*Conceptual Reverse-Reductionism*
    - Any concept of "art" (e.g., fine art, cave art, outsider art, junk art) is deemed acceptable in postmodernism (as already addressed in Ch.4 of *FHC*).

(continued on next page)

Table 4.2. On Reductionism and Reverse-Reductionism (Part V)

- **Against the Varieties of Reverse-Reductionism *(cont'd)***
  —*Theoretical Reverse-Reductionism*
  - There are numerous art and literary theories co-exist. Take the case of literary studies, as there are now Literary Structuralism, Marxist Literary Criticism, New Criticism, Phenomenology, Hermeneutics, Language-Game Literary Criticism, Feminist Literary Criticism, Reception Theory, Reader Response Criticism, Poststructuralism, Semiotics, Pyschoanalytic Literary Criticism, just to cite some well-known ones, with no one being said to be better than any others (as detailedly analyzed in Ch.4 of *FHC*). (S. Raman 1997) In *BNN*, I even introduced "the compromise fallacy" as another good illustration of theoretical reverse-reductionism, in misleadingly treating both genetic and environmental approaches as equally valid.

  —*Methodological Reverse-Reductionism*
  - There is the "anything-goes" mentality in postmodernism (e.g., doing art without praxis, doing art with praxis, and doing art by sublation), as analyzed in Ch.4 of *FHC*. And in *FPHFS*, I also exposed the problems of reverse-reductionism in the context of formal science in relation to systems theory.

  —*Ontological Reverse-Reductionism*
  - There are likewise no privileged ontology, and the door is open for anything in postmodernism (e.g., the equal status of the ontology of Being vs. that of Becoming, as already addressed in Ch.4 of *FHC*—and also in *FPHC)*. In *FAE*, I also introduced another version of reverse-reductionism, that is, "the pluralist fallacy," in the context of understanding aesthetic experience, for instance—although this fallacy has been committed not exclusively in relation to the ontological level (but also at the conceptual, theoretical, and methodological ones).

Sources: From my previous books.

Table 4.3. The Concepton of Existential Dialectics (Part I)

- **Sets and Elements**
  - —Sets
    - Ex: the Same
    - Ex: the Others
  - —Elements
    - Ex: whites in 20th century America (in the set of "the Same")
    - Ex: Iraq during the U.S. invasion in 2003 (in the set of "the Others")

- **Relations, Operations, Functions**
  - —Relations (e.g., "belongs," "equals to," "is greater than")
    - Ex: symmetric interactions within the Same (or the Others)
    - Ex: asymmetric interactions between the Same and the Others
  - —Operations (e.g., "and," "or," "not," "if…then")
    - Ex: if the Same oppresses the Others, it will also oppress itself.
    - Ex: the Same is not the Others.
  - —Functions (e.g., goals)
    - Ex: the Same is hegemonic in relation to the Others.

- **Truth Values**
  - —"1" if True (in Symbolic Logic)
    - Ex: the proposition that imperial Japan was hegemonic to China during WWII
  - —"0" if False (in Symbolic Logic)
    - Ex: the proposition that Grenada invaded France in 2003
  - —"1" & "0" if Both True and False (in Dialectic Logic)
    - Ex: the proposition that the rabbit-duck picture refers to a duck
  - —"~1" & "~0" if Neither True Nor False (or N/A, in Dialectic Logic)
    - Ex: the proposition that God really exists

(continued on next page)

Table 4.3. The Conception of Existential Dialectics (Part II)

- **Axioms and Postulates**
  - —Axioms
    - Ex: the reflexive axiom—"any quantity is equal to itself"
  - —Postulates
    - Ex: the SSS postulate—"if the three sides of a triangle are congruent to their corresponding parts, then the triangles are congruent"

- **Theorems and Principles**
  - —Theorems (and Principles) in Existential Dialectics
    - In Relation to Method
      - –#24: The formalness-informalness principle
      - –#23: The absoluteness-relativeness principle
      - –#22: The partiality-totality principle
      - –#21: The predictability-unpredictability principle
      - –#20: The explicability-inexplicability principle
      - –#19: The fiction-reality principle
    - In Relation to Structure
      - –#18: The finiteness-transfiniteness principle
      - –#17: The preciseness-vagueness principle
      - –#16: The simpleness-complicatedness principle
      - –#15: The openness-hiddenness principle
      - –#14: The denseness-emptiness principle
    - In Relation to Process
      - –#13: The change-constancy principle
      - –#12: The order-chaos principle
      - –#11: The slowness-quickness principle
      - –#10: The expansion-contraction principle

(continued on next page)

Table 4.3. The Conception of Existential Dialectics (Part III)

- **Theorems and Principles (*cont'd*)**
  —Theorems (and Principles) in Existential Dialectics
  - <u>In Relation to Agency</u>
    –#9: The theory-praxis principle
    –#8: The convention-novelty principle
    –#7: The evolution-transformation principle
    –#6: The symmetry-asymmetry principle
    –#5: The softness-hardness principle
    –#4: The seriousness-playfulness principle
  - <u>In Relation to Outcome</u>
    –#3: The regression-progression principle
    –#2: The same-difference principle
    –#1: The stability-reaction principle

Notes: The categories and examples in each are solely illustrative (not exhaustive). The comparison is also relative (not absolute), nor are they mutually exclusive. As generalities, they allow exceptions.
Sources: Starting from Ch.6 of *BCPC* and also from other books of mine

Table 4.4. The Syntax of Existential Dialectics I: The Principles (Part I)

- **In Relation to Method**

—*The Formalness-Informalness Principle*
(On the Formal Requirements of Logical Systems)

- The formal requirements of a logical system (e.g., consistency, soundness, and completeness) have both usefulness and non-usefulness, to the extent that, if there are formal systems requiring them, there are alternative ones which do not. Thus, it does *not* exclude classical logics but simply goes beyond both classical and non-classical logics, while learning something from each. There is no formalness without informalness—and vice versa.
- For instance, existential dialectics can make use of both classical logics under certain conditions (e.g., especially, though not exclusively, when they are clear-cut, etc.) and non-classical logics under alternative conditions (especially, though not exclusively, when they are "unknown," "irrelevant," "ambiguous," "possible," with "different degrees of truth," empirically inconsistent in a desirable way, etc.). (WK 2008u)
- Family resemblance: e.g., logicalness-nonlogicalness, rationality-nonrationality, etc.
- Sources: From *FPHML*, *FPHH*, *FPHGAM*, and my later books.

—*The Explicability-Inexplicability Principle*
(On the Underlying Mechanisms of Things)

- Both explicability and inexplicability are part of the understanding of things. There is no explicability without inexplicability—and vice versa.
- This principle tells us the duality of the research dilemma, in that, if reality can be explained in some ways, it also has its other ways which are not quite explainable, at a given point in time.
- Family resemblance: e.g., underlyingness-regularness, causation-regularness, causation-correlation, explanation-description, explanation-narration, explanation-interpretation, etc.
- Sources: Especially from *FPHU*—and also *FPHC*. See also my later books.

(continued on next page)

Table 4.4. The Syntax of Existential Dialectics I: The Principles (Part II)

- **In Relation to Method** ***(cont'd)***

  —The Predictability-Unpredictability Principle
  (On the Occurrence of Events)

  - Both predictability and unpredictability have a major role to play in the occurrence of things, so that neither determinism nor indeterminism wins the centuries-old fight. There is no predictability without unpredictability—and vice versa.
  - There are events which are predictable, just as there are those which are not. Or what is regarded as unpredictable at one point in time may turn out to be predictable later, and, conversely, what is deemed as predictable may turn out to not be so predictable. Even in predictability, outcomes are subject to uncertainty, the degree of which varies from case to case.
  - Works on complexity theory (as in *FC*), quantum mechanics (as in *FPHST* and *FPHG*), frequency probability, subjective (or Bayesian) probability, and "special causes" (as in *FPHGAM*), just to name a few of them, have contributed to the understanding of uncertainity.
  - Family resemblance: e.g., sureness-arbitrariness, certainty-uncertainty, sureness-doubtfulness, deterministicness-indeterministicness, deterministicness-randomness, likeliness-riskiness, etc.
  - Sources: Especially from *FC* & *FPHGAM.* See also my later books.

(continued on next page)

Table 4.4. The Syntax of Existential Dialectics I: The Principles (Part III)

- **In Relation to Method *(cont'd)***

—The Partiality-Totality Principle
(On the Relationships between Whole and Parts)

- The whole is not the sum of the parts. There is no partiality without totality—and vice versa.
- Any inquiry about a phenomenon in the work is to guard against the varieties of (a) reductionism and (b) reverse-reductionism.
- Reductionism and reverse-reductionism can be (i) conceptual, (ii) theoretical, (iii) methodological, and (iv) ontological.
- Works on emergence (as in *FC*), gestalt psychology (as in *FPHGEOL*), and the hermeneutic circle (as in *FPHDA*), just to name a few of them, have contributed to the understanding of whole and parts.
- Family resemblance: e.g., individualisticness-holisticness, partness-wholeness, analysis-synthesis, etc.
- Sources: Especially from *FC* & *FPHGEOL*. See also my later books.

—The Fiction-Reality Principle
(On the different faces of Reality)

- If there is reality, there is also fiction, especially for any agent with some kind of intelligence life, be it about humans, animals, or, later, post-humans. There is no reality without fiction—and vice versa.
- The distinction between reality and fiction is not absolute, since there can be a mixture of both. Fiction can have non-fictional elements, just as non-fiction can have fictional ones. So, the distinction is in degree, not in kind.
- Family resemblance: e.g., fiction-nonfiction, fiction-fact, imagination-fact, imagination-nonfiction, subjectivity-objectivity, etc.
- Sources: First from *FPHLIT*.

(continued on next page)

Table 4.4. The Syntax of Existential Dialectics I: The Principles (Part IV)

• **In Relation to Method** ***(cont'd)***

—*The Absoluteness-Relativeness Principle*

(On the Multiplicity of Things)

- There is the multiplicity of things in reality, be they about entities, qualities (or properties), and relationships, such that what is acceptable from one standpoint may not be so from another. For instance, if there is something absolute, there is likewise something relative. There is no absoluteness without relativeness—and vice versa.
- Both absoluteness and relativeness here are also relevant to different modalities often cited in the literature on ontology, such as possibility (e.g., something "can" happen) and its opposite (e.g., impossibility), probability (e.g., something "will" happen) and its opposite (e.g., improbability), and necessity (e.g., something "should" happen) and its opposite (e.g., contingency).
- Family resemblance: e.g., uniformity-diversity, internalness-externalness, immanence-transcendence, independence-dependence, isolatedness-interconnectiveness, etc.
- Sources: From *FPHK*. See also my later books.

(continued on next page)

Table 4.4. The Syntax of Existential Dialectics I: The Principles (Part V)

- **In Relation to Structure**

—*The Finiteness-Transfiniteness Principle*

(On the Nature of Numbers)

- If there are finite things, there are likewise transfinite ones. There is no finiteness without transfiniteness—and vice versa.
- To avoid confusion, my usage of the word "transfinite" here differs radically from the one used by Cantor (and other mathematicians) for "relative" infinity—and is more limited, in light of the problems confronting any attempt to understand the idea of infinity, be it by intuition, imagination, and conception (as detailedly analyzed in *Sec. 2.2.3* of *FPHG*).
- Instead, by "transfinity," I allow numbers which can be many times larger—or smaller, for that matter—than the finite things that we encounter in daily life, but they do not have to be related to the idea of infinity at all (which may not exist).
- Of course, there may be some *borderline* cases, in which it is not clear whether the number in question is transfinite (in my usage) or simply a mathematical convenience. A good example of a borderline case is the Planck unit of length for "the smallest space possibly measured in nature," which is "less than billionths of trillionths of trillionths of an inch" (or something like 1.6×10^{-35} meters).
- That said—my usage of "transfinity" can also resolve (or better, dissolve) an age-old problem in philosophy known as "Zeno's paradoxes" (as already explained in *Sec. 2.2.3* of *FPHG*).
- Family resemblance: e.g., boundedness-quasiunboundedness, smallness-largeness, microness-macroness, boundedness-unboundedness, etc.
- Source: From *FPHG*. See also my later books.

(continued on next page)

Table 4.4. The Syntax of Existential Dialectics I: The Principles (Part VI)

- **In Relation to Structure *(cont'd)***
  —*The Simpleness-Complicatedness Principle*
  (On the Interconnection among Things)
  - Both simpleness and complicatedness are vital, without favoring one over the other, in that each is utilized, depending on the basis of the perspectives of nature, the mind, culture, and society. And even when a combination of them is preferred, the dilemma is only shifted to a combinational degree of concern. There is no simpleness without complicatedness—and vice versa.
  - In relation to taxonomy, simpleness has its heuristic usefulness, just as complicatedness has its realistic representation, for instance. And in relation to network, simpleness has its economical attractiveness, just as complicatedness has its practical reliability, for instance.
  - Family resemblance: e.g., inflexibility-flexibility, standardization-specialization, imperfectness-perfectness, superficiality-depth, shallowness-deepness, economicalness-elaboratedness, plainness-circumspection, onesidedness-multisidedness, etc.
  - Sources: Especially from *FIA*. See also my later books.

  —*The Denseness-Emptiness Principle*
  (On the Distribution of Entities in Space)
  - Both density and void are needed, in relation to the mind, nature, culture, and society, albeit in different ways. There is no denseness without emptiness—and vice versa.
  - For clarity, the term "void" is used here only as an approximation of emptiness (depending on the degree of the lack of density), since, in physics, it is well known that "empty" space is not really empty all the way, because it can be full of energy (e.g., random quantum fluctuations at the sub-atomic level, and, for that matter, dark energy in the universe) and matter (e.g., different versions of sub-atomic particles, and, for that matter, dark matter in the universe).
  - Family resemblance: e.g., fullness-voidness, nearness-farness, concentration-dispersion, denseness-nothingness, etc.
  - Sources: First worked out in *FPHUP*. See also my later books.

(continued on next page)

Table 4.4. The Syntax of Existential Dialectics I: The Principles (Part VII)

- **In Relation to Structure** ***(cont'd)***
  —*The Preciseness-Vagueness Principle*
  (On the Refinement of Things)
  - Both preciseness and vagueness are important, not that one is better than the other, but that both are used, in different degrees of preference, in accordance to the contextual application from the perspectives of nature, the mind, culture, and society. Even when both are used in a combination, the dilemma is shifted instead to one of combinational concern. There is no preciseness without vagueness—and vice versa.
  - In relation to taxonomy, preciseness has its taxonomic clarity, just as vagueness has its classificatory flexibility, for instance. And in relation to network, vagueness has its explorative liberty, just as preciseness has its conceptual definitiveness, for instance.
  - Family resemblance: e.g., clarity-ambiguity, directness-indirectness, quantitativeness-qualitativeness, describability-nondescribability, specificity-obscurity, specificity-generality, thickness-thinness, concretness-abstractness, translatability-untranslatability, clearcutness-transitionalness, etc.
  - Sources: Especially from *FIA*. See also my later books.

(continued on next page)

Table 4.4. The Syntax of Existential Dialectics I: The Principles (Part VIII)

- **In Relation to Structure** ***(cont'd)***
  —*The Openness-Hiddenness Principle*
  (On the Detection of Things)
  - Reality has its hidden face, just as it is open to outside view in some other ways. There is no openness without hiddenness—and vice versa.
  - For instance, in the context of anomalous experience, certain aspects of reality can be open for examination (e.g., the experiment with SPECT images, the transmission of telepathemic bit, and the use of metonymies and metaphors). Other aspects, however, remain hidden, and examples include the elusive deeper nature of spiritual reality in the intangible realm, the undetectability of different branched universes, the underlying mechanisms of signals traveling between different folds of physical space-time, and the dependence on language and the bias for science and logic).
  - Family resemblance: e.g., overtness-covertness, publicness-privateness, openness-closedness, transparency-secrecy, openness-biasedness, participation-obstruction, exposition-illusion, etc.
  - Sources: Especially from *FPHU*. See also my later books.

- **In Relation to Process**
  —*The Change-Constancy Principle*
  (On the Alteration of Things)
  - Change occurs over time, although constancy is also allowed. There is no change without constancy—and vice versa.
  - Asymmetry undergoes changes over time, so does symmetry.
  - Old players fade away, and new ones emerges, with ever new causes and ever new forms.
  - Family resemblance: e.g., dynamicness-staticness, instability-stability, etc.
  - Sources: First named in *BCPC*. Especially from *FHC*, *FCD*, and *FPHC*. See also other books of mine (from that point on).

(continued on next page)

Table 4.4. The Syntax of Existential Dialectics I: The Principles (Part IX)

- **In Relation to Process *(cont'd)***

—*The Expansion-Contraction Principle*
(On the Growth of Things)

- Entities in the world can both expand in some ways and contract in other ones, as part of their nature. There is no expansion without contraction—and vice versa.
- For instance, in the context of anomalous experience, one of its most fascinating expressions is none other than the analysis of different causes of the slow but steady expansion (and, furthermore, if put in a different way, unfolding) of unconsciousness.
- Family resemblance: e.g., conquest-autarky, rise-fall (or up-down), spread-shrink, extendingness-shorteningness, etc.
- Sources: Especially from *FPHU*. See also my later books.

—*The Slowness-Quickness Principle*
(On the Speed of Change)

- Both slowness and quickness co-exist, with their own internal tension, to the extent that each fights for its own relevance with the other, in accordance to the perspectives of nature, the mind, culture, and society, without one being the victor and the other being the vanquished in the long haul. Even when both are chosen in other cases, this dilemma is only transferred into something else with a combinational character. There is no slowness without quickness—and vice versa.
- In relation to taxonomy, quickness has its efficient usability, just as slowness has its aesthetic appeal, for instance. And in relation to network, quickness has its adventurous readiness, just as slowness has its risk-adverse convenience, for instance.
- Family resemblance: e.g., inconvenience-convenience, passiveness-activenessness, gradualness-abruptness, deceleration-acceleration, etc.
- Sources: Especially from *FIA*. See also my later books.

(continued on next page)

Table 4.4. The Syntax of Existential Dialectics I: The Principles (Part X)

- **In Relation to Process** ***(cont'd)***
  —*The Order-Chaos Principle*
  (On the Pattern of Things)
  - Both order and chaos are vital in the process of change in the world. The preference for order is biased, since it does not give sufficient attention to the vital role of chaos in the transformation of the world (without somehow reducing it for the understanding of order). There is no order without chaos—and vice versa.
  - The scientific search for order in the world is often a hidden bias in its ontological obsession with order, since chaos is often treated as the "bad" guy, with order as the "good" guy (for the end goal of science).
  - Neither order nor chaos is the final end of the world, and one is not to be treated as the means for the other in the transformation of things. Both are fundamental in their recurrent dialectical interactions with each other over time, without reducing one for the other.
  - Family resemblance: e.g., lawfulness-disorder, order-disorder, linearity-nonlinearity, linearity-irreversalness, pattern-chaos, etc.
  - Sources: Especially from *FC*. See also my later books.

- **In Relation to Agency**
  —*The Softness-Hardness Principle*
  (On the Force of Change)
  - This has to do with *the force of change* under the category about agency in existential dialectics, in that any change by an agent, be it organic (like humans) or non-organic (like natural objects), can occur in a forceful (aggressive) or gentle (pacific) way, which can come in all shapes and sizes, of course. There is no softness without hardness—and vice versa.
  - Family resemblance: e.g., peacefulness-violence, cooperation-competition, dovishness-hawkishness, reward-punishment, peace-war, pacificity-aggression, acceptance-discrimination, etc.
  - Sources: Especially from *ALD*. See also my later books.

(continued on next page)

Table 4.4. The Syntax of Existential Dialectics I: The Principles (Part XI)

- **In Relation to Agency *(cont'd)***
  —*The Symmetry-Asymmetry Principle*
  (On the Relationships among Existents)
  - There is no symmetry without asymmetry—and vice versa.
  - For instance, the Same can be symmetric and asymmetric towards the Others. But in case of asymmetry, oppression and self-oppression can occur. So, when the Same is asymmetric towards the Others, the Same can also be relatively asymmetric towards itself in self-oppression, just as the Others can be likewise towards themselves.
  - The subsequent oppressiveness is dualistic, as much by the Same against the Others and itself, as by the Others against the Same and themselves.
  - Both oppression and self-oppression can be achieved by way of downgrading differences between the Same and the Others and of accentuating them.
  - This is true, even though not all forms of asymmetry have to be about oppression and self-oppression.
  - In addition, from Chapter Three of *FPHG*, symmetry is not perfect, to be understood in an approximate sense under many life circumstances.
  - Family resemblance: e.g., balance-extremity, harmoniousness-unharrmoniousness, reflexiveness-unreflexiveness, equality-inequality, center-periphery, compatibility-incompatibility, etc.
  - Sources: From all my books, starting with *FHC*. First named in *BCPC*. See, for instance, *FCD* for more details on the Same and the Others.

(*continued on next page*)

Table 4.4. The Syntax of Existential Dialectics I: The Principles (Part XII)

- **In Relation to Agency** ***(cont'd)***

  —*The Evolution-Transformation Principle*
  (On the Multiple Kinds of Agency)

  - This principle (and the symmetry-asymmetry principle) are both about *the agency of change*. The word "agency," in a formal definition, refers to "a person or thing through which power is exerted or an end is achieved." (MWD 2007b) It therefore does not have to necessarily involve an intelligent lifeform.
  - Because of this dual meaning in agency, the evolution-transformation principle is more concerned with *the multiple kinds of agency*, that is, both about the *evolution* in the state of nature (e.g., an object of natural beauty) and the *transformation* in the world of intelligent lifeforms (e.g., a work of art, an air-conditioner, etc.). There is no evolution without transformation—and vice versa.
  - And the transformative part of the principle precisely refers to the other dimension in the dual meaning of agency, in giving technology a major role to play in the change of the world, which is something that I extensively analyzed in *FHC* in the context of the technophilic lifeworld, espeically though not exclusively since modern times. The same theme is also elaborated in other books of mine.
  - Family resemblance: e.g., nonwillingness-willingness, naturalness-technologicalness, naturalness-nonnaturalness, inorganicness-volition, inorganicness-motivation, evolution-creation, naturalness-artificialness, etc.
  - Sources: Especially from *FAE*. See also other books of mine.

(*continued on next page*)

Table 4.4. The Syntax of Existential Dialectics I: The Principles (Part XIII)

- **In Relation to Agency** ***(cont'd)***
  - *—The Convention-Novelty Principle (On the Nature of Creative Thinking)*
    - If there is conventional wisdom, there is likewise novel challenge, to the extent that both convergent and divergent thinking are part of life. There is no convention without novelty—and vice versa.
    - In addition, in this context of the convention-novelty principle, there are, in the absence of better words, what I want to call (a) *creative techniques* and (b) *creative traits*, which, when satisfied—in relation to the larger context of the mind, nature, society, and culture—can be used to enhance creative works. The list of creative techniques and traits (as summarized in *Table 4.1* of *FPHCT*) are not exhaustive, of course, but the examples here constitute a great beginning of understanding the structure of creative thinking.
    - Yet, lest the reader gets carried away by the euphoria about creative thinking in our time, it should be reminded that creative thinking has its own possibilities and limits (as shown in *Table 2.1* of *FPHCT* on invention), just as it has its own promises and pitfalls (as shown in *Table 3.1* of *FPHCT* on innovation). In the end, it should also be stressed that creative thinking has its own desirability and dark sides (as shown in *Table 4.2* of *FPHCT*).
    - This has important daily implications for an average individual, and the word "average" here allows different degrees of variation from one individual to another in a population, at any given point in history.
    - Family resemblance: e.g., normalness-nonnormalness, conformity-nonconformity, convergence-divergence, etc.
    - Sources: First from *FPHCT*. See also my later books.

(*continued on next page*)

Table 4.4. The Syntax of Existential Dialectics I: The Principles (Part XIV)

• **In Relation to Agency** ***(cont'd)***
—*The Seriousness-Playfulness Principle*
(On the Extent of Seriousness)
• If there is seriousness, there is also playfulness, especially for any agent with some kind of intelligence life, be it about humans, animals, or, later, post-humans. There is no seriousness without playfulness—and vice versa.
• Lest any misunderstanding occurs, the word "playfulness" here should not be confused with other terms like "laughter," "smile," "joke," "tickleness," and "laugh-like vocalization" (in some animals), for instance.
• Being playful can come in all shapes and sizes and therefore is not necessarily tied up with any of these terms, alhough it can result from any of them, or a combination of them, or something else altogether.
• This principle, together with some other principles (especially though not exclusively, like the formalness-informalness principle), have important implications for what constitutes rationality, be it about "rationality of application" at the practical level or "rationality of knowledge" at the meta-theoretical level (as summarized in *Sec. 4.20* of *FPHH*).
• Family resemblance: e.g., seriousness-jokingness, seriousness-humorousness, seriousness-wittiness, seriousness-nonseriousness, seriousness-restingness, seriousness-relaxedness, etc.
• Sources: First from *FPHH* & *FPHGAM* See also my later books.

(*continued on next page*)

Table 4.4. The Syntax of Existential Dialectics I: The Principles (Part XV)

- **In Relation to Agency** ***(cont'd)***
  —*The Theory-Praxis Principle*
  (On the Duality of Knowledge))
  - This has to do with the duality of knowledge in existential dialectics, in that, if there is theoretical construction, there is likewise its practical application, both technical and normative. There is no theory without praxis—and vice versa.
  - Family resemblance: e.g., discovery-application, knowledge-action, invention-innovation, theory-intentionality, discourse-action, knowledge-motivation, universality-particularity, knowledge-interest, etc.
  - Source: Especially from *FPHE*. See also my later books.

- **In Relation to Outcome**
  —*The Regression-Progression Principle*
  (On the Direction of History)
  - Neither the cyclical nor the linear views are adequate for explaining many phenomena at all levels. There is regression without progression—and vice versa.
  - History progresses to more advanced forms, but with a regressive touch. Examples include no freedom without unfreedom, no equality without inequality, and no civilization without barbarity.
  - This is not an inevitable law, but merely a highly likely empirical trend.
  - Family resemblance: e.g., cost-benefit, undesirability-desirability, badness-goodness, risk-opportunity, etc.
  - Sources: From all my books, starting with *FHC*. First named in *BCPC*.

(*continued on next page*)

Table 4.4. The Syntax of Existential Dialectics I: The Principles (Part XVI)

- **In Relation to Outcome *(cont'd)***
  - *—The Same-Difference Principle (On the Metamorphosis of Change)*
    - An entity, as it evolves over time, can be both different from and similar to its opposing alternatives and does not have to be solely more different from them over time. There is no similarity without difference—and vice versa.
    - Opposites are not absolute in a black-or-white fashion; so, an entity can become relatively more similar to (or more different from) its opposite over time.
    - Family resemblance: e.g., homogeneity-heterogeneity, we-they, etc.
    - Sources: Especially from ALD. See also my later books.
  - *—The Stability-Reaction Principle (On the Feedback Mechanism for Further Changes)*
    - Once an outcome becomes stabilized over time (which varies from case to case), it will undergo further changes in its multifaceted interactions with the outside world (e.g., with other entities, energy, space, and so on).
    - There is no stability without reaction for further changes—and vice versa.
    - Family resemblance: e.g., stability-forwardness, balance-reaction, balance-forwardness, consolidation-modification, normalcy-feedback, normalcy-feedback, stability-breakup, etc.
    - Sources: First from *FPHCHEM.*

Notes: The features in each principle are solely illustrative (not exhaustive). The comparison is also relative (not absolute), nor are they mutually exclusive. As generalities, they allow exceptions.

Table 4.5. The Syntax of Existential Dialectics II: The Principles as Short Cuts (Part I)

• The principles should be treated with caution, lest misunderstanding occurs, since they do not constitute rigid dualities (or dichotomies).

• The reason is that each pair in an ontological principle consists of two opposites, which are, however, merely short cuts both for multiple variations and degrees, as well as for different interactions with multiple other entities unlike them. In this light, each pair can end up having hundred (or even thousand, if not more) different versions, which interact with hundred (or even thousand, if not more) other entities. There are two clarifications here.

• Firstly, this conception of shortcuts is not mutually exclusive nor absolute, in that the opposites can come in all shapes and sizes, with different degrees. For instance, by analogy, just as there are different degrees of the two colors "white" and "black"—there are likewise different degrees of the opposites in each ontological pair, to the extent that there can be multiple entities (not only two) interacting with each other in each pair.

• And secondly, this conception of shortcuts do not ignore other possible entities in interacting with the two opposites (with their different versions) in each pair. For instance, by analogy, there are not only the two colors "white" and "black" as opposites, since there are other colors too besides them like "yellow," "green," "purple," or else. The same logic can be applied to each ontological pair, in that they also interact with other entities, not with only two of them (with their different versions).

• Therefore, with these two clarifications in mind—each pair in an ontological principle serves only as an abbreviation for something more complicated and, therefore, although it contains two opposites, it should not be confused as a duality (dualism).

(continued on next page)

Table 4.5. The Syntax of Existential Dialectics II: The Principles as Short Cuts (Part II)

• Existential dialectics rejects any dualism (or dichotomy) as too rigid and instead allows the multiplicity of entities, to the extent that between the two opposites in each pair exist many other alternatives to choose from. They are named in that short form for aesthetic elegance, instead of listing all possible entities between the two opposites in the title.

• To be dialectic is to go beyond any rigid dichotomy and transcend into something different altogether in the long haul. One may be tempted to call the dialectic logic here with a different name like existential "multilectics" (instead of "dialectics"), but this naming is incorrect (or even misleading), for the two reasons aforestated.

• After all, each ontological pair come in all shapes and sizes (with different degrees) and do not exist by themselves but also interact with other entities unlike them. The virtue of revealing an ontological pair is to show how they relate within themselves (in multiple versions) and also interact with others unlike them (also in multiple versions).

• This will be clear in the table on "the dialectic constraints imposed by the principles."

Source: From Ch.1 of *FPHK*

Table 4.6. The Syntax of Existential Dialectics III: The Principles as Family Resemblances

• Each ontological principle is generic, with some other comparable ontological pairs to be put in the same family (like a *family resemblance*). Ludwig Wittgenstein (1953) once suggested the idea of "family resemblance" in explaining different games classified under the same family called *games*.

• Why should, for instance, playing football and chess as playing "games," when it is well understood that football is not the same as chess? The answer is that, although each game is different and has different rules, many of them (though not all) share, more or less, some commonalites (e.g., scoring as necessary for winning).

• And this is so, even though some games share more than some others in any given selection of critieria, and no two games are exactly identical. So, his point here is that there is no essential core which is common to all games, and the best that one can look for is some characteristics which are common to many (but not all) games. (A. Biletzki 2006)

• By the same logic—in the previous section on selection criteria, the flexibility-inflexibility pair can be put in the family resemblance of the simpleness-complicatedness principle, although the two pairs are not exactly identical. Likewise, the directness-indirectness pair can be put in the family resemblance of the preciseness-vagueness principle, although, again, the two pairs are not exactly identical.

• In this sense, which specific pair in a family should be used to designate the name of the family can be at times a bit arbitrary, but with good reason.

• For illustration, in the context of method, the partiality-totality principle can take the different form like individualisticness-holisticness, just as the explicability-inexplicability principle can take the different form like underlyingness-regularness—although each two pairs are not exactly identical and have slightly different meanings and usages. A more comprehensive listing of this family resemblance for all other principles is shown in the table on the syntax of existential dialectics.

Source: From Ch.1 of *FPHK*

Table 4.7. The Syntax of Existential Dialectics IV: The Dialectic Constraints Imposed by the Principles (Part I)

• **Co-Existent and Asymmetric**

—The principles, as they constitute the syntax of existential dialectics, are dialectic in character, such that, when they are applied, they impose dialectic constraints on how reality is to be understood. Consider, say, the symmetry-asymmetry principle as an illustration here, in order to summarize two main characters of the dialectic constraints in question.

—Firstly, to be dialectic here is to go beyond the narrow dichotomies (and, for that matter, any rigid multi-dimensional classificatory scheme), be they about "self" vs. "world," "freedom" vs. "unfreedom," "barbarity" vs. "civilization," "individuality" vs. "communality," and so on.

—One way to do so (to go beyond) is to consider them all in terms of co-existence (without favoring one over the rest). For instance, my theory of "post-civilization" (to be summarized later in the section on the pragmatics of existential dialectics) is to go beyond barbarity and civilization in terms of understanding barbarity and civilization as being co-existent. And the same logic can be said in relation to my theories of "post-democracy," "post-capitalsm," and others (also to be introduced later in the section on the pragmatics of existential dialectics), in regard to freedom vs. unfreedom, equality vs. inequality, communality vs. individuality, spirituality vs. materiality, and so on.

—But to consider them all (in the dichotomies—and, for that matter, in any rigid multi-dimensional classifactory schme) as co-existent is not the same as to imply that the opposites in any classificatory scheme are all equal, since, in accordance to the symmetry-asymmetry principle (as an illustration here), if they are equal in terms of being considered as co-existent, they are asymmetric in terms of being unequal in dialectic interaction (e.g., X can be more dominant than Y in case A, or Y is more dominant than X in case B).

(continued on next page)

Table 4.7. The Syntax of Existential Dialectics IV: The Dialectic Constraints Imposed by the Principles (Part II)

—For this reason, there are different versions of "post-democracy" and "post-capitalism" in my theories. As an illustration, in version I of the theory of post-democracy, freedom is more dominant than equality, whereas in version II of the theory of post-democracy, equality is more so than freedom.

—But this "X more than Y" has to be understood in the context of dialectic logic (not in conventional logic), in that both "X" and "Y" are important in post-democracy (in the context of dialectic logic), but in an asymmetry way. By contrast, in conventional logic, it often favors one over the other—be it in regard to privileging freedom over equality in Fascism, favoring freedom relatively more than equality in Liberal Democracy, or favoring equality relatively more than freedom in Socialist Democracy. In the latter two cases (about Liberal Democracy and Socialist Democracy), the difference between dialectic logic and conventional logic can be one in degree, not in kind—in this sense, albeit not in other senses.

—The same logic can be said about the relatonships between individuality and communality, between spirituality and materiality, and between formal legalism and informal legalism in the different versions of my theory of post-capitalism.

• **Transcendent**

—Secondly, to be dialectic is to go beyond the narrow dichotomies (and, for that matter, any rigid multi-dimensional classificatory scheme) in another way, this time, in a transcendent way, that is, in exploring other possibilities or even other issues not considered within the narrow confines of narrow dichotomies (and, for that matter, any rigid multi-dimensional classificatory scheme).

(continued on next page)

Table 4.7. The Syntax of Existential Dialectics IV: The Dialectic Constraints Imposed by the Principles (Part III)

—As an analogy, to go beyond the narrow color dichotomy of "black" and "white" is not just to choose both "black" and "white" (as in the first meaning) but also to explore other color options (e.g., "green," "purple," "blue," etc.—and, alternatively, "shade," "line," "curve," etc.). By the same logic, to go beyond "democracy" is to transcend democracy (as in version III of the theory of "post-democracy") and to explore other possibilties of lifeforms (e.g., floating consciousness, hyper-spatial consciousness, etc., to live beyond the narrow obsession with freedom and equality).

—This dialectic character of the principles in existential dialectics has important implications for the pragmatics of existential dialectics (as will be clear shortly, in the section on the pragmatics of existential dialectics).

Notes: The examples here are solely illustrative (not exhaustive). The comparison is also relative (not absolute), nor are they mutually exclusive. As generalities, they allow exceptions.

Sources: First explicitly stated in *ALD*. Also from all other books of mine.

Table 4.8. The Syntax of Existential Dialectics V: Further Clarifications (Part I)

- **Pioneering**
  —Firstly, the total number of ontological principles is unknown, to be discovered later, as our knowledge of the world becomes more advanced.
  —Consequently, the principles as introduced in my books are not exhaustive, with new ones being added, whenever more of them are discovered in later research.
  —At least, future generations can pick up where I leave off and continue the discovery.
  —In this sense, my work should be treated as a pioneering effort for the development of a systematic, comprehensive analysis of a new general ontology for the future of knowledge.

- **Flexible**
  —Secondly, the principles are not rigidly classified, as they can be reclassified in a different way.
  —For instance, the preciseness-vagueness principle is classified under the category of "structure" but can be reclassified under the category of "method," although in so doing, it has a different meaning in the context of method.
  —The same logic applies to the same-difference principle under the category of "outcome," which can be reclassified under the category of "structure," although in so doing, once more, it has a different meaning in the context of structure.

- **Mutually Constraining**
  —Thirdly, the principles are to be understood together, not that each prinicple is to be analyzed independently of others. After all, the principles are mutually constraining, in that they work together as a whole.

(continued on next page)

Table 4.8. The Syntax of Existential Dialectics V: Further Clarifications (Part II)

- **Selectively Useful**
  —Fourthly, the principles are relevant to all subject matters, but some principles are more useful to some subject matters than others—as implied in the symmetry-asymmetry principle. This is true, even if different studies of the same kind can yield different views about the degree of relevance for each ontological principle, depending on the specific nature of a research in question, needless to say.

- **Anti-Reductionistic**
  —And fifthly, the principles constitute only two levels of analysis, this time, at the ontological and methodological levels—while other levels of analysis (from the perspectives of the mind, nature, society, and culture) are also needed, in order to understand reality in its totality (as explained in the section on "sophisticated methodological holism," so as to avoid the dual dangers of reductionism and reverse-reductionism).

Notes: The examples here are solely illustrative (not exhaustive). The comparison is also relative (not absolute), nor are they mutually exclusive. As generalities, they allow exceptions.
Sources: From my previous books like *FHC*, *FCD*, *FPHC*, *BDPD*, *BCPC*, *BNN*, *FC*, *FAE*, and so on.

Table 4.9. The Syntax of Existential Dialectics VI: The Dilemma of Specific vs. General Ontology (Part I)

• Ontology is often more complicated and imposes some seemingly insurmountable difficulties concerning what constitutes a kind of ontology which can be valid enough to be accepted by the wider intellectual community in question. An excellent example concerns what I want to call *the dilemma of ontology* in relation to the relationship between "specific ontology" and "general ontology." (M. Bunge 1999; R. Corazzon 2007)

• On the one hand, "general ontology," as Mario Bunge (1999) put it, "studies all existents." But, on the other hand, "special ontology studies one genus of thing or process—physical, chemical, biological, social, etc." (M. Bunge 1999; R. Corazzon 2007)

• Consequently, the relationship between the former and the latter is that, as an illustration, "whereas general ontology studies the concepts of space, time, and event, the ontology of the social investigates such general sociological concepts as those of social system, social structure, and social change." (M. Bunge 1999; R. Corazzon 2007)

• With this formal definition in mind, the relationship between the two forms of ontology only poses *the dilemma of ontology*, in that general ontology is possible to the extent that its general contructs must be broad enough to serve as the "lowest common denominator" for all disciplines, but this character of the lowest common denominator is not only hard to find but also its usefulness is limited. (WK 2008b)

• But specific ontology fares no better either, in that its specific constructs primarily serve a specific discipline in question, although other disciplines can appropriate them for their own purposes, but in a different context (especially, though not exclusively, when used as a metaphor from one field to another). For instance, the Kantian category of understanding under the heading of "quantity" constitutes a type of specific ontology, which is specific to the discipline of mathematics, although other disciplines can make use of the categories in their own context.

(continued on next page)

Table 4.9. The Syntax of Existential Dialectics VI: The Dilemma of Specific vs. General Ontology (Part II)

• A solution lies in my proposal of "sophisticated methodological holism," which requires all levels of analysis in any subject matter. See the tables on sophisticated methodological holism, and on reductionism and reverse-reductionism, for more details.

Source: From *Sec. 1.4* of *FPHK*

Table 4.10. The Syntax of Existential Dialectics VII: Types of Inappropriate Family Resemblances (Part I)

- **Inadequate Family Resemblances**
  —Ex: consistency-inconsistency: the formalness-informalenss principle
  —Ex: cognition-noncognition: the formalness-informalness principle

- **Compound Family Resemblances**
  —Ex: reality-nonreality: the formalness-informalness principles, the absoluteness-relativeness principle
  —Ex: normality-abnormality: the symmetry-asymmetry principle, the density-emptiness principle, the convention-novelty principle, the same-difference principle
  —Ex: structure-context: the simpleness-complicatedness principle, the theory-praxis principle
  —Ex: freedom-unfreedom: the symmetry-asymmetry principle, the expansion-contraction principle
  —Ex: morality-immorality: the denseness-emptiness principle, the symmetry-asymmetry, the same-difference principle
  —Ex: particularness-universalness: the preciseness-vagueness principle, the simpleness-complicatedness principle
  —Ex: pureness-mixedness: the absoluteness-relativeness principle, the simpleness-complicatedness principle, the same-difference principle
  —Ex: shortsightedness-foresightedness: the partiality-totality principle, the simpleness-complicatedness principle
  —Ex: shorttermness-longtermness: the partiality-totality principle, the simpleness-complicatedness principle, the slowness-quickness principle

- **One-Sided Family Resemblances**
  —Ex: sound-silence: asymmetry (in the symmetry-asymmetry principle)
  —Ex: vision-blindness: asymmetry (in the symmetry-asymmetry principle)

(continued on next page)

Table 4.10. The Syntax of Existential Dialectics VII: Types of Inappropriate Family Resemblances (Part II)

- **Vague Family Resemblances**
  —Ex: shortness-longness, narrowness-broadness, smallness-largeness, smallness-greatness, microness-macroness, and littleness-muchness: three different possibilities, namely, (a) finiteness (in the finiteness-transfiniteness principle), if both opposites are defined on a finite scale, (b) transfiniteness (in the finiteness-transfiniteness principle) if both opposites are defined on a transfinite scale, and (c) the finiteness-transfiniteness principle if one opposite is defined on a finite scale and the other opposite is defined on a transfinite scale.

- **One-Sided and Compound Family Resemblances**
  —Ex: extroversion-introversion: one-sided, in being part of the side (category) of "normality" in the normality-abnormality pair. And compounded, because the normality-abnormality pair is in turn related to the symmetry-asymmetry principle, the density-emptiness principle, the convention-novelty principle, the same-difference principle
  —Ex: organizedness-spontaneousness: one-sided, in being part of the side (category) of "normality" in the normality-abnormality pair. And compounded, because the normality-abnormality pair is in turn related to the symmetry-asymmetry principle, the density-emptiness principle, the convention-novelty principle, the same-difference principle
  —Ex: risk-caution: one-sided, in being part of the side (category) of "normality" in the normality-abnormality pair. And compounded, because the normality-abnormality pair is in turn related to the symmetry-asymmetry principle, the density-emptiness principle, the convention-novelty principle, the same-difference principle

Notes: The examples here are solely illustrative (not exhaustive). The comparison is also relative (not absolute), nor are they mutually exclusive. As generalities, they allow exceptions.
Source: First summarized in *BEPE* (and updated in later books)

Table 4.11. The Semantics of Existential Dialectics (Part I)

• **Abstract Structure vs. Specific Meanings**

—The syntax of existential dialectics so understood in terms of ontological principles only gives us the structure of ontology in the world, in an abstract (general) sense. These principles by themselves do not tell us the specific meanings in a given context.

—In order to grasp the specific meanings of the principles in a given context, it is necessary to study the semantics of existential dialectics. And the analysis of the ontological principles as family resemblances in the section on syntax is only a starting point (and thus overlaps a bit with the semantics of existential dialectics here).

—With this caveat in mind, the reason that I have often gone in great lengths in my previous books on different subjects is to explain the specific meanings of the principles when applied in different contexts.

• **Specific Meanings in Specific Fields**

—For instance, in *FPHST*, I used the first three principles (i.e., the change-constancy principle, the regression-progression principle, and the symmetry-asymmetry principle) to propose "the perspectival theory of space-time," for a better way to understand space and time—especially, though not exclusively, in relation to future post-human history (as summarized in *Table 3.6*, *Table 3.7*, *Table 3.8*, and *Table 3.9*). In so doing, I had to introduce concepts and theories specific to the field of physics and other related fields (e.g., "absolute space" and "absolute time" in "classical mechanics" and "relative space-time" in "the theory of relativity").

—In *BNN*, I also exploited the three principles to propose the "transcendent" approach to the study of genes and memes as a new way to understand the interaction between nature and nurture. In so doing, I had to explore concepts and theories in the world of evolutionary theory (e.g., "mutation," "variation," "adaptation," "selection," and "inheritance" in Darwinian evolutionary theory) and neural biology (e.g., "chromosome," "gene," "DNA," "RNA," "protein," "neuron," "neural network," and "behavior").

(continued on next page)

Table 4.11. The Semantics of Existential Dialectics (Part II)

- **Two Types of Translation (Interpretation)**
  —In the technical jargons of semantics, these exercises in my books perform two essential functions of translation (or interpretation), namely, (a) "intensive" interpretation (about the properties or relations of the principles in existential dialectics), and (b) "extensive" interpretation (about the classes or cases to which the properties and relations of the principles in existential dialectics are applied). (WK 2011oo)

- **Limits of Translation (Interpretation)**
  —But there are limits of translation (or interpretation), and an excellent example is that "there is a problem with the 'objectivity' of translation, as it is relative to different cultural (or sub-cultural) contexts. For instance, one of the still unresolved problems in linguistics is: 'Is there an objective gauge for the quality of translation?'" (*FPHCOM* & WK 2011pp)
  —This is one serious problem about the logic of translation and interpretation that I already addressed in *Sec. 3.5.1* of *FPHCOM (*and also, for that matter, in *FHC*, in the context of hermeneutics).

Notes: The categories and examples in each are solely illustrative (not exhaustive). The comparison is also relative (not absolute), nor are they mutually exclusive. As generalities, they allow exceptions.
Sources: First explicitly stated in *ALD* (and later revised in *FPHCOM*). Also from all other books of mine.

Table 4.12. The Pragmatics of Existential Dialectics (Part I)

- **The Two-Way Street Connecting Theory and Meta-Theory**
  - —The pragmatics of existential dialectics is not a one-way street (that is, using the ontological principles for theoretical insights in praxis) but a two-way one, that is, (a) from meta-theory to theory, and (b) from theory to meta-theory.
  - —(a) On one side of the street, the ontological principles can inspire some theoretical insights in praxis, that is, in relation to some specific fields.
  - —(b) On the other (opposing) side of the street, however, the study of a subject matter in the specific fields in turn reveals some more hitherto unknown ontological principles to be discovered and identified. For this reason, three new principles were added in *FC*, one in *FAE*, and two in *ALD*, on top of the original three in *BCPC*—after some research on the specific subject matters.

- **Direct and Indirect Applications**
  - —*Direct*
    - The logic of existential dialectics can shed some theoretical insights on diverse phenomena in the world, and good instances are the pertinent use of the principles of existential dialectics for the theoretical insights on the freedom/unfreedom dialectics, the equality/inequality dialectics, and the wealth/poverty dialectics in my previous works.
    - My latest books like *FPHST* and *BNN* also use the principles to reveal some theoretical insights on the perspectives of space and time (as in *FPHST*) and of nature and nurture (as in *BNN*).
  - —*Indirect*
    - The theoretical insights can further be used to reveal other phenomena directly from them (viz., the theoretical insights) and therefore indirectly from the principles themselves. A good instance is the use of the theoretical insights on the freedom/unfreedom and equality/inequality dialectics for the understanding of the civilization/barbarity dialectics.

(continued on next page)

Table 4.12. The Pragmatics of Existential Dialectics (Part II)

- **Direct and Indirect Applications (*cont'd*)**
    - Even in indirect applications, however, a phenomenon under study can still be directly related back to the principles themselves. In the example as cited above, the civilization/barbarity dialectics can be directly related to the principles of existential dialectics without the intermediate role of the freedom/unfreedom and equality/inequality dialectics.

- **Multiple Levels of Application**

—The theoretical insights can be applied to different levels of analysis, even though in a given example, it may refer to one level only. For instance, in the example concerning the freedom/unfreedom dialectics, it can be used at the structural level (e.g., in relation to the theory of cyclical progression of hegemony), but it can be exploited as well for other levels (e.g., the theory of post-capitalism at the institutional level).

Notes: The categories and examples in each are solely illustrative (not exhaustive). The comparison is also relative (not absolute), nor are they mutually exclusive. As generalities, they allow exceptions.
Sources: From Ch.6 of *BCPC*. See also other books of mine.

Table 4.13. The Freedom-Unfreedom Dialectics (Part I)

- **On Having**
  - *—In Relation to the Technological*
    - (1) if freer from submission to Nature, then less free from ecological degradation (Deep and Social Ecology), even if in a hi-tech form
    - (2) if freer from technological inconvenience / backwardness, then less free from technological control and the loss of privacy
    - (3) if freer from technological (material) backwardness, then less free from the abusive (barbaric) maltreatment of the primitive Others
  - *—In Relation to the Everyday*
    - (1) if freer from abject poverty, then less free from artificial needs/discontents (Frankfurt School)
    - (2) if freer from sensual suppression, then less free from violent sublimation (Freud)
    - (3) if freer from the snobbishness of high culture, then less free from the shabbiness (leveling-off effect) of mass culture (Tocqueville)
    - (4) if freer from the inefficiency of traditional "compassionate economy," then less free from the bondage of a "ruthless [competitive] economy" (Keynes)
    - (5) if freer from anarchy in the state of nature (system fragmentation), then less free from government regulations and controls in system integration

---

(continued on next page)

Table 4.13. The Freedom-Unfreedom Dialectics (Part II)

• **On Belonging**

—*In Relation to the Good and the Just*

(1) if freer from disciplinary society, then less free from society of control (Foucault)

(2) if freer from the tyranny of one or a few, then less free from the tyranny of the majority (or sometimes, minority veto)

(3) if freer from elitist decision making, then less free from political gridlock/cleavage

(4) if freer from arbitrary (discretionary) administration, then less free from bureaucratic irrationality (Weber) and legal trickery (loopholes)

• **On Being**

—*In Relation to the True*

(1) if freer from unscientific dogmas, then less free from instrumental abyss (nihilism). Or conversely, if freer from meaninglessness, then less free from dogmas.

(2) if freer from the bondage of partiality/partisanship (e.g., prejudice, discrimination), then less free from the danger of impartiality and neutrality (e.g., opportunism, unrealisticness, lack of compassion, inaction)

(3) if freer from making generalizations, then less free from being unable to understand much of anything

—*In Relation to the Holy*

(1) if freer from collective conscience, then less free from social loneliness

(2) if freer from religious absoluteness, then less free from spiritual emptiness

—*In Relation to the Beautiful/Sublime*

(1) if freer from artistic non-autonomy, then less free from aesthetic disillusion (deconstruction)

(continued on next page)

Table 4.13. The Freedom-Unfreedom Dialectics (Part III)

Notes: The examples in each category are solely illustrative (not exhaustive), and the comparison is relative (not absolute), nor are they necessarily mutually exclusive. And some can be easily re-classified elsewhere. As generalities, they allow exceptions.
Sources: A reconstruction from Ch.10 of *FCD*, based on *FHC*

Table 4.14. The Equality-Inequality Dialectics (Part I)

- **On Having**

  —*In Relation to the Technological*

  (1) if more equal in treating Nature with spiritual unity, then less equal in suppressing the dominant drive to transcend it altogether

  —*In Relation to the Everyday*

  (1) if more equal in building social plurality, then less equal in leveling-off effects (e.g., the subsequent relative intolerance of high/intellectual ethos in mass culture industry)

  (2) if more equal in socioeconomic distribution beyond a certain point, then less equal in efficiency (e.g. resentment, the erosion of work ethics)

  (3) if more equal in urging an affirmative action program, then less equal in creating victim mentality (in oneself), stigma (from others), reverse discrimination (against the once privileged), and mediocracy (against the more able)

- **On Belonging**

  —*In Relation to the Good and the Just*

  (1) if more equal in banning monarchic/oligarchic exclusion, then less equal in producing "the tyranny of the majority" or of "minority veto"

  (2) if more equal in encouraging participatory decision making, then less equal in inducing political divisiveness (gridlock/cleavage in power blocs) and organizational oligarchy

  (3) if more equal in institutionalizing a decentralized bureaucracy, then less equal in falling into more territorial/turf politics (intrigues)

(continued on next page)

Table 4.14. The Equality-Inequality Dialectics (Part II)

- **On Being**
  - —In Relation to the Beautiful / Sublime
    - (1) if more equal in accepting diverse styles ("anything goes" mentality), then less equal in artistic good quality (in leveling-off effects against the best)
  - —In Relation to the True
    - (1) if more equal in tolerating multiple viewpoints (no matter how extreme), then less equal in epistemic standards
  - —In Relation to the Holy
    - (1) if more equal in celebrating any cults and sects (no matter how questionable), then less equal in spiritual depth and authenticity

---

Notes: The examples in each category are solely illustrative (not exhaustive), and the comparison is relative (not absolute), nor are they mutually exclusive. And some can be easily reclassified elsewhere. As generalities, they allow exceptions.
Sources: A reconstruction from Ch.10 of *FCD*, based on *FHC*

Table 4.15. The Duality of Oppression in Existential Dialectics: Oppression and Self-Oppression (Part I)

- **From the Same to the Others and Itself**
  —The Oppression by the Same against the Others
  - *By way of downgrading differences*
    –Ex: on judiciary caprice for corporate crimes (*Sec.2.2.1.2.1*)
    –Ex: on the deceptive politics of liberation (*Sec.3.5*)
    –Ex: on the humanitarian mystique (*Sec.4.4*)
    –Ex: on the fad of emotional intelligence (*Sec.5.3*)
  - *By way of accentuating differences*
    –Ex: on the legal sophistry of self-defense (*Sec.2.3*)
    –Ex: on the legal semantics of proportionality (*Sec.2.4*)
    –Ex: on the tricky politics of external threat (*Sec.3.4*)
    –Ex: on the appeal of the Far Right for democracy (*Sec.5.4*)
    –Ex: on the democratic axis of evil (*Sec.5.5*)
    –Ex: on the democratic way of brutality and revenge (*Sec.5.6*)
    –Ex: on democratic autocracy (*Sec.6.4*)
  —The Oppression by the Same against Itself
  - *By way of downgrading differences*
    –Ex: on the politics of fear (*Sec.2.2*)
    –Ex: on the trickery of compassionate conservatism (*Sec.3.2*)
    –Ex: on the deceptive politics of patriotism (*Sec.3.3*)
  - *By way of accentuating differences*
    –Ex: on the caprice of due process on domestic suspects (*Sec.2.2*)
    –Ex: on the false security/freedom dilemma (*Sec.6.5.2*)

(continued on next page)

Table 4.15. The Duality of Oppression in Existential Dialectics: Oppression and Self-Oppression (Part II)

• **From the Others to the Same and Themselves**
—The Oppression by the Others against the Same
• *By way of downgrading differences*
–Ex: on judiciary caprice in the reverse direction (*Sec.2.2.1.2.2*)
–Ex: on equal pay (*Sec.6.2.1.1*)
–Ex: on equal representation (*Sec.6.2.1.2*)
–Ex: on affirmative action program (*Sec.6.3.1.1*)
–Ex: on same-sex marriage (*Sec.6.3.1.2*)
• *By way of accentuating differences*
–Ex: on sexual harassment (*Sec.6.2.2.1*)
–Ex: on physical violence (*Sec.6.2.2.2*)
–Ex: on sexual exploitation (*Sec.6.2.2.3*)
—The Oppression by the Others against Themselves
• *By way of downgrading differences*
–Ex: on the reverse-class mystique (*Sec.4.2*)
–Ex: on the reverse-black mystique (*Sec.4.3*)
–Ex: on self-discrimination by downgrading (*Sec.6.3.2.2*)
• *By way of accentuating differences*
–Ex: on self-discrimination by accentuating (*Sec.6.3.2.1*)

Notes: The examples are solely illustrative (not exhaustive), nor are they mutually exclusive. As generalities, they allow exceptions. Also, both forms of oppression co-exist in all of the examples, so the listing of them are only meant in a relative, not absolute, sense.
Source: A summary of the sections (as cited) in Chs.2-6 of *BDPD*. See text for more info and references.

Table 4.16. The Structure of Existential Dialectics I: The Freedom/Unfreedom and Equality/Inequality Dialectics

- Each freedom and equality produces its own unfreedom and inequality, regardless of whether the pair occurs in political society (with the nation-state), in civil society (with some autonomy from the state), or elsewhere (e.g., in the private sphere of individual homes)—and regardless of whether freedom and equality are understood as "negative" or "positive."

- Oppression is dualistic, as much by the Same against the Others and itself, as by the Others against the Same and themselves.

- Both forms of oppression and self-oppression can be achieved by way of downgrading differences (between the Same and the Others) and of accentuating them.

- The relationships are relatively asymmetric between the Same and the Others and relatively symmetric within them. This is true, even when the Same can be relatively asymmetric towards itself in self-oppression, just as the Others can be likewise towards themselves.

- Symmetry and asymmetry change over time, with ever new players, new causes, and new forms.

Notes: The examples in each category are solely illustrative (not exhaustive) nor necessarily mutually exclusive, and the comparison is relative (not absolute). As generalities, they allow exceptions. "Negative" freedom is freedom "from" (e.g., freedom from poverty), whereas "positive" freedom is freedom "to" (e.g., freedom to the state of enlightenment). "Negative" equality is "procedural" equality (e.g., equality of opportunity), while "positive" equality is "substantive" equality (e.g., equality of outcome). Existential dialectics impose constraints on freedom and equality in democracy, non-democracy, and post-democracy. There is no utopia, in the end; even should there be one, dystopia would exist within it.
Sources: From *Table 1.5* of *BDPD*—and also from *FHC*, *FCD*, and *FPHC*

Table 4.17. The Structure of Existential Dialectics II: The Wealth/Poverty Dialectics

---

• There is no wealth without poverty, just as there is no poverty without wealth.

• The wealth/poverty dialectics occurs in the realms of having, belonging, and being, in relation to the material, relational, and spiritual.

• The wealth/poverty dialectics also expresses itself at the multiple levels of analysis in accordance to methodological holism, be they about the micro-physical, the chemical, the biological, the psychological, the organizational, the institutional, the structural, the systemic, the cultural, and the cosmological.

• The wealth/poverty dialectics is a different manifestation of existential dialectics in general, subject to the principles in its logic of ontology—just as the freedom/unfreedom and equality/inequality dialectics are likewise.

• There is no economic utopia, in the end; even should there be one, dystopia would exist within it.

Notes: The main points here are solely illustrative (not exhaustive) nor necessarily mutually exclusive, and the comparison is relative (not absolute). As generalities, they allow exceptions.
Sources: From *BCPC*. See also *FCD* and *FHC*.

Table 4.18. The Structure of Existential Dialectics III: The Civilization/Barbarity Dialectics

• There is no civilization without barbarity.

• The civilization/barbarity dialectics applies in the four civilizing processes (e.g., the rationalizing process, the pacifying process, the stewardizing process, and the subliming process).

• The civilization/barbarity dialectics is another (different) manifestation of existential dialectics in general, subject to the principles in its logic of ontology—just as the freedom/unfreedom and equality/inequality dialectics and the wealth/poverty dialectics are likewise.

• There is no utopia, in the end; even should there be one, dystopia would exist within it.

Notes: The main points here are solely illustrative (not exhaustive) nor necessarily mutually exclusive, and the comparison is relative (not absolute). As generalities, they allow exceptions.
Sources: From *BCIV*. See also *FCD*, *FHC*, and *BDPD*.

Table 4.19. The Double Sides of Virtual Organizations (Part I)

- **Psychological**
  - —Ex: virtual psychosis
  - —Ex: impersonality and loneliness in quaternary social relations
- **Organizational**
  - —Ex: the race for power and interests
  - —Ex: the world of unequal successes
  - —Ex: the bureaucratic life of its own
  - —Ex: lesser accountability and transparency
- **Economic**
  - —Ex: the economic divides
  - —Ex: the erosive impact of commercialization
- **Political**
  - —Ex: the anti-authoritarian myth
  - —Ex: different power struggles among groups (e.g., Conservative, Reformist, Radical)
- **Structural**
  - —Ex: different localities (e.g., climates, scenery)
  - —Ex: divided domains (e.g., core, peripheral, sub-peripheral)
  - —Ex: substitution effect of social capital
- **Systemic**
  - —Ex: uneven advances in transportation and communications, and geographical migration
- **Cultural**
  - —Ex: conflicting civilizational fabrics (e.g., Confucian, Islamic, Western)
  - —Ex: the bias of the Liberal Democratic agenda
  - —Ex: the addiction to moral fanaticism

(continued on next page)

Table 4.19. The Double Sides of Virtual Organizations (Part II)

Notes: These categories and examples are solely illustrative (not exhaustive), and some of the items can be reclassified somewhere else. Nor are they mutually exclusive. Since they are generalities, exceptions are expected.
Source: From Ch.7 of *FCD*

Table 4.20. Beyond the World of Titans, and the Remaking of the World Order

- **Hyper-Empires**
  —Ex: The Chinese Union
  —Ex: The Indian Union

- **Meso-Empires**
  —Ex: The European Union
  —Ex: The North American Union

- **Micro-Empires**
  —Ex: The Latin American Union
  —Ex: The Middle Eastern Union

- **The Rest of the World**
  —<u>Odd Powers</u>
    - Ex: Japan
    - Ex: Russia

  —<u>The Poor Club</u>
    - Ex: The African Union

  —<u>Ambivalent Regions</u>
    - Ex: Southeast Asia
    - Ex: Oceania
    - Ex: South Asia
    - Ex: Central Asia
    - Ex: Southern/Eastern Europe
    - Ex: North Africa
    - Ex: Central America
    - Ex: Others (e.g., the Korean peninsula)

Source: A summary of Chs.2-5 (of *BWT*)

Table 4.21. The Origins of Authoritarian Liberal Democracy

- **The Geopower of Nature (Ch.4 of *ALD*)**
  —Ex: Power Character and Geographical Strategy
  —Ex: Living Space and Territorial Expansion
  —Ex: Strategic Heartland and Containment

- **The Biopsychology of the Mind (Ch.5 of *ALD*)**
  —Ex: The Bell Curve and Mass Intelligence
  —Ex: Group Analysis and Mass Knowledge
  —Ex: Groupthink and Elite (Mis)calculation

- **The Disciplinary Control of Society (Ch.6 of *ALD*)**
  —Ex: Social Organizations of Ruthlessness
  —Ex: Social Institutions of Greed
  —Ex: Social Structure of Exclusion
  —Ex: Social Systems of Violence

- **The Molding Force of Culture (Ch.7 of *ALD*)**
  —Ex: The Tradition of Conquest
  —Ex: The Rationalization of Unreason

Notes: The examples in each category are solely illustrative (not exhaustive), and they are also relative (not absolute), nor are they mutually exclusive. As generalities, they allow exceptions.
Source: A summary of Chs.4-7 of *ALD*

Table 4.22. The Theory of Post-Democracy I: The Priority of Freedom over Equality (Part I)

• **Differences**

—*For the aggressive Lions (the strong Elitists)*

•Setting up rank distinctions among unequals (e.g., between inferior humans and superior post-humans, or later among inferior post-humans and superior ones, relatively speaking)

•Yearning for being not only distinguished from unequals, but also the first among equals (the best of the very best)

•Soul-searching for a high spiritual culture (not the trashy one for the masses). Mass culture is a dirty joke for them.

—*For the manipulative Foxes (the weak Counter-Elitists)*

•Seeking a gentle hegemony by way of more communitarian concerns (for inferior humans and, later, inferior post-humans)

•Being more sympathetic to less formal-legalistic institutions and values

• **Similarities**

—*For both Lions and Foxes*

•Exploring different spheres of non-human consciousness in the cosmos (something vastly superior than the human one)

•Recognizing the democratic illusions (e.g., no freedom without unfreedom, no equality without inequality, or simply no justice without injustice, and vice versa)

(continued on next page)

Table 4.22. The Theory of Post-Democracy I: The Priority of Freedom over Equality (Part II)

Notes: The two callings and examples in each category are solely illustrative (not exhaustive), since there will be many different post-human value ideals in the distant future of post-human civilization. The comparison is also relative (not absolute) towards post-democracy, so this is not just a version of free-market democracy (nor Fascism/Nazism, as shown in the table later on democracy, non-democracy, and post-democracy). Nor are they mutually exclusive. As generalities, they allow exceptions. And the specific forms of post-human post-democratic ideals need to be further developed in future after-postmodern history, as they will be different from the ones we now know. The point here is to solely give an extremely rough picture of a small part of the world to come that we have never known.
Source: From Ch.10 of *FCD*. Refer to text for more info and references.

Table 4.23. The Theory of Post-Democracy II: The Priority of Equality over Freedom

• **Hybrid Versions of**
—Ex: the Trans-Feminine Calling
—Ex: the Trans-Sinitic Calling
—Ex: the Trans-Islamic Calling
—Ex: the Trans-Outerspace Calling

• **Qualifications**
—These four versions of post-capitalist value ideals need not automatically be post-democratic, just as capitalism does not necessarily mean democracy. They are two different entities—though closely related.
—But up to a certain threshold of elevating equality at the farther expense of freedom, the democratic ideals will be overcome and cease to exist.
—The overcome will not be socialist or communist, but post-democratic with no freedom without unfreedom and no equality without inequality, subject to the constraints of existential dialectics.

Notes: The callings are solely illustrative (not exhaustive), since there will be many different post-human value ideals in the distant future of post-human lifeforms. The comparison is also relative (not absolute), nor are they mutually exclusive. As generalities, they allow exceptions. And the specific forms of post-human post-democratic ideals need to be further developed in future after-postmodern history, as they will be different from the ones we now know. The point here is to solely give an extremely rough picture of a small part of the world to come that we have never known.
Source: From Ch.10 of *FCD*. Refer to text for more info and references.

Table 4.24. The Theory of Post-Democracy III: The Transcendence of Freedom and Equality (Part I)

- **Transcending Freedom in Floating Existence**
  —*Freedom*: seeking an ultimate elimination of the body. Being without the body. The aim is to transcend freedom in the end into a metaphysical state (i.e., beyond the physique).
  —*Unfreedom*: yet facing difficult trade-offs. The sacrifice of bodily existence and its joyfulness. An eternal boredom in floating existence in dark deep space, though with alternative pleasures. There is no free lunch even in the state of transcending freedom.

- **Transcending Equality in the Rivalry of Cosmic Hegemony**
  —*Inequality*: competing to outlast other lifeforms in floating existence, or just marginalizing them for one's hegemonic expansiveness in the rest of the cosmos (and even beyond). Universalism is only for the mediocre.
  —*Equality*: accepting only those of one's rank as equal partners in the vast spacetime for cosmic supremacy. Even here, the aim is to transcend equality into a metaphysical state.

(continued on next page)

Table 4.24. The Theory of Post-Democracy III: The Transcendence of Freedom and Equality (Part II)

Notes: Do not confuse this transcendence of freedom and equality (as one version of post-democracy) with the naïve temptation to transcend the freedom/unfreedom and equality/inequality dialectics. Existential dialectics hold true for freedom and equality in all cultures and societies—past, present, or future (i.e., democracy, non-democracy, and post-democracy), regardless of whether freedom and equality are conventionally understood as "negative" or "positive."
Also, the two features and examples in each are solely illustrative (not exhaustive), since there will be many different post-human value ideals in the distant future of post-human lifeforms. The comparison is also relative (not absolute), nor are they mutually exclusive. As generalities, they allow exceptions. And the specific forms of post-human ideals even for these radically alien floating lifeforms (and others unknown to us) need to be further developed in future after-postmodern history, as they will likely be different from the ones herein illustrated. The point here is to solely give a very rough picture of a small part of the extremely alien world to come that we have never known.
Source: From Ch.10 of *FCD*. Refer to text for more info and references.

Table 4.25. Democracy, Non-Democracy, and Post-Democracy (Part I)

- **Democracy**
  —*Theoretical Constructs*
  - The pursuit of freedom and equality (in various degrees), regardless of whether freedom and equality can be understood as "negative" or "positive"
    (1) more equality than freedom: The relative priority of the good over the right
    (2) more freedom than equality: The relative priority of the right over the good

  —*Types*
  - Only (1): Different versions of communitarian moral universalism
  - Only (2): Different versions of liberal moral universalism
  - (1) or (2): Different versions of anarchic (non-nation-state) moral universalism
  - (1) or (2): Different versions of postmodern moral localism

- **Non-Democracy**
  —*Theoretical Constructs*
  - The focus on (1') equality or (2') freedom, but not both, regardless of whether freedom and equality can be understood as "negative" or "positive"

  —*Types*
  - Only (1'): Different versions on the Far Left (e.g., Stalinism, Robespierrianism)
  - Only (2'): Different versions on the Far Right (e.g., Nazism, absolute monarchism)

(continued on next page)

Table 4.25. Democracy, Non-Democracy, and Post-Democracy (Part II)

- **Post-Democracy**

  —*Theoretical Constructs*

  •The priority of (1'') equality over freedom, or (2'') freedom over equality, or (3'') the transcendence of freedom and equality, regardless of whether freedom and equality are "negative" or "positive." In degree, (1'') or (2'') is less than (1') or (2') but more than (1) or (2)—respectively.

  •Like democracy and non-democracy, post-democracy is also subject to the freedom/unfreedom and equality/inequality dialectics (or existential dialectics in general). Unlike them, post-democracy acknowledges the constraints of existential dialectics and no longer value freedom and equality as sacred virtues. There is no utopia, in the end; even were there one, dystopia would exist within it.

  —*Types*

  •(1''): Different versions of trans-Sinitic value ideals
  •(1''): Different versions of trans-feminine value ideals
  •(1''): Different versions of trans-Islamic value ideals
  •(1''): Different versions of trans-outerspace value ideals
  •(2''): Different versions of post-human elitist value ideals
  •(3''): Different versions of the value ideals of floating consciousness (etc.)

(continued on next page)

Table 4.25. Democracy, Non-Democracy, and Post-Democracy (Part III)

Notes: The examples are solely illustrative (not exhaustive), nor are they mutually exclusive. As generalities, they allow exceptions. "Negative" freedom is freedom "from" (e.g., freedom from poverty), whereas "positive" freedom is freedom "to" (e.g., freedom to the state of enlightenment). "Negative" equality is "procedural" equality (e.g., equality of opportunity), while "positive" equality is "substantive" equality (e.g., equality of outcome). Existential dialectics impose constraints on freedom and equality in democracy, non-democracy, and post-democracy, regardless of whether freedom and equality can be understood as "negative" or "positive" in conventional discourse. Therefore, do not confuse the transcendence of freedom and equality in (3'') with the naïve temptation to transcend existential dialectics. There is no utopia, in the end; even should there be one, it would not exist without dystopia embedded within it.
Sources: A summary, based on my previous works, especially Ch.5 of *FHC*, Chs.5-10 of *FCD*, Chs.2-4 of *FPHC*, and Chs.1 & 7 of *BDPD*. The reader should consult the books for more analysis, as this is only a summary here.

Table 4.26. Multiple Causes of the Emergence of Post-Democracy (Part I)

- **At the Micro-Physical Level**
  —Ex: intelligent life without the human physical-chemical system
  —Sources: Ch.7 of *FHC*; Chs.9-10 of *FCD*; Ch.1 of *FPHC*

- **At the Chemical Level**
  —Ex: space radiation and toxins
  —Sources: Ch.7 of *FHC*; Chs.9-10 of *FCD*

- **At the Bio-Psychological Level**
  —Ex: exo-biological evolution in deep space
  —Ex: genetic engineering of new beings
  —Ex: limits of cognitive partiality
  —Ex: illusions of emotional neutrality
  —Ex: human biological inequality
  —Ex: the rise of unfolding unconsciousness
  —Sources: Ch.2 & Chs.9-10 of *FCD*; Ch.7 of *FHC*; Ch.4 of *BCPC*; *FPHU*

- **At the Institutional Level**
  —Ex: the flawed logic of equality
  —Ex: the conflicting nature of governance
  —Sources: Ch. 5 of *FHC*; Chs. 6 & 10 of *FCD*; Ch. 3 of *FPHC*; Chs.2-5 of *BDPD*

- **At the Organizational Level**
  —Ex: e-civic alienation
  —Ex: the dark sides of formal-legalistic routines
  —Sources: Ch.3 of *FHC*; Ch.7 of *FCD*; Ch.3 of *FPHC*

- **At the Structural Level**
  —Ex: ever new forms of inequities, at home and abroad
  —Ex: the emergence of China, women, and Islam as major actors
  —Sources: Chs.5-6 of *FHC*; Chs.7, 9 & 10 of *FCD*; Chs.4-5 of *BDPD*

(continued on next page)

Table 4.26. Multiple Causes of the Emergence of Post-Democracy (Part II)

- **At the Cultural Level**
  —Ex: freedom/unfreedom dialectics
  —Ex: equality/inequality dialectics
  —Ex: system fragmentation and integration
  —Sources: Ch.5 of *FHC*; Chs. 3, 9 & 10 of *FCD*; Ch.4 of *FPHC*; Ch.1 of *BDPD*; Ch.4 of *BCPC*

- **At the Systemic Level**
  —Ex: space habitats (in zero-gravity) and colonization
  —Ex: ultra advanced future info systems
  —Ex: qualitative demography
  —Sources: Ch.7 of *FHC*; Chs.9 &10 of *FCD*

- **At the Cosmological Level**
  —Ex: the colonization of multiverses
  —Ex: the alteration of space-time and the creation of new matter-energy
  —Ex: the expansion of floating consciousness
  —Ex: the spread of hyper-spatial consciousness
  —Sources: Ch.7 of *FHC*; Chs.9 &10 of *FCD*; Ch.4 of *FPHC*; *FPHST*

Notes: The examples in each category are solely illustrative (not exhaustive), and some of the items can be reclassified somewhere else. Nor are they always mutually exclusive. Since they are generalities, exceptions are expected.
Sources: Especially from *FHC*, *FCD*, *FPHC*, *BCPC*, *BDPD*, *FPHST*, and *FPHU*. See also other books and my perspectives on civilizational holism.

Table 4.27. Some Clarifications about Post-Capitalism and Post-Democracy (Part I)

• The prefix "trans-" in the first category of post-capitalism (with its four versions) refers to something "going beyond" (not "uniting" or "combining"). Ex: *Sec.10.3.3* of *FCD*; *Sec.2.4* & *Sec.4.4* of *FPHC*; *Sec.7.2* of *BCPC*

• Such terms like "post-democracy," "post-capitalism," "post-human elitist," "trans-feminine calling," and the like as used in my works are more for our current intellectual convenience than to the liking of future humans and post-humans, who will surely invent more tasteful neologisms to call their own eras, entities, and everything else, for that matter. But the didactic point here is to use the terms to foretell what the future might be like, not that its eras and entities must be called so exactly and permanently.
Ex: *Sec.11.1* of *FCD*; *Sec.7.2* of *BCPC*

• The four versions in the first category of post-capitalist value ideals need not automatically be post-democratic, just as capitalism does not necessarily mean democracy. They are two different entities—though closely related. But up to a certain threshold of elevating equality at the farther expense of freedom, the democratic ideals will be overcome and cease to exist. The same is true for the post-human elitist calling in the second category of post-capitalism in relation to post-democracy, depending on the extent to which freedom is elevated at the expense of equality.
Ex: *Sec.10.4.3.3* of *FCD*; *Table 3.9* of *FPHC*; *Table 7.6* of *BDPD*

• The comparison in each of the three realms of existence in all forms of post-capitalism is not absolute, but relative. Examples include "communal" vs. "individualistic," and the like.
Ex: Notes in *Table 10.8*, *Table 10.9*, *Table 10.10*, & *Table 10.11* of *FCD*; Chs.2-4 of *FPHC*; *Sec.7.2* of *BCPC*

(continued on next page)

Table 4.27. Some Clarifications about Post-Capitalism and Post-Democracy (Part II)

• The emergence of post-capitalism (and post-democracy, for that matter) has multiple causes (to not be reduced to one or only a few).
Ex: Ch.10 of *FCD*, Chs.2-4 of *FPHC*; *Sec.1.3* & *Sec.7.2* of *BCPC* (or *Table 1.8* & *Table 7.11*)

• The specific forms of post-capitalism (and post-democracy, for that matter) need to be further developed in future after-postmodern history, as they will be different from the ones we now know. The point here is to solely give an extremely rough sketch of a world to come that we have never known.
Ex: *Sec.10.3.3* & *Sec.10.4.3.3* of *FCD*; *Table 10.14* & *Table 10.15* of *FCD*; *Sec.7.2* of *BCPC*

• All forms of post-capitalism are not part of a "teleological law," but of "historical trends" only. The same is also true for all forms of post-democracy.
Ex: *Sec.7.1* of *FHC*; *Sec.9.5.3.2* & *Sec.10.3.4.2* of *FCD*; *Sec.7.2* of *BCPC*

• Post-capitalism is not better than capitalism in an "absolute" sense but only fits in better, on the basis of the historical contingency of culture, society, nature, and the mind in some future eras. The same is true for post-democracy in relation to democracy. The term "better" is historically relative.
Ex: *Sec.10.3.3* of *FCD*; *Sec.1.7* of *BDPD*; *Sec.1.5* of *BCPC*

(continued on next page)

Table 4.27. Some Clarifications about Post-Capitalism and Post-Democracy (Part III)

• All forms of post-capitalism and post-democracy are subject to the constraints of existential dialectics. In the process, the dialectic direction is to go beyond the conventional "either-or" dichotomies (e.g., freedom vs. unfreedom, equality vs. inequality, freedom vs. equality, individuality vs. communality, spirituality vs. materiality, formal legalism vs. informal legalism, etc.). As is true in post-civilization, to go beyond the dichotomies is to acknowledge the co-existence of both in each dichotomy, although the degree of scaling one over the over varies from case to case (e.g., the theory of post-capitalism I, the theory of post-capitalism II, the theory of post-democracy I, the theory of post-democracy II, etc.)—but is not to be extreme in largely favoring one over the other, *on average* (all things considered). There is no utopia to be had in the end; even should there be one, dystopia would exist within it.
Ex: Ch.5 of *FHC*; *Sec.10.4.4.2* of *FCD*; *Sec.1.5* of *BDPD*; *Sec.1.3* of *BCPC*; *BCIV*

• All forms of post-capitalism, however different from each other though they are, share one common feature, in that they all inspire for a higher spiritual culture. The same is also true for post-democracy.
Ex: *Sec.10.3*, *Sec.10.4* & *Sec.10.5* of *FCD*; *Chs.2-4* of *FPHC*; *Sec.7.2* of *BCPC*

• All forms of post-capitalism try to avoid the excess in capitalist consumerism by favoring more basic than artificial needs in having, but the quality and quantity of these "basic" needs will be measured by future standards, not by our current ones. Standards are historically relative.
Ex: *Sec.10.3*, *Sec.10.4* & *Sec.10.5* of *FCD*; Ch.2 of *FPHC*; *Sec.7.2* of *BCPC*

(continued on next page)

Table 4.27. Some Clarifications about Post-Capitalism and Post-Democracy (Part IV)

• All forms of post-capitalism make use of a different degree of political authority with advanced info systems in future history and strives for higher spiritual cultures (especially in the post-human age), while acknowledging the constraints of existential dialectics and no longer valuing free market (as in capitalism) and economic control (as in non-capitalism) as sacred virtues.
Ex: *Sec.10.3.4.2*, *Sec.10.3*, *Sec.10.4* & *Sec.10.5* of *FCD*; Chs.2-4 of *FPHC*; *Sec.1.5* of *BDPD*; *Sec.7.2* of *BCPC*

Notes:: The main points here are solely illustrative (not exhaustive) nor necessarily mutually exclusive, and the comparison is relative (not absolute). As generalities, they allow exceptions. The sections as cited are only illustrative (not exhaustive).
Sources: From *FHC*, *FCD*, *FPHC*, and *BDPD*

Table 4.28. The Theory of Post-Capitalism I.1: By Group— Ex: Spiritual/Communal in the Trans-Feminine Calling

- **More Communal Than Individual**
  —*Sharing*: learning from others, as different ideas mutually enrich
  —*Cooperative*: encouraging a sense of shared leadership and teamwork

- **More Informal-Legalistic Than Formal-Legalistic**
  —*Specific*: listening more from the heart than from the head, to know a person as a concrete, not as an abstract, unit
  —*Affective*: thinking and acting with others on a more affective tone. Business can mix with an emotional touch.
  —*Ascriptive*: hiring (or firing) can be done on the basis of merit (or lack of it), but deep solidarity (sisterhood) is important too.
  —*Particularistic*: making decisions on the basis of cost-benefit analysis, but a given group relationship is vital

- **More Spiritual Than Secular**
  —*Long-Term Looking*: sharing for a long-term relationship (e.g., love, friendship), not just for a short-term gain
  —*Loving/Caring*: showing compassion for the sufferings of others, without quickly blaming and pre-judging
  —*Respectful*: showing acceptance about others' feelings (and thoughts)

Notes: The categories and examples are solely illustrative, since there can be different versions, and the comparison is relative (not absolute), nor are they mutually exclusive. As generalities, they allow exceptions. The specific forms of the trans-feminine version need to be further developed in future after-postmodern history, as they will be different from the ones we now know, since the prefix "trans-" here means going beyond or deconstructing the feminine values, while using them as the inspirational point at the beginning.
Source: From Ch.10 of *FCD*. Refer to text for more info and references.

Table 4.29. The Theory of Post-Capitalism I.2: By Nation-State— Ex: Spiritual/Communal in the Trans-Sinitic Calling

- **More Communal Than Individualistic**
  —*Centralized*: being more top-down in management
  —*Collective*: encouraging more group cooperation
  —*Social*: investing in trust and connection

- **More Informal-Legalistic Than Formal-Legalistic**
  —*Specific*: knowing more of those related or connected
  —*Affective*: behaving in a paternalistic, hierarchical way
  —*Ascriptive*: favoring family members and those related
  —*Particularistic*: building connection (guanxi) as imperative

- **More Spiritual Than Secular**
  —*Expansionist*: diffusing civilizational values (e.g., the superiority complex of civilizationalism)
  —*Holistic*: synthesizing things into a panoramic horizon
  —*Historical*: learning from the lessons of the ancient past
  —*Respectful*: deferential to elders and superiors

Notes: The categories and examples are solely illustrative, since there can be different versions, and the comparison is relative (not absolute), nor are they mutually exclusive. As generalities, they allow exceptions. The specific forms of the trans-Sinitic version need to be further developed in future after-postmodern history, as they will be different from the ones we now know, since the prefix "trans-" here means going beyond or deconstructing the Sinitic values, while using them as the inspirational point at the beginning.
Source: From Ch.10 of *FCD*. Refer to text for more info and references.

Table 4.30. The Theory of Post-Capitalism I.3: By Region— Ex: Spiritual/Communal in the Trans-Islamic Calling

- **More Communal Than Individualistic**
  —*Collective*: building the webs of relationships to bind individuals
  —*Sharing*: cultivating the established "wisdom" through common experience
  —*Cooperative*: stressing harmony, solidarity, and commonality

- **More Informal-Legalistic Than Formal-Legalistic**
  —*Specific*: making efforts to know well the participants (family and larger community) in matters of common concern
  —*Affective*: mixing work with language and ritual on explicit religious (Islamic) ideals, texts, stories, and examples
  —*Ascriptive*: privileging local history and custom on relationships among kinship groups
  —*Particularistic*: preferring an unbiased insider with ongoing connections to all parties

- **More Spiritual Than Secular**
  —*Historical*: learning from the lessons of the past as a source of stability and guidance
  —*Deferential*: showing respect for age, experience, status, and leadership in communal affairs
  —*Honorable*: emphasizing face, dignity, prestige, and fairness
  —*Compassionate*: giving mercy and charity ("Zahah") to others

Notes: The categories and examples are solely illustrative (not exhaustive), and the comparison is relative (not absolute), nor are they mutually exclusive. As generalities, they allow exceptions. The specific forms of the trans-Islamic version need to be further developed in future after-postmodern history, as they will be different from the ones we now know, since the prefix "trans-" here means going beyond or deconstructing the Islamic values, while using them as the inspirational point at the beginning.
Sources: From Ch.10 of *FCD*. Refer to text for more info and references, especially from the works by George Irani (2000) and C. Murphy (September 19, 2001).

Table 4.31. The Theory of Post-Capitalism I.4: By Universe— Ex: Spiritual/Communal in the Trans-Outerspace Calling

- **More Communal Than Individual**
  —*Cooperative*: requiring teamwork in small space habitats
  —*Sharing*: learning from, and enjoying being with, each other in a small group in outer space

- **More Informal-Legalistic Than Formal-Legalistic**
  —*Specific*: knowing more about each other to facilitate living and working together in space, both as fellow astronauts and space-mates
  —*Affective*: being friendly and social to each other as vital to working and living in small space quarters
  —*Ascriptive*: nurturing comaraderie among fellow astronauts as if they are family members over time
  —*Particularistic*: building work relationship with enduring memory in a space mission

- **More Spiritual Than Secular**
  —*Long-Term*: looking beyond selfish materialistic concerns in a precarious space environment with potential life or death
  —*Loving/Caring*: cultivating deep bondage for the success of a long term space mission
  —*Transcendent*: searching for life meaning in outer space

Notes: The calling and examples in each category are solely illustrative (not exhaustive), since there will be many different outer-space value ideals in the distant future of space colonization. The comparison is also relative (not absolute), nor are they mutually exclusive. As generalities, they allow exceptions. And the specific forms of trans-outer-space calling need to be further developed in future after-postmodern history, as they will be different from the ones we now know, since the prefix "trans-" here means going beyond or deconstructing the current outer-space values, while using them as the inspirational point at the beginning. The point here is to solely give an extremely rough picture of a small part of the world to come that we still do not know much about.
Source: From Ch.10 of *FCD*. Refer to text for more info and references.

Table 4.32. The Theory of Post-Capitalism II: Spiritual/Individualistic in the Post-Human Elitist Calling (Part I)

- **More Individualistic Than Communal**
  —Setting up rank distinctions among unequals (e.g., between inferior humans and superior post-humans, or later among inferior post-humans and superior ones, relatively speaking)
  —Yearning for being not only distinguished from unequals, but also the first among equals (the best of the very best)
  —Recognizing the constraints of equality/inequality dialectics (or existential dialectics in general)

- **More Spiritual Than Secular**
  —Soul-searching for a high spiritual culture (not the trashy one for the masses). Mass culture is a dirty joke for them.
  —Exploring different spheres of non-human consciousness in the cosmos (something vastly superior than the human one)
  —Recognizing the constraints of freedom/unfreedom dialectics (or existential dialectics in general)

- **Qualifications**
  —Although post-human elitist post-democracy is comparable to post-human elitist post-capitalism in some respects, the former does not necessarily imply the latter (post-human elitist post-capitalism), just as democracy does not have to entail capitalism. They are two different (though related) entities.
  —But up to a certain threshold of incorporating government intervention with advanced info systems in future civilizations for higher spiritual concerns at the expense of the free market and materialist pursuit, the capitalist ideal will be overcome.
  —The overcome will not be Fascist or feudalistic, but post-capitalist, subject to the constraints of existential dialectics.

(continued on next page)

Table 4.32. The Theory of Post-Capitalism II: Spiritual/Individualistic in the Post-Human Elitist Calling (Part II)

Notes: The calling and examples in each category are solely illustrative (not exhaustive). The comparison is also relative (not absolute), nor are they mutually exclusive. As generalities, they allow exceptions. And the specific forms of post-human elitist post-capitalism need to be further developed in future after-postmodern history, as they will be different from the ones we now know, while using them as the inspirational point at the beginning. The point here is to solely give an extremely rough picture of a small part of the world to come that we still do not know much about.
Sources: From Ch.10 of *FCD* (and also *FPHC*, *BDPD*, and *BCPC*). Refer to the text for more info and references.

Table 4.33. Capitalism, Non-Capitalism, and Post-Capitalism (Part I)

- **Capitalism**
  - —*Theoretical Constructs*
    - •Allocation of scarce resources among alternative wants largely by free market for competition (whose characteristics in its ideal form include, for instance, no barrier to entry or exit, homogeneity, perfect information, a large number of buyers/sellers, and perfect factor mobility)
    - •More formal-legalistic than informal-legalistic, more individualistic than communal, and more material (secular) than spiritual
    - •Either (1) minimal government or (2) relatively active government
  - —*Types*
    - •Only (1): Different versions of market capitalism (e.g., the U.S.)
    - •Only (2): Different versions of welfare capitalism (e.g., Sweden)

- **Non-Capitalism**
  - —*Theoretical Constructs*
    - •Allocation of scarce resources among alternative wants mainly by political authority for policies (which can be regulative, redistributive, symbolic, and participatory)
    - •More informal-legalistic than formal-legalistic
    - •Either (1') more individualistic (for the elites), often (though not always) for material (secular) concerns, or (2') more communal (for the masses), often (though not always) for spiritual concerns
  - —*Types*
    - •Only (1'): Different versions on the Right (e.g., Fascist corporate-state economy for the glory of the new Rome, medieval lord-vassal-serf economy for the power of the feudalistic order)
    - •Only (2'): Different versions on the Left (e.g., Soviet command economy for the creation of the New Socialist Man)

(continued on next page)

Table 4.33. Capitalism, Non-Capitalism, and Post-Capitalism (Part II)

• **Post-Capitalism**

—*Theoretical Constructs*

•Allocation of scarce resources among alternative wants largely by political authority with advanced info systems in future civilizations, subject to existential dialectics. In degree of allocating by authority, post-capitalism is more than capitalism but less than non-capitalism.

•More spiritual than secular (material)

•Either (1'') more individualistic or (2") more communal

•Like capitalism and non-capitalism, post-capitalism is also subject to the freedom/unfreedom and equality/inequality dialectics (or existential dialectics in general). There is no utopia, in the end; even were there one, dystopia would exist within it.

•Unlike capitalism and non-capitalism, post-capitalism makes use of a different degree of political authority with advanced info systems in future civilizations and strives for higher-spiritual cultures (especially in the post-human age), while acknowledging the constraints of existential dialectics and no longer valuing free market (as in capitalism) and economic control (as in non-capitalism) as sacred virtues.

—*Types*

•Only (1"): Different versions of post-human elitist value ideals

•Only (2''): Different versions of trans-Sinitic value ideals

•Only (2''): Different versions of trans-feminine value ideals

•Only (2''): Different versions of trans-Islamic value ideals

•Only (2''): Different versions of trans-outerspace value ideals

---

(continued on next page)

Table 4.33. Capitalism, Non-Capitalism, and Post-Capitalism (Part III)

Notes: The calling and examples in each category are solely illustrative (not exhaustive). The comparison is also relative (not absolute), nor are they mutually exclusive. As generalities, they allow exceptions. And the specific forms of each calling need to be further developed in future after-postmodern history, as they will be different from the ones we now know, while using them as the inspirational point at the beginning. The point here is to solely give an extremely rough picture of a small part of the world to come that we still do not know much about.
Source: From Ch.10 of *FCD*. Refer to the text for more info and references.

Table 4.34. Multiple Causes of the Emergence of Post-Capitalism (Part I)

• **At the Micro-Physical Level**
—Ex: intelligent life without the human physical-chemical system
—Ex: mastering of quantum mechanics, electromagnetism, and other fields for the understanding of a broad range of anomalous experiences and the application for artificial intelligence for spiritual quest
—Sources: Ch.7 of *FHC*; Chs.9-10 of *FCD*; Ch.1 of *FPHC*

• **At the Chemical Level**
—Ex: space radiation and toxins
—Sources: Ch.7 of *FHC*; Chs.9-10 of *FCD*

• **At the Bio-Psychological Level**
—Ex: exo-biological evolution in deep space
—Ex: genetic engineering of new beings
—Ex: limits of human cognition
—Ex: the rise of unfolding unconsciousness
—Sources: Ch.2 & Chs.9-10 of *FCD*; Ch.7 of *FHC*; *FPHU*

• **At the Institutional Level**
—Ex: the flawed logic of the free market
—Ex: the need of a post-autistic economics
—Sources: Ch.10 of *FCD*

• **At the Organizational Level**
—Ex: the dark sides of formal-legalistic routines
—Sources: Ch.3 of *FHC*; Ch.7 of *FCD*; Ch.3 of *FPHC*

• **At the Structural Level**
—Ex: ever new forms of inequities, at home and abroad
—Ex: the emergence of China, women, and Islam as major actors
—Sources: Chs.5-6 of *FHC*; Chs.7, 9 & 10 of *FCD*; Chs.4-5 of *BDPD*

(continued on next page)

Table 4.34. Multiple Causes of the Emergence of Post-Capitalism (Part II)

- **At the Cultural Level**
  —Ex: freedom/unfreedom dialectics
  —Ex: equality/inequality dialectics
  —Sources: Ch.5 of *FHC*; Chs.3 & 10 of *FCD*; Ch.4 of *FPHC*; Ch.1 of *BDPD*

- **At the Systemic Level**
  —Ex: space habitats (in zero-gravity) and colonization
  —Ex: ultra advanced future info systems
  —Ex: qualitative demography
  —Sources: Ch.7 of *FHC*; Chs. 9 & 10 of *FCD*

- **At the Cosmological Level**
  —Ex: the colonization of multiverses
  —Ex: the alteration of space-time and the creation of new matter-energy
  —Ex: the expansion of floating consciousness
  —Ex: the spread of hyper-spatial consciousness
  —Sources: Ch.7 of *FHC*; Chs. 9 & 10 of *FCD*; Ch.4 of *FPHC*; *FPHST*

Notes: The examples in each category are solely illustrative (not exhaustive), and some of the items can be reclassified somewhere else. Nor are they always mutually exclusive. Since they are generalities, exceptions are expected.
Sources: Especially from *FHC*, *FCD*, *FPHC*, *BCPC*, *BDPD*, *FPHST*, and *FPHU*. See also other books and my perspectives on civilizational holism.

Table 4.35. The Theoretical Debate on Civilization

- **The Progressive Theory of Civilization**
  —*Thesis*: The "civilizing" process is "good," as opposed to the "barbarizing" process as something "bad," relatively speaking.
  —*Discourse*: Especially, though not exclusively, in the Enlightenment era and a bit before. Example: Thomas Hobbes—in that the tribes in primitive societies were "savages."

- **The Romantic Theory of Barbarity**
  —*Thesis*: The "civilizing" process is "bad," as opposed to the "barbarizing" process as something "good," relatively speaking.
  —*Discourse*: Especially, though not exclusively, in the Counter-Enlightenment circle. Example: Jean-Jacques Rousseau—in that civilization "corrupts" men, and the "savages" are in fact "noble."

- **The Moderate Theory of Civilization**
  —*Thesis*: The "civilizing" process is "good," but there is a price to pay, especially in systematic (compulsive) self-control.
  —*Discourse*: Especially, though not exclusively, in some late modern and postmodern circles. Example: Norbert Elias—in that social manners become more refined in the civilizing process, but self-control also becomes more systematic.

- **The Theory of Post-Civilization**
  —*Thesis*: The civilizing process is as evil and good as barbarity, and each cannot exist without the other, to be eventually superseded by post-civilization unto the post-human age.
  —*Discourse*: Proposed by Peter Baofu. See the rest of *BCIV* for more analysis.

Source: From *BCIV* on the theoretical debate

Table 4.36. No Freedom Without Unfreedom in the Civilizing Processes (Part I)

- **The Rationalizing Process (at the Level of Culture)**
  —if freer from the dominance of unreason (as in barbarism) in the civilizing process, then less free from the rationalizing process (be it in the form of the principle of either transcendence or immanence)
  —if freer from the principle of immanence in the rationalizing process, then less free from the inclination to commit terror in the name of reason and the relative underdevelopment of non-reason (e.g., in relation to yoga and meditation)
  —if freer from the principle of transcendence in the rationalizing process, then less free from the relative underdevelopment of reason (e.g., in relation to systematic methodology) and the occurrence of oppression in the name of non-reason

- **The Pacifying Process (at the Level of Society)**
  —if freer from the dominance of pillage (as in savagery) in the civilizing process, then less free from the pacifying process (be it in the form of external control or self-control)
  —if freer from self-control in the pacifying process, then less free from the temptation of expansionist oppression and rebellious mindset in external control
  —if freer from external control in the pacifying process, then less free from the gruesome psychological self-torture and conformism in self-control

- **The Stewardizing Process (at the Level of Nature)**
  —if freer from the dominance of nature (as in the state of nature) in the civilizing process, then less free from the stewardizing process (be it in the form of the stewardship of creation or the covenant with nature)
  —if freer from the stewardship of creation in the stewardizing process, then less free from material underdevelopment, relatively speaking, and spiritual exclusion in the covenant with nature
  —if freer from the covenant with nature in the stewardizing process, then less free from ecological degradation and spiritual disconnection from nature in the stewardship of creation

---

(continued on next page)

Table 4.36. No Freedom Without Unfreedom in the Civilizing Processes (Part II)

- **The Subliming Process (at the Level of the Mind)**
  —if freer from the dominance of spontaneity (as in the wild state of the mind) in the civilizing process, then less free from the subliming process, be it in the form of (cyclical-centric) self-refinement or (linear-centric) self-discipline
  —if freer from (cyclical-centric) self-refinement in the subliming process, then less free from the (linear-centric) self-regimen (as a form of neurosis)
  —if freer from (linear-centric) self-discipline in the subliming process, then less free from the (cyclical-centric) self-torture (equally as a form of neurosis)

Notes: The examples in each category are solely illustrative (not exhaustive), and the comparison is relative (not absolute), nor are they necessarily mutually exclusive. And some can be easily re-classified elsewhere. As generalities, they allow exceptions.
Sources: From *BCIV*. See also *FHC*, *FCD*, *FPHC*, *BDPD*, and *BCPC*.

Table 4.37. No Equality Without Inequality in the Civilizing Processes (Part I)

- **The Rationalizing Process (at the Level of Culture)**
  —if more equal for the role of rationalization in the rationalizing process (of civilizational making), then less equal for that of mythicization (as in barbarism)
  —if more equal for the principle of transcendence in (linear-centric) rationalizing process, then less equal for the principle of immanence
  —if more equal for the principle of immanence in (cyclical-centric) rationalizing process, then less equal for the principle of transcendence

- **The Pacifying Process (at the Level of Society)**
  —if more equal for pacification in civilizational making, then less equal for the institution of pillaging and others (as in savagery)
  —if more equal for external control, relatively speaking, in pacifying process, then less equal for self-control
  —if more equal for self-control, relatively speaking, in pacifying process, then less equal for external-control

- **The Stewardizing Process (at the Level of Nature)**
  —if more equal for stewardship in the stewardizing process (of civilizational making), then less equal for reverent (submissive) existence (as in barbarism)
  —if more equal for the stewardship of creation in (linear- centric) stewardizing process, then less equal for the (cyclical-centric) covenant with nature for harmonious co-existence
  —if more equal for the (cyclical-centric) covenant with nature in the stewardizing process, then less equal for the (linear-centric) stewardship of nature for domination

---

(continued on next page)

Table 4.37. No Equality Without Inequality in the Civilizing Processes (Part II)

- **The Subliming Process (at the Level of the Mind)**
  —if more equal for the role of reason in the subliming process, then less equal for that of unreason (as in the natural state of wildness)
  —if more equal for the primacy of reason in (linear-centric) subliming process, then less equal for other faculties (e.g., intuition, existential feelings, and analogous thinking) in cyclical-centric one
  —if more equal for the exercise of other faculties (e.g., intuition, existential feelings, and analogous thinking) in cyclical-centric subliming process, then less equal for the role of reason in linear-centric counterpart

Notes: The examples in each category are solely illustrative (not exhaustive), and the comparison is relative (not absolute), nor are they mutually exclusive. And some can be easily reclassified else-where. As generalities, they allow exceptions.
Sources: From *BCIV*. See also *FHC*, *FCD*, *FPHC*, *BDPD*, and *BCPC*.

Table 4.38. Five Theses on Post-Civilization

• Post-civilization no longer treats civilization as good and barbarity as evil (relatively speaking), nor does it nostalgically regard barbarity as good and civilization as evil (relatively speaking again). Civilization is as evil and good as barbarity.

• Post-civilization also no longer accepts the dichotomy between civilization and barbarity. Civilization cannot exist without barbarity. It is no longer necessary to preserve civilization, any more than it is imperative to destroy barbarity. To go beyond civilization and barbarity is to acknowledge the co-existence of both, although the degree of scaling one over the over varies from case to case—but is not to be extreme in largely favoring one over the other, *on average* (subject to the constraints of existential dialectics).

• Post-civilization is thus subject to the constraints of existential dialectics. There is no freedom without unfreedom, and no equality without inequality, for instance. There will be no utopia; even should there be one, there would be dystopia embedded within it.

• Post-civilization will eventually replace civilization (as a form of life settlement), to be dominated by post-capitalist and post-democratic lifeforms here on earth and in deep space (besides other alien lifeforms that we have never known), unto the post-human age in multiverses. Those few post-humans who keep civilization will live in a "post-human civilization," while the rest (the majority), who choose post-civilization, will evolve towards the state of "post-human post-civilization." One therefore should not confuse "post-human civilization" with "post-human post-civilization," as the two are not the same.

• Post-civilization will confront psychosis as a primary problem in the culture of virtuality unto the post-human age, just as civilization has neurosis as a primary one of its own (although both neurosis and psychosis are major problems in both).

Notes: The comparison in each category is relative (not absolute), nor are they necessarily mutually exclusive. And some can be easily re-classified elsewhere. As generalities, they allow exceptions.
Sources: From *BCIV*. See also *FHC*, *FCD*, *FPHC*, *BDPD*, and *BCPC*.

Table 4.39. Barbarity, Civilization, and Post-Civilization

- **The Rationalizing Process (at the Level of Culture)**
  —*Barbarity*
    - More mythicizing than rationalizing, relatively speaking

  —*Civilization*
    - More rationalizing than mythicizing, relatively speaking

  —*Post-Civilization*
    - Beyond the dichotomy, subject to existential dialectics

- **The Pacifying Process (at the Level of Society)**
  —*Barbarity*
    - More pillaging than pacifying, relatively speaking

  —*Civilization*
    - More pacifying than pillaging, relatively speaking

  —*Post-Civilization*
    - Beyond the dichotomy, subject to existential dialectics

- **The Stewardizing Process (at the Level of Nature)**
  —*Barbarity*
    - More revering than stewardizing, relatively speaking

  —*Civilization*
    - More stewardizing than revering, relatively speaking

  —*Post-Civilization*
    - Beyond the dichotomy, subject to existential dialectics

- **The Subliming Process (at the Level of the Mind)**
  —*Barbarity*
    - More impulsing than subliming, relatively speaking

  —*Civilization*
    - More subliming than impulsing, relatively speaking

  —*Post-Civilization*
    - Beyond the dichotomy, subject to existential dialectics

Notes: The comparison in each category is relative (not absolute), nor are they necessarily mutually exclusive. And some can be easily re-classified elsewhere. As generalities, they allow exceptions.
Sources: From *BCIV*. See also *FHC*, *FCD*, *FPHC*, *BDPD*, and *BCPC*.

Table 4.40. Types of Super Civilization in the Cosmos (Part I)

- **Type I**
  —a civilization which gains control of and uses the total energy output "falling on its planet from its sun for interstellar communication" (or, in general, space colonization). For N. Kardashev, who proposed the first three types, human civilization is currently Type Zero (Type O), which is below even Type I, since its present energy consumption for all purposes, let alone for interstellar communication, is still 10,000 times less.

- **Type II**
  —a civilization which gains control of and uses directly the total energy output of its sun for interstellar communication (or, in general, space colonization).

- **Type III**
  —a civilization which gains control of and uses the total energy output of its galaxy for interstellar communication (or, in general, space colonization).

- **Type IV**
  —a civilization which gains control of and uses the total energy output of its cluster of galaxies for interstellar communication (or, in general, space colonization).

- **Type V**
  —a civilization which gains control of and uses the total energy output of its supercluster of galaxies for interstellar communica-tion (or, in general, space colonization).

(continued on next page)

Table 4.40. Types of Super Civilization in the Cosmos (Part II)

- **Type…n**
  —So continues the series in what I call the cyclical progression of hegemony in the cosmos and beyond.

Notes: The Russian astrophysicist Nikolai Kardashev proposed the first three types of super civilization in terms of total energy out-put for interstellar communication. (CSM 1979) I extend his argument further to propose Type IV, Type V, Type VI, and Type…n, in the context of my claim about the cyclical progression of he-gemony in the cosmos and beyond.
Sources: From *Table 9.4* of *FCD*. See *FHC*, *FCD*, and *FPHC* for more info.

Table 4.41. The Civilizational Project from Pre-Modernity to After-Postmodernity (Part I)

	Pre-Modern	*Modern*	*Postmodern*	*After-Postmodern*
Main narratives	•Sacralness •Courtliness •Vitalism •Animism	•Freedom •Equality •Fraternity	•Multiplicity •Hybridization	•Naked contingency •Cyclical progression of hegemony
Main institutions	•Monarchy •Aristocracy •Feudalism •Holy order •Primitivism	•Capitalism •Liberalism •Socialism •Nazism •Fascism	•Capitalism •Liberalism •Postmodern politics of difference	•Post-Capitalism •Post-Democracy •Others
Main technological and economic revolutions	•Agricultural	•Service •Industrial	•Informational	•Biological •Material •Energy •Space •Others

(continued on next page)

Table 4.41. The Civilizational Project from Pre-Modernity to After-Postmodernity (Part II)

	Pre-Modern	*Modern*	*Postmodern*	*After-Postmodern*
Main agents	•Males •Upper strata •Mini-states	•Males •Upper strata •Whites •Empires	•Males •Upper strata •Whites •Others •Supra-states •IO's	•Post-humans •Humans •Others
Main impacts	•Local	•Inter-national	•Global	•Outer-space •Multiverse
Main outcomes	•Towards moderntiy •Rise of linear- & cyclical-centric civiliza-tions	•Towards post-modernity •Dominance of linear-centric civilization	•Towards after-post-moderntiy •Linear-centric civiliza-tion in crisis	•Towards human (& maybe post-human) extinction •Rise of post-civiliza-tion, especially in post-human forms of space-time

Notes: The examples in each category are solely illustrative (not exhaustive) nor necessarily mutually exclusive, and the comparison is relative (not absolute). As generalities, they allow exceptions.

Sources: From *Table 10.16* of *FCD*—and also from *BCIV* on post-civilization (and *FPHST*)

Table 4.42. Civilizational Holism
(Part I)

- **At the Micro-Physical Theoretical Level**
  —Ex: Mastering of quantum mechanics, electromagnetism, and other fields for the understanding of a broad range of anomalous experiences and theapplication for artificial intelligence (*Sec.1.4.1* of *FPHC*)

- **At the Chemical Theoretical Level**
  —Ex: Unprecedented expansion of (and violence to) the mind through ever new forms of drugs (and virtual technologies, for that matter) (Ch.9 of *FCD*)

- **At the Biological Theoretical Level**
  —Ex: Humans are not biologically equal, on the basis of race, gender, ethnicity, age, and whatnot. (*Sec.2.6* & Ch.10 of *FCD*; *BNN*) And post-humans will experience the same fate, in an even more amazing way.

- **At the Psychological Theoretical Level**
  —Ex: Human cognitive impartiality and emotional neutrality are quite limited. (*Secs.2.4-2.5* of *FCD*)
  —Ex: Rise of Floating Consciousness (Ch.10 of *FCD*; Chs.1 & 4 of *FPHC*) and Unfolding Unconsciousness (*FPHU*)

- **At the Organizational Theoretical Level**
  —Ex: Administrative colonization of deep space, with less legal-formalism in some corners. (Chs.9-10 of *FCD*)

- **At the Institutional Theoretical Level**
  —Ex: Both capitalism and democracy will not last, to be superseded by different versions of post-capitalism and post-democracy in after-postmodernity. (Ch.10 of *FCD*)

(continued on next page)

Table 4.42. Civilizational Holism (Part II)

- **At the Structural Theoretical Level**
  —Ex: Social stratification reappears in ever new forms, also with new causes and new players in the cyclical progression of hegemony. (Chs.8-10 of *FCD*)
  —Ex: The world of hyper-empires, and the union of the unions (*BWT*)

- **At the Systemic Theoretical Level**
  —Ex: Outerspace expansion: local → regional → global → solar → galactic → clustery → multiversal (Ch.9 of *FCD*)
  —Ex: Demographic transition: human extinction, and the rise of post-humans (e.g., cyborgs, thinking machines, thinking robots, genetically altered superior beings, floating consciousness, hyper-spatial consciousness) (Ch.4 of *FPHC*; Ch.10 of *FCD*; & Ch.7 of *FHC*)
  —Ex: New technological forces in material sciences, electronic and communication sciences, energy sciences, biosciences, manufacturing and engineering sciences, and space sciences (Ch.10 of *FCD* & Ch.7 of *FHC*)
  —Ex: Systematic dominance towards nature for space colonization (Chs.9-10 of *FCD*; Chs.2 & 7 of *FHC*)

- **At the Cultural Theoretical Level**
  —Ex: The post-human transcendence of freedom and equality (Ch.10 of *FCD*)
  —Ex: Methodological Holism (Ch.1 of *FCD*; Ch.1 of *FPHC*; *Sec.2.1* & *Sec.2.5* of *BCPC*)
  —Ex: The Evolution from Barbarity to Post-Civilization (*BCIV*)

- **At the Cosmological Theoretical Level**
  —Ex: Mastering of dark matter and dark energy, and the exploration of multiverses (Ch.4 of *FPHC*; Ch.10 of *FCD*; & Ch.7 of *FHC*)
  —Ex: Alternation of space-time (*FPHST*)
  —Ex: The emergence of hyper-spatial consciousness (*FPHC*)

(continued on next page)

Table 4.42. Civilizational Holism (Part III)

- **At Other Levels**
  —Ex: Historical: pre-modernity → modernity → postmodernity → after-postmodernity (human distinction, and the rise of post-humans, including floating consciousness, hyper-spatial consciousness, and unfolding unconsciousness) (Ch.7 of *FHC*; Ch.10 of *FCD*; *FPHC*; *FPHU*)

Notes: These examples are solely illustrative (not exhaustive), and some of the items can be reclassified somewhere else. Nor are they always mutually exclusive. Since they are generalities, exceptions are expected. And the comparison is relative, not absolute.
Sources: From *Table 5.1* of *FPHC*—with details from *FHC*, *FCD*, and the rest of my books. Not every aspect in each category as discussed in all my books are presented here, since there are too many issues. For more info, also consult the table on theories on civilizational holism and, of course, the books themselves.

Table 4.43. Theories on Civilizational Holism (Part I)

I. Theories in Relation to Nature

—*At the Chemical Theoretical Level*

- 53. Creational Theory of Chemistry (Peter Baofu) (*FPHCHEM*)

—*At the Macro-Physical (Cosmological) Theoretical Level*

- 52. Multilateral Theory of Acoustics (Peter Baofu) (*FPHA*)
- 51. Resettlement Theory of Geology (Peter Baofu) (*FPHGEOL*)
- 50. Theory of Post-Cosmology (Peter Baofu) (*BCOS*)
- 49. Theory of Hyper-Spatial Consciousness (Peter Baofu) (Ch.4 of *FPHC*; *FPHG*)
- (• 47). Selective Theory of Geometry (Peter Baofu) (*FPHG*)
- (• 46). Perspectival Theory of Space-Time (Peter Baofu) (*FPHST*)
- (• 45). Dialectic Theory of Complexity (Peter Baofu) (*FC*)
- (• 26). Theory of Floating Consciousness (Peter Baofu) (Ch.10 of *FCD*; Chs.1 & 4 of *FPHC*)
- 48. Theory of the Geopower of Nature (Peter Baofu) (Ch.4 of *ALD*)

—*At the Micro-Physical Theoretical Leve*

- 47. Selective Theory of Geometry (Peter Baofu) (*FPHG*)
- 46. Perspectival Theory of Space-Time (Peter Baofu) (*FPHST*)
- 45. Dialectic Theory of Complexity (Peter Baofu) (*FC*)

(continued on next page)

Table 4.43. Theories on Civilizational Holism (Part II)

II. Theories in Relation to Culture
—At the Cultural Theoretical Level

- • 44. Comparative-Impartial Theory of Literature (Peter Baofu) (*FPHLIT*)
- • 43. Mediative-Variative Theory of Chess (Peter Baofu) (*FPHCHESS*)
- • 42. Theory of Post-Ethics (Peter Baofu) (*BEPE*)
- • 41. Dualistic Theory of Mass Culture (Peter Baofu) (Ch.2 of *FHC*)
- • 40. Comparative Theory of Religion—also known as the Comparative-Substitutive Theory of Religion (Peter Baofu) (Ch.3 of *FHC*; Ch.9 of *FCD*; Ch.1 of *FPHK*; *FPHR*)
- • 39. Theory of Post-Civilization (Peter Baofu) (*BCIV*)
- • 38. Theory of the Trinity of Modernity to Its After-Postmodern Counterpart (Peter Baofu) (*FHC*; Ch.10 of *FCD*)
- • 37. Transformative Theory of Aesthetic Experience (Peter Baofu) (*FAE*)
- (• 17). Theory of Post-Capitalism (Peter Baofu) (Ch.10 of *FCD*; Chs.2 & 4 of *FPHC*; *BCPC*)
- (• 16). Theory of Post-Democracy (Peter Baofu) (Ch.10 of *FCD*; Chs.3 & 4 of *FPHC*; *BDPD*)
- (• 5). Theory of Existential Dialectics, or the Holistic Theory of Knowledge (Peter Baofu) (*FHC*; *FCD*; *FPHC*; *BDPD*; *FC*; *FAE*; *ALD; FIA; FPHK*; etc,)
- (• 4). Contrastive Theory of Rationality (Peter Baofu) (*FPHML*)
- (• 2). Theory of Methodological Holism (Peter Baofu) (Ch.1 of *FCD*; Ch.1of *FPHC*; *Sec.2.1* & *Sec.2.5* of *BCPC*; *FC*; *FPHK*; etc.)

(continued on next page)

Table 4.43. Theories on Civilizational Holism (Part III)

III. Theories in Relation to the Mind

—*At the Biological Theoretical Level*

- 36. Theory of Contrastive Advantages (Peter Baofu) (*Sec.2.6* & Ch.10 of *FCD; BNN*)
- (• 26). Theory of Floating Consciousness (Peter Baofu) (Ch.10 of *FCD*; Chs.1 & 4 of *FPHC*)

—*At the Psychological Theoretical Level*

- 35. Complex Theory of Data Analysis (Peter Baofu) (*FPHDA*)
- 34. Supersession Theory of Computing (Peter Baofu) (*FPHCOM*)
- 33. Metamorphic Theory of Humor (Peter Baofu) (*FPHH*)
- 32. Contrarian Theory of Personality (Peter Baofu) (*FPHP*)
- 31. Theory of Virtual Sexuality (Peter Baofu) (*FPHS*)
- 30. Expansive-Contractive Theory of Martial Arts (Peter Baofu) (*FPHMA*)
- 29. Multilogical Theory of Learning (Peter Baofu) (*FPHL*)
- 28. Comprehensive Theory of Creative Thinking (Peter Baofu) (*FPHCT*)
- 27. Theory of Unfolding Unconsciousness—also known as the Unfolding Theory of Anomalous Experience (Peter Baofu) (*FPHU*)
- 26. Theory of Floating Consciousness (Peter Baofu) (Ch.10 of *FCD*; Chs.1 & 4 of *FPHC*)
- 25. Theory of Cognitive Partiality (Peter Baofu) (*Sec.2.4* of *FCD*; *Sec.4.5.1.1* of *BCPC*)
- 24. Theory of Emotional Non-Neutrality (Peter Baofu) (*Sec.2.5* of *FCD*; *Sec.4.5.2* of of *BCPC*)
- 23. Theory of Behavioral Alteration (Peter Baofu) (*Sec.4.5.3* of *BCPC*)

Table 4.43. Theories on Civilizational Holism (Part IV)

IV. Theories in Relation to Society

—At the Organizational Theoretical Level

- 22. Theory of E-Civic Alienation (Peter Baofu) (Ch.7 of *FCD*)
- 21. Combinational Theory of Organization (Peter Baofu) (*FPHO*; Ch.6 of *ALD*)

—At the Institutional Theoretical Level

- 20. Detached of Theory of Gambling (Peter Baofu) (*FPHGAM*)
- 19. Heterodox of Theory of Education (Peter Baofu) (*FPHEDU*)
- 18. Reconstruction of Theory of Law (Peter Baofu) (*FPHLAW*)
- 17. Theory of Post-Capitalism (Peter Baofu) (Ch.10 of *FCD*; Chs.2 & 4 of *FPHC*; *BCPC*)
- 16. Theory of Post-Democracy (Peter Baofu) (Ch.10 of *FCD*; Chs.3 & 4 of *FPHC*; *BDPD*)
- 15. Dynamic Theory of Comparative Political Systems (Peter Baofu) (*ALD*)

—At the Systemic Theoretical Level

- 14. Contingent Theory of Urban Planning (Peter Baofu) (*FPHUP*)
- 13. Totalistic Theory of Communication (Peter Baofu) (*FPHMM*; *FCD*; *FHC*)
- 12. Ambivalent Theory of Technology (Peter Baofu) (*FPHE*; *FCD*; *FHC*)
- 11. Multifaceted Theory of War and Peace (Peter Baofu) (Ch.9 of *FCD*; Ch.1 of *FPHK*)
- 10. Theory of Post-Humanity (Peter Baofu) (Ch.7 of *FHC*; Chs.3, & 10 of *FCD*; Chs.1, 3 & 4 of *FPHC*; and other books of mine)
- 9. Theory of the Cyclical Progression of System Integration and Fragmentation (Peter Baofu) (Chs.9-10 of *FCD*)
- 8. Synthetic Theory of Information Architecture (Peter Baofu) (*FIA*)

(continued on next page)

Table 4.43. Theories on Civilizational Holism (Part V)

IV. Theories in Relation to Society (*cont'd*)
—*At the Structural Theoretical Level*
- • 7. Theory of the Cyclical Progression of Hegemony (Peter Baofu) (Chs.9-10 of *FCD*; Chs.1, 3 & 4 of *FPHC*; *BDPD*)
- • 6. Theory of the Cyclical Progression of Empire-Building (Peter Baofu) (*BWT*)

V. Meta-Theories (in Relation to Theories)
—*At the Ontological Meta-Theoretical Level*
- • 5. Theory of Existential Dialectics, or the Holistic Theory of Knowledge (Peter Baofu) (*FHC*; *FCD*; *FPHC*; *BDPD*; *FC*; *FAE*; *ALD; FIA; FPHK*; etc,)
- • 4. Contrastive Theory of Rationality (Peter Baofu) (*FPHML*)

—*At the Methodological Meta-Theoretical Level*
- (• 35). Complex Theory of Data Analysis (Peter Baofu) (*FPHDA*)
- • 3. Critical-Dialectic Theory of Formal Science (Peter Baofu) (*FPHFS*)
- • 2. Theory of Methodological Holism (Peter Baofu) (Ch.1 of *FCD*; Ch.1of *FPHC*; *Sec.2.1* & *Sec.2.5* of *BCPC*; *FC*; *FPHK*; etc.)

VI. Theories in Relation to the Rest
—*At Other Levels (Historical)*
- • 1. Theory of the Evolution from Pre-Modernity to After-Postmodernity (Peter Baofu) (*FHC*; Ch.9-10 of *FCD*; *FPHC*)

Notes: All these theories are my constructions, as some of the main contributions of my grant project on civilization and its future. These examples are solely illustrative (not exhaustive), and some of the items can be reclassified somewhere else. Nor are they always mutually exclusive. Since they are generalities, exceptions are expected.
Sources: From my previous books.

Table 4.44. Three Great Future Transformations of Mind Games

- **Virtual Games**
  - —Ex: online chess
  - —Ex: virtual experience
- **Novel Games**
  - —Ex: new chess variants
  - —Ex: new chess engines
- **Post-Human Mind Games**
  - —Ex: the quest for broader/deeper mental benefits of chess playing
  - —Ex: games designed for the evolution of the mind into different body-less forms (e.g., "floating consciousness," "hyper-spatial consciousness," "unfolding unconsciousness")

Notes: The examples in the categories are solely illustrative (not exhaustive), and the comparison is relative (not absolute), nor are they necessarily mutually exclusive. And some can be easily re-classified elsewhere. As generalities, they allow exceptions.
Sources: From *Sec. 4.16* of *FPHCHESS*. See text for more info. Also, consult *FCD* for strategy and tactics in warfare and *FPHO* for strategy and tactics in organization.

BIBLIOGRAPHY

Aanstoos, C.; I. Serlin; & T. Greening. 2000. "A History of Division 32 (Humanistic Psychology) of the American Psychological Association." D. Dewsbury, ed.*Unification through division: Histories of the divisions of the American Psychological Association*, Vol. V. Washington, DC: American Psychological Association.

Adèr, H. J.; & G.J. Mellenbergh, eds. 2008. *Advising on Research Methods: A Consultant's Companion*. Huizen, the Netherlands: Johannes van Kessel Publishing.

Adorno, T. W.; & Max Horkheimer. 2002. *Dialectic of Enlightenment*. Edmund Jephcott, ed. Stanford, CA: Stanford University Press.

Allan, G. 2003. "A Critique of Using Grounded Theory as a Research Method." *Electronic Journal of Business Research Methods*, 2 (1): 1-10.

American Association for the Advancement of Science (AAAS). 2006. "Statement on the Teaching of Evolution." <http://www.aaas.org/news/releases/2006/pdf/0219boardstatement.pdf>.

Archer, Margaret. 2000. *Being Human: The Problem of Human Agency*. Cambridge, England: Cambridge University Press.

Arkin. A.; J. Ross; H. McAdams. 1998. "Stochastic Kinetic Analysis of Developmental Pathway Bifurcation in Phage Lambda-Infected Escherichia Coli Cells." *Genetics*, 149 (4): 1633–48. <http://www.pubmedcentral.nih.gov/articlerender.fcgi?tool=pmcentrez&artid=1460268>.

Ashley, D.; & D. Orenstein. 2005. *Sociological Theory: Classical Statements*. Boston, MA: Pearson Education.

Baofu, Peter. 2011. *The Future of Post-Human Chemistry: A Preface to a New Theory of Substances and their Changes*. Cambridge, England: Cambridge Scholars Publishing.

_____.2011a. *The Future of Post-Human Literature: A Preface to a New Theory of Fiction and Non-Fiction*. Cambridge, England: Cambridge International Science Publishing, Ltd.

_____.2011b. *The Future of Post-Human Computing: A Preface to a New Theory of Hardware, Software, and the Mind*. Cambridge, England: Cambridge International Science Publishing, Ltd.

_____.2011c. *The Future of Post-Human Gambling: A Preface to a New Theory of*

Risk and Caution. Cambridge, England: Cambridge International Science Publishing, Ltd.

_____.2011d. *The Future of Post-Human Acoustics: A Preface to a New Theory of Sound and Silence*. Cambridge, England: Cambridge International Science Publishing, Ltd.

_____.2011e. *The Future of Post-Human Humor: A Preface to a New Theory of Joking and Laughing*. Cambridge, England: Cambridge International Science Publishing, Ltd.

_____.2011f. *Beyond Ethics to Post-Ethics: A Preface to a New Theory of Morality and Immorality*. Charlotte, NC: Information Age Publishing.

_____.2011g. *The Future of Post-Human Education: A Preface to a New Theory of Teaching and Learning*. Cambridge, England: Cambridge International Science Publishing, Ltd.

_____.2010. *The Future of Post-Human Religion: A Preface to a New Theory of Spirituality*. Cambridge, England: Cambridge International Science Publishing, Ltd.

_____.2010a. *The Future of Post-Human Chess: A Preface to a New Theory of Tactics and Strategy*. Cambridge, England: Cambridge International Science Publishing, Ltd.

_____.2010b. *The Future of Post-Human Geology: A Preface to a New Theory of Statics and Dynamics*. Cambridge, England: Cambridge International Science Publishing, Ltd.

_____.2010c. *Beyond Cosmology to Post-Cosmology: A Preface to a New Theory of Different Worlds*. Cambridge, England: Cambridge International Science Publishing, Ltd.

_____.2010d. *The Future of Post-Human Personality: A Preface to a New Theory of Normality and Abnormality*. Cambridge, England: Cambridge International Science Publishing, Ltd.

_____.2010e. *The Future of Post-Human War and Peace: A Preface to a New Theory of Aggression and Pacificity*. Cambridge, England: Cambridge Scholars Publishing.

_____.2010f. *The Future of Post-Human Formal Science: A Preface to a New Theory of Abstraction and Application*. Cambridge, England: Cambridge Scholars Publishing.

_____.2010g. *The Future of Post-Human Law: A Preface to a New Theory of Necessity, Contingency, and Justice*. Cambridge, England: Cambridge Scholars Publishing.

_____.2010h. *The Future of Post-Human Sexuality: A Preface to a New Theory of the Body and Spirit of Love Makers*. Cambridge, England: Cambridge Scholars Publishing.

______,2009. *The Future of Post-Human Martial Arts: A Preface to a New Theory of the Body and Spirit of Warriors*. Cambridge, England: Cambridge Scholars Publishing.

______.2009a. *The Future of Post-Human Organization: A Preface to a New Theory of Communication, Decision-Making, and Leadership*. Cambridge, England: Cambridge Scholars Publishing.

______.2009b. *The Future of Post-Human Language: A Preface to a New Theory of Structure, Context, and Learning*. Cambridge, England: Cambridge Scholars Publishing.

______.2009c. *The Future of Post-Human Urban Planning: A Preface to a New Theory of Density, Void, and Sustainability*. Cambridge, England: Cambridge Scholars Publishing.

______.2009d. *The Future of Post-Human Geometry: A Preface to a New Theory of Infinity, Symmetry, and Dimensionality*. Cambridge, England: Cambridge Scholars Publishing.

______.2009e. *The Future of Post-Human Creative Thinking: A Preface to a New Theory of Invention and Innovation*. Cambridge, England: Cambridge Scholars Publishing.

______.2009f. *The Future of Post-Human Mass Media: A Preface to a New Theory of Communication*. Cambridge, England: Cambridge Scholars Publishing.

______.2009g. *The Future of Post-Human Engineering: A Preface to a New Theory of Technology*. Cambridge, England: Cambridge Scholars Publishing.

______.2008. *The Future of Post-Human Mathematical Logic: A Preface to a New Theory of Rationality*. Cambridge, England: Cambridge Scholars Publishing.

______.2008a. *The Future of Post-Human Knowledge: A Preface to a New Theory of Methodology and Ontology*. Oxford, United Kingdom: Chandos Publishing (Oxford) Ltd.

______.2008b. *The Future of Post-Human Unconsciousness: A Preface to a New Theory of Anomalous Experience*. Cambridge, England: Cambridge Scholars Publishing.

______.2008c. *The Future of Information Architecture: Conceiving a New Way to Understand Taxonomy, Network, and Intelligence*. Oxford, United Kingdom: Chandos Publishing (Oxford) Ltd.

______.2007. *The Rise of Authoritarian Liberal Democracy: A Preface to a New Theory of Comparative Political Systems*. Cambridge, England: Cambridge Scholars Publishing.

______.2007a. *The Future of Aesthetic Experience: Conceiving a Better Way to Beauty, Ugliness, and the Rest*. Cambridge, England: Cambridge Scholars Publishing.

______.2007b. *The Future of Complexity: Conceiving a Better Way to Understand*

Order and Chaos. London, United Kingdom: World Scientific Publishing Co.

_____.2007c. *Beyond the World of Titans, and the Remaking of World Order: A Preface to a New Logic of Empire-Building*. Cambridge, England: Cambridge Scholars Publishing.

_____.2006. *Beyond Nature and Nurture: Conceiving a Better Way to Understand Genes and Memes*. Cambridge, England: Cambridge Scholars Publishing.

_____.2006a. *The Future of Post-Human Space-Time: Conceiving a Better Way to Understand Space and Time*. New York: Peter Lang Publishing, Inc.

_____.2006b. *Beyond Civilization to Post-Civilization: Conceiving a Better Model of Life Settlement to Supersede Civilization*. New York: Peter Lang Publishing, Inc.

_____.2005. *Beyond Capitalism to Post-Capitalism: Conceiving a Better Model of Wealth Acquisition to Supersede Capitalism*. New York: The Edwin Mellen Press.

_____.2004. *The Future of Post-Human Consciousness*. New York: The Edwin Mellen Press.

_____.2004a. *Beyond Democracy to Post-Democracy: Conceiving a Better Model of Governance to Supersede Democracy*. Volume 1. New York: The Edwin Mellen Press.

_____.2004b. *Beyond Democracy to Post-Democracy: Conceiving a Better Model of Governance to Supersede Democracy*. Volume 2. New York: The Edwin Mellen Press.

_____.2002. *The Future of Capitalism and Democracy*. Maryland: The University Press of America.

_____.2000. *The Future of Human Civilization*. Volume 1. New York: The Edwin Mellen Press.

_____.2000a. *The Future of Human Civilization*. Volume 2. New York: The Edwin Mellen Press.

Baron, Jonathan. 1994. *Thinking and Deciding*. UK: Cambridge University Press.

Barrow, John D. 1998. *Impossibility: The Limits of Science and the Science of Limits*. Oxford: Oxford University Press.

Bentz, V. M.; J. Shapiro. 1998. *Mindful Inquiry in Social Research*. Thousand Oaks, CA: Sage.

Bernard, H. Russell, ed. 1998. *Handbook of Methods in Cultural Anthropology*. Walnut Creek, CA: AltaMira Press.

Biletzki, Anat; & Anat Matar. 2006. "Ludwig Wittgenstein" (Nov. 17). *Stanford Encyclopedia of Philosophy*. <http://plato.stanford.edu/entries/wittgenstein/>.

Blake, W.J.; M. KAErn; C. Cantor; & J. Collins. 2003. "Noise in Eukaryotic Gene Expression." *Nature*, 422 (6932): 633–7. <http://www.bu.edu/abl/pdf/

blake2003.pdf?file=/nature/journal/v422/n6932/full/nature01546_fs.html&co ntent_filetype=PDF>.

Boaz, Noel T.; & Alan J. Almquist. 1997. *Biological Anthropology: A Synthetic Approach to Human Evolution.* Upper Saddle River, NJ: Prentice Hall.

Bornstein, Robert F.; & Catherine Crave-Lemley. 2004. "Mere Exposure Effect." *Cognitive Illusions: A Handbook on Fallacies and Biases in Thinking, Judgement and Memory.* Rüdiger Pohl, ed. Hove, UK: Psychology Press.

Bourdieu, Pierre. 2003. "Participant Objectivation." *Journal of the Royal Anthropological Institute*, 9 (2): 281–294.

_____; & Loic Wacquant. 1992. *An Invitation to Reflexive Sociology.* Cambridge: Polity Press.

Bowler, Peter J. 2003. *Evolution:The History of an Idea.* CA: University of California Press.

Box, George E.P.; & N. R. Draper. 1987. *Empirical Model-Building and Response Surfaces.* Wiley.

Bull, J.J., & H. A. Wichman. 2001. "Applied Evolution." *Annu Rev Ecol Syst*, 32 (1): 183–217.

Bunge, Mario. 1999. *Dictionary of philosophy.* Amherst, MA: Prometheus Books.

Byrne, R. M. J. 2005. *The Rational Imagination: How People Create Counterfactual Alternatives to Reality.* Cambridge, MA: MIT Press.

Chandra, Siddharth; & William Shadel. 2007. "Crossing Disciplinary Boundaries: Applying Financial Portfolio Theory to Model the Organization of the Self-Concept." *Journal of Research in Personality*, 41 (2): 346–373.

Chase, Scott. 1993. "Tachyons" (March). *Physics FAQ.* <http://math.ucr.edu/home/baez/physics/ParticleAndNuclear/tachyons.html>.

Chavez, Leo. 1997. *Shadowed Lives: Undocumented Workers in American Society.* Fort Worth, TX: Harcourt Brace College Publishers.

Churchman, C. West. 1968. *The Systems Approach.* New York: Dell publishing.

Colman, Andrew M., ed. 2009.. "Humanistic Psychology." *A Dictionary of Psychology.* England: Oxford University Press.

Collingwood, R.1940. *An Essay on Metaphysics.* Oxford: Clarendon Press.

Comte, Auguste. 1853. *The Positive Philosophy of Auguste Comte.* Volume 1. Harriet Martineau, tr. London: J. Chapman.

Corazzon, Raul. 2007. "Definitions of Ontology: From Nicolai Hartmann to the Present Time." <http://www.formalontology.it/section_4_two.htm>.

Corbett, Dan. 2008. "Introduction to String Theory" (November 23). *ThinkQuest.* <http://library.thinkquest.org/27930/stringtheory1.htm>.

Cuizon, Gwendolyn. 2007. "Aristotle's Four Causes" (March 19). <http://socyberty.com/philosophy/aristotles-four-causes/>.

Dodge, Y. 2003. "Statistics." *The Oxford Dictionary of Statistical Terms.* NY:

Oxford University Press.

Edwards, W. 1968. *Conservatism in Human Information Processing. Formal Representation of Human Judgment.* B. Kleinmutz, ed. New York: John Wiley and Sons.

Eiben, A. E., et al 1994. "Genetic Algorithms with Multi-Parent Recombination." *PPSN III: Proceedings of the International Conference on Evolutionary Computation. The Third Conference on Parallel Problem Solving from Nature*: 78–87.

Elowitz, M. B.; A. Levine; E. D. Siggia; & P. S. Swain. 2002. "Stochastic Gene Expression in a Single Cell." *Science*, 297 (5584): 1183–6. <http://www.sciencemag.org/cgi/pmidlookup?view=long&pmid=12183631>.

_____; & S. Leibler. 2000. "A Synthetic Oscillatory Network of Transcriptional Regulators." *Nature*, 403 (6767): 335–8.

Elton, Edwin; & Martin J. Gruber. 1997. "Modern Portfolio Theory, 1950 to Date." *Journal of Banking & Finance*, 21: 1743-1759.

Ember, Carol; & Melvin Ember. 2006. *Cultural Anthropology*. NY: Prentice Hall.

Entwistle, N. J., & T. Brennan. 1971. "Types of Successful Students." *British Journal of Educational Psychology,* 41: 268-276.

Epp, Susanna S. 2004. *Discrete Mathematics with Applications*. Belmont, CA: Wadsworth Pub. Co.

Everett, Hugh. 1957. "'Relative State' Formulation of Quantum Mechanics" (July). *Reviews of Modern Physics*, vol. 29, no.3. <http://www.univer.omsk.su/omsk/Sci/Everett/paper1957.html>.

Feldman, Ronen. 2004. "Text Analytics: Theory and Practice." <http://www.ir.iit.edu/cikm2004/tutorials.html#T2>.

Ferguson, Sinclair B.; David F Wright; & James Innell Packer. 1988. "Hermeneutics." *New Dictionary of Theology*. Downers Grove, IL: InterVarsity Press.

Festinger, L. 1957. *A Theory of Cognitive Dissonance*. Stanford, CA: Stanford University Press.

Fine, Gary Alan. 1993. "Ten Lies of Ethnography." *Journal of Contemporary Ethnography*, 22 (3): 267-294.

Firehammer, Reginald. 2005. "Basic Principles of Ontology" (January 05). <http://usabig.com/autonomist/philosophy/ontology2.html>.

Fishwick, P. A. 1995. *Simulation Model Design and Execution: Building Digital Worlds*. Upper Saddle River, NJ: Prentice Hall.

Fisk, John E. 2004. "Conjunction Fallacy." *Cognitive Illusions: A Handbook on Fallacies and Biases in Thinking, Judgement and Memory*. Rüdiger Pohl. Hove, UK: Psychology Press.

Flyvbjerg, B. 2008. "Curbing Optimism Bias and Strategic Misrepresentation in Planning: Reference Class Forecasting in Practice." *European Planning*

Studies, 16 (1): 3-21.

_____.2001. *Making Social Science Matter: Why Social Inquiry Fails and How It Can Succeed Again*. England: Cambridge University Press.

Fogel, David B. 2000. *Evolutionary Computation: Towards a New Philosophy of Machine Intelligence*. New York: IEEE Press.

Forster, Michael. 2002. "Friedrich Daniel Ernst Schleiermacher" (April 17). *Stanford Encyclopedia of Philosophy*. <http://plato.stanford.edu/entries/schleiermacher/>.

Fraser. A. S. 1958. "Monte Carlo Analyses of Genetic Models." *Nature*, 181 (4603): 208–9.

Freudenthal, Hans, ed. 1951. *The Concept and the Role of the Model in Mathematics and Natural and Social Sciences*. Dordrecht, Holland: D. Reidel.

Frisvad, Jens. 2011. "Cluster Analysis" (May 09). <http://www2.imm.dtu.dk/courses/27411/doc/Lection10/Cluster%20analysis.pdf>.

Gadamer, Hans-Georg. 1975. "Hermeneutics and Social Science." *Philosophy Social Criticism/Cultural Hermeneutics*, 2: 307-316.

Gardner, T. S.; C. Cantor; & J. Collins. 2000. "Construction of a Genetic Toggle Switch in Escherichia Coli." *Nature*, 403 (6767): 339–42.

Garson, G. David. 2008. "Ethnographic Research: Statnotes, from North Carolina State University, Public Administration Program." <http://faculty.chass.ncsu.edu/garson/PA765/ethno.htm>.

Gasking, D. 1955. "Causation and Recipes." *Mind*, 64: 479-487.

Geard, N., & J. Wiles. 2005. "A Gene Network Model for Developing Cell Lineages." *Artif. Life*, 11 (3): 249–67. <http://www.mitpressjournals.org/doi/abs/10.1162/1064546054407202?url_ver=Z39.88-2003&rfr_id=ori:rid:crossref.org&rfr_dat=cr_pub%3dncbi.nlm.nih.gov>.

Giorgi, Amedeo. 2009. *The Descriptive Phenomenological Method in Psychology: A Modified Husserlian Approach*. Pittsburgh, PA: Duquesne University Press.

Gjesdal, Bjorn Ramberg Kristin. 2007. "Hermeneutics." *Stanford Encyclopedia of Philosophy*. <http://plato.stanford.edu/entries/hermeneutics/>.

Glaser, B. G. 1998. *Doing Grounded Theory—Issues and Discussions*. Mill Valley, CA: Sociology Press.

_____. 1978. *Theoretical Sensitivity: Advances in the Methodology of Grounded Theory*. Mill Valley, CA: Sociology Press.

_____; & A. Strauss. 1967. *Discovery of Grounded Theory. Strategies for Qualitative Research*. Mill Valley, CA: Sociology Press.

Gott, J. Richard. 2002. *Time Travel in Einstein's Universe*. Boston : Houghton Mifflin.

Gouldner, Alwin W. 1980. *Marxism and Critique in the Two Marxisms*. New York: Oxford University Press.

Griffiths, Emily. 2009. "What is a Model?" (October). BBC Radio 4's Today Programme. <http://www.emily-griffiths.postgrad.shef.ac.uk/models.pdf>.

Groves, Robert; Floyd J. Fowler Jr.; Mick P. Couper; James M. Lepkowski; Eleanor Singer; & Roger Tourangeau. 2009. *Survey Methodology*. New York: Wiley.

Guth, Alan. 1997. *The Inflationary Universe: The Quest for a New Theory Of Cosmic Origins*. New York: Helix Books.

Habermas, Jürgen. 1987. *The Philosophical Discourse of Modernity*: Twelve Lectures. Frederick Lawrence. Cambridge, MA: MIT Press.

Hardman, David. 2009. *Judgment and Decision Making: Psychological Perspectives*. Malden, MA: Wiley-Blackwell.

Harrison, John. 1994. "Do You Suffer from Paradigm Paralysis?" 1st World Congress on Fluency Disorders in Munich, Germany (Aug. 1-5). <http://www.mnsu.edu/comdis/kuster/Infostuttering/Paradigmparalysis.html>.

Heidegger, Martin. 1962. *Being and Time*. New York: Harper & Row.

Heider, Karl. 2001. *Seeing Anthropology*. NY: Prentice Hall.

Hicks, Stephen R.C. 2004. *Explaining Postmodernism: Skepticism and Socialism from Rousseau to Foucault*. Tempe, AZ: Scholargy Press.

Holland, John H. 1975. *Adaptation in Natural and Artificial Systems*. MI: University of Michigan Press.

Hoorens, Vera. 1993. "Self-Enhancement and Superiority Biases in Social Comparison."*European Review of Social Psychology*, 4 (1): 113–139.

Hsee, Christopher K.; & Reid Hastie. 2006. "Decision and Experience: Why Don't We Choose what Makes Us Happy?" *Trends in Cognitive Sciences,* 10 (1): 31–37.

_____.& Jiao Zhang. 2004. "Distinction Bias: Misprediction and Mischoice due to Joint Evaluation." *Journal of Personality and Social Psychology*, 86 (5): 680–695.

Huberty, C. J.; E. Jordan; & W. C. Brandt. 2005. "Cluster Analysis in Higher Education Research." *Higher Education: Handbook of Theory and Research*. J. C. Smart, ed. Great Britain: Springer.

Husserl, Edmund. 1970. *The Crisis of the European Sciences and Transcendental Phenomenology*. Evanston, IL: Northwestern University Press.

Interacademy Panel on International Issues (IPII). 2006. "IAP Statement on the Teaching of Evolution." <http://web.archive.org/web/20070422234034/http://www.interacademies.net/Object.File/Master/6/150/Evolution+statement.pdf>.

Jamshidi, M. 2003. "Tools for Intelligent Control: Fuzzy Controllers, Neural Networks and Genetic Algorithms." *Philosophical Transactions. Series A,*

Mathematical, Physical, and Engineering Sciences, 361 (1809): 1781–808.

Jaynes, Edwin T. 2003. *Probability Theory: The Logic of Science*. Cambridge, UK: Cambridge University Press.

Jeng, M. 2006. "A Selected History of Expectation Bias in Physics." *American Journal of Physics*, 74: 578-583.

Johnson, Richard Arvid. 1976. *Management, Systems, and Society: An Introduction*. Pacific Palisades, CA: Goodyear Pub. Co..

Kahneman, Daniel; Alan B. Krueger, David Schkade, Norbert Schwarz, & Arthur A. Stone. 2006. "Would You Be Happier if You Were Richer? A Focusing Illusion." *Science*, 312 (5782): 1908–10.

_____.& Shane Frederick. 2002. "Representativeness Revisited: Attribute Substitution in Intuitive Judgment." *Heuristics and Biases: The Psychology of Intuitive Judgment*. Thomas Gilovich, Dale Griffin, and Daniel Kahneman, eds. UK: Cambridge University Press.

_____.& A. Tversky. 1972. "Subjective Probability: A Judgment of Representativeness." *Cognitive Psychology*, 3: 430–454.

Kaku, Michio. 1994. *Hyperspace: A Scientific Odyssey Through Parallel Universes, Time Warps, and The Tenth Dimension*. Illustrations by Robert O'Keefe. Oxford: Oxford University Press.

Kantardzic, Mehmed. 2003. *Data Mining: Concepts, Models, Methods, and Algorithms*. NJ: John Wiley & Sons.

Kauffman, Stuart. 1993. *The Origins of Order: Self-Organization and Selection in Evolution*. NY: Oxford University Press.

_____. 1969. "Metabolic Stability and Epigenesis in Randomly Constructed Genetic Nets." *Journal of Theoretical Biology* (22): 437–467.

Kelle, U. 2005. "'Emergence' vs. 'Forcing' of Empirical Data? A Crucial Problem of 'Grounded Theory' Reconsidered." *Forum: Qualitative Social Research* (On-line Journal), 6 (2), Art. 27, paragraphs 49-50.

Kish, L. 1965. *Survey Sampling*. New York: Wiley.

Klauer, K. C.; J. Musch, & B. Naumer. 2000. "On Belief Bias in Syllogistic Reasoning." *Psychological Review*, 107 (4): 852–884.

Knabe, J.F., M. J. Schilstra; & C. Nehaniv. 2008. "Evolution and Morphogenesis of Differentiated Multicellular Organisms: Autonomously Generated Diffusion Gradients for Positional Information." *Artificial Life XI: Proceedings of the Eleventh International Conference on the Simulation and Synthesis of Living Systems*. Cambridge, MA: MIT Press. <http://panmental.de/papers/FlagPottsGRNALife11.pdf>.

_____; C. Nehaniv; M. Schilstra; T. Quick. 2006. "Evolving Biological Clocks using Genetic Regulatory Networks." *Proceedings of the Artificial Life X Conference (Alife 10)*. Cambridge, MA: MIT Press. <http://citeseerx.ist.psu.edu/viewdoc/download?doi=10.1.1.72.5016&rep=rep1&type=pdf>.

_____; C. Nehaniv; & M. Schilstra. 2006a. "Evolutionary Robustness of Differentiation in Genetic Regulatory Networks." *Proceedings of the 7th German Workshop on Artificial Life 2006 (GWAL-7)*. Berlin: Akademische Verlagsgesellschaft Aka. <http://citeseerx.ist.psu.edu/viewdoc/download?doi=10.1.1.71.8768&rep=rep1&type=pdf>.

Koza, John R. 1992. *Genetic Programming*. Cambridge, MA: MIT Press.

Kuhn, Thomas. 1977. *The Essential Tension: Selected Studies in Scientific Tradition and Change*. Chicago, IL: University of Chicago Press.

Kutschera, U.; & K. Niklas. 2004. "The Modern Theory of Biological Evolution: An Expanded Synthesis." *Naturwissenschaften*, 91 (6): 255–76. <http://www.ncbi.nlm.nih.gov/pubmed/15241603>.

Laplace, P. S. 1951. *A Philosophical Essay on Probabilities*. New York: Dover Publications Inc.

Legewie, Heiner; & Barbara Schervier-Legewie. 2004. "Anselm Strauss interviewed by Heiner Legewie and Barbara Schervier-Legewie." *Forum: Qualitative Social Research* (On-line Journal), 5 (3), Art. 22.

Lewin, K. 1958. *Group Decision and Social Change*. New York: Holt, Rinehart and Winston.

_____.1946. "Action Research and Minority Problems." *J Soc. Issues*, 2(4): 34-46.

MacKay, D.J. 2003. *Information Theory, Inference, and Learning Algorithms*. England: Cambridge University Press.

Madhu, P. 2005. "Towards a Praxis Model of Social Work: A Refelexive Account of 'Praxis Intervention' With the Adivasis of Attappady." PhD Dissertation, Kottayam: Mahatma Gandhi University. <http://www.mgutheses.in/page/?q=T%201357&search=&page=&rad=#>.

Mandelbrot, B., & R. Hudson. 2004. *The (Mis)Behaviour of Markets: A Fractal View of Risk, Ruin, and Reward*. London: Profile Books.

Mann, P. S. 1995. *Introductory Statistics*. NJ: Wiley.

Markoviç, Mahilo. 1974. *From Affluence to Praxis*. Boston, MA: Beacon Press.

Markowitz, H. M. 1959. *Portfolio Selection: Efficient Diversification of Investments*. New York: John Wiley & Sons.

_____.1952. "Portfolio Selection." *The Journal of Finance*, 7 (1): 77-91.

Marks, Gary; & N. Miller. 1987. "Ten Years of Research on the False-Consensus Effect: An Empirical and Theoretical Review." *Psychological Bulletin*, 102 (1): 72–90.

Martin, Patricia Yancey; & Barry A. Turner. 1986. "Grounded Theory and Organizational Research." *The Journal of Applied Behavioral Science*, 22 (2): 141.

Marx, Karl. 2008. "Theses on Feuerbach" (August 22). Marxists Internet Archive. <http://www.marxists.org/archive/marx/works/1845/theses/theses.htm>.

Maynard, M.; & J. Purvis. 1994. *Researching Women's Loves from a Feminist Perspective*. London: Taylor & Frances.

McAleer, Neil. 1987. *The Mind-Boggling Universe*. New York: Doubleday & Co.

Menzies, P.; & H. Price. 1993. "Causation as a Secondary Quality." *British Journal for the Philosophy of Science*, (44): 187-203.

Merriam-Webster's Collegiate Dictionary Online (MWD). 2011. "Paradigm" (May 03). <http://www.merriam-webster.com/dictionary/paradigm>.

_____.2007. "Agency" (March 20). <http://www.m-w.com/dictionary/agency>.

Miller, G. & P. Johnson-Laird. 1976. *Language and Perception*. Cambridge, MA: Harvard University Press.

Natanson, M. 1973. *Edmund Husserl: Philosopher of Infinite Tasks*. Evanston, IL: Northwestern University Press.

Nickles, Thomas. 2002. *Thomas Kuhn*. NY: Cambridge University Press.

Nova Online (NO). 2000. "Time Can Vary." <http://www.pbs.org/wgbh/nova/time/thinktime.html>.

Oswald, Margit E.; Stefan Grosjean. 2004. "Confirmation Bias." *Cognitive Illusions: A Handbook on Fallacies and Biases in Thinking, Judgement and Memory*. Rüdiger F. Pohl, ed. Hove, UK: Psychology Press.

Outhwaite, William. 1988 *Habermas: Key Contemporary Thinkers*. UK: Polity Press.

Palace, Bill. 1996. "Data Mining: What is Data Mining?" <http://www.anderson.ucla.edu/faculty/jason.frand/teacher/technologies/palace/datamining.htm>.

Pearl, Judea. 2000. *Causality: Models, Reasoning, and Inference*. NY: Cambridge University Press.

Philosophy Professor (PP). 2011. "Aristotle's Four Causes" (May 09). <http://www.philosophyprofessor.com/philosophies/aristotles-four-causes.php>.

Plous, Scott. 1993. *The Psychology of Judgment and Decision Making*. New York: McGraw-Hill.

Pohl, Rüdiger F. 2004. "Hindsight Bias." *Cognitive Illusions: A Handbook on Fallacies and Biases in Thinking, Judgement and Memory*. Rüdiger Pohl, ed. Hove, UK: Psychology Press.

Pronin, Emily; & Matthew B. Kugler. 2007. "Valuing Thoughts, Ignoring Behavior: The Introspection Illusion as a Source of the Bias Blind Spot." *Journal of Experimental Social Psychology*, 43 (4): 565–578.

_____.; J. Kruger; K. Savitsky; & L. Ross. 2001. "You Don't Know Me, but I Know You: The Illusion of Asymmetric Insight." *Journal of Personality and Social Psychology*, 81 (4): 639–656.

Pullan, Wendy. 2000. *Structure. Cambridge, England: Cambridge University Press.*

Pyke, G. H. 1984. "Optimal Foraging Theory: A Critical Review." *Annual Review of Ecology and Systematics*, 15: 523-575. <http://www.annualreviews.org/doi/abs/10.1146/annurev.es.15.110184.002515?journalCode=ecolsys>.

Quotes.net (*QN*). 2011. "Maitri Upanishadshttp" (May 19). <http://www.quotes.net/authors/Maitri+Upanishads>.

Raman, Shankar. 1997. Lectures on Literary Theory. Cambridge, MA: MIT (spring).

Ramsey, Frank. 1926. "Truth and Probability." <http://www.newschool.edu/nssr/het/texts/ramsey/ramsess.pdf>.

Raser, J. M.; & E. O'Shea. 2005. "Noise in Gene Expression: Origins, Consequences, and Control." *Science*, 309 (5743): 2010–3. <http://www.sciencemag.org/cgi/pmidlookup?view=long&pmid=16179466>.

Rennie, David L. 2007. "Hermeneutics and Humanistic Psychology." *The Humanistic Psychologist*, 35 (1). <http://www.apa.org/divisions/div32/pdfs/hermeneutics.pdf>.

Ribeiro, A.; &,J. Lloyd-Price. 2007. "SGN Sim, a Stochastic Genetic Networks Simulator." *Bioinformatics*, 23 (6): 777–9. <http://bioinformatics.oxfordjournals.org/cgi/pmidlookup?view=long&pmid=17267430>.

_____; R. Zhu; & S. Kauffman. 2006. "A General Modeling Strategy for Gene Regulatory Networks with Stochastic Dynamics." *J. Comput. Biol.*, 13 (9): 1630–9.

Richardson, L. 2000. "Evaluating Ethnography." *Qualitative Inquiry*, 6 (2): 253-255

Rollinger, Robin. 1999. *Husserl's Position in the School of Brentano*. London: Kluwer.

Roussel, M. R.; & R. Zhu. 2006. "Validation of an Algorithm for Delay Stochastic Simulation of Transcription and Translation in Prokaryotic Gene Expression." *Phys Biol*, 3 (4): 274–84. <http://stacks.iop.org/1478-3967/3/274>.

Rubin, R. B.; A. Rubin; & L. Piele. 2005. *Communication Research: Strategies and Sources*. Belmont, CA: Thomson Wadworth.

Russell, B. 1948. *Human Knowledge*. Simon and Schuster.

Safranski, R. 1998. *Martin Heidegger: Between Good and Evil*. Cambridge, MA: Harvard University Press.

Salmon, W. 1984. *Scientific Explanation and the Causal Structure of the World*. NJ: Princeton University Press.

Sanna, Lawrence J.; & Norbert Schwarz. 2004. "Integrating Temporal Biases: The Interplay of Focal Thoughts and Accessibility Experiences." *Psychological Science*, 15 (7): 474–481.

Schilstra, M. J., & H. Bolouri. 2002. "Modelling the Regulation of Gene Expression in Genetic Regulatory Networks." Biocomputation group, University of Hertfordshire. <http://strc.herts.ac.uk/bio/maria/NetBuilder/

Theory/NetBuilderModelling.htm>.

Schokel, Luis Alonso; & Jose Maria Bravo. 1998. *A Manual of Hermeneutics (Biblical Seminar)*. Liliana M. Rosa, tr. Brook W. R. Pearson, ed. Sheffield: Sheffield Academic Press.

Schunk, Dale H. 2004. *Learning Theories: An Educational Perspective*. Upper Saddle River, NJ: Pearson/Merrill/Prentice Hall.

Shafir, Eldar; Diamond, Peter; & Amos Tversky. 2000. "Money Illusion." *Choices, Values, and Frames*. UK: Cambridge University Press.

Shklar, Judith N. 1986. "Squaring the Hermeneutic Circle." *Social Research*, 71 (3): 657.

Shlain, Leonard. 1991. *Art and Physics: Parallel Visions in Space,Time, and Light*. New York: William Morrow and Co., Inc.

Slattery, Martin. 2003. *Key Ideas in Sociology*. Cheltenham: Nelson Thornes.

Smith, David Woodruff. 2007. *Husserl*. NY: Routledge.

SociologyGuide.com (SGC). 2011. "Auguste Comte." <http://www.sociology guide.com/thinkers/Auguste-Comte.php>.

Sokolowski, J. A.; & C. Banks. 2009. *Principles of Modeling and Simulation*. Hoboken, NJ: John Wiley and Sons.

Sokolowski, Robert. 2000. *Introduction to Phenomenology*. England: Cambridge University Press.

Sowa, John. 2006. "Processes and Causality" (Dec. 5). <http://www.jfsowa.com/ ontology/causal.htm>.

Stanford Encyclopedia of Philosophy (SEP). 2006. "Interpretations of Probability" (December 23). <http://plato.stanford.edu/entries/probability-interpret/>.

Sutherland, Stuart. 2007. *Irrationality*. England: Pinter & Martin.

Syswerda, G. 1989. "Uniform Crossover in Genetic Algorithms." *Proceedings of the Third International Conference on Genetic Algorithms*. J. D. Schaffer, ed. Morgan Kaufmann.

ThinkExist.com (TEC). 2011. "Aaron Levenstein" (May 03) <http://thinkexist.com/quotation/statistics_are_like_a_bikini-what_they_re veal_is/222844.html>.

Thompson, Suzanne C. 1999. "Illusions of Control: How We Overestimate Our Personal Influence." *Current Directions in Psychological Science*, 8 (6): 187–190. <http://www.jstor.org/stable/20182602>.

Ting, Chuan-Kang. 2005. "On the Mean Convergence Time of Multi-parent Genetic Algorithms Without Selection." *Advances in Artificial Life*: 403–412.

Torbert, W. & et al. 2004. *Action Inquiry: The Secret of Timely and Transforming Leadership*. San Francisco, CA: Berrett-Koehler.

Tversky, Amos; & Daniel Kahneman. 1974. "Judgment under Uncertainty:

Heuristics and Biases." *Science*, 185 (4157): 1124–1131.

Vohradsky, J. 2001. "Neural Model of the Genetic Network." *J. Biol. Chem.*, 276 (39): 36168–73. <http://www.jbc.org/cgi/pmidlookup?view=long&pmid=11395518>.

von Wright, G. 1971. *Explanation and Understanding*. Ithaca, NY: Cornell University Press.

Weber, Max. 1930. *The Protestant Ethic and the Spirit of Capitalism*. Talcott Parsons, tr. New.York: Scribner.

Weisberg, Herbert F. 2005. *The Total Survey Error Approach*. Chicago, IL: *University of Chicago Press*.

Whitehead, J. & J. McNiff. 2006. *Action Research Living Theory*. London, England: Sage.

_____. 1989. "Living Theories." <http://www.actionresearch.net>.

Wikipedia (WK). 2011. "Postpositivism" (April 30). <http://en.wikipedia.org/wiki/Postpositivism>.

_____.2011a. "Data Analysis" (April 30). <http://en.wikipedia.org/wiki/Data_analysis>.

_____.2011b. "Qualitative Research" (April 30). <http://en.wikipedia.org/wiki/Qualitative_research>.

_____.2011c. "Qualitative Data" (April 30). <http://en.wikipedia.org/wiki/Qualitative_data>.

_____.2011d. "Quantitative Research" (April 30). <http://en.wikipedia.org/wiki/Quantitative_research>.

_____.2011e. "Numerical Data" (April 30). <http://en.wikipedia.org/wiki/Numerical_data>.

_____.2011f. "Quantity" (April 30). <http://en.wikipedia.org/wiki/Quantity>.

_____.2011g. "Qualitative Psychological Research" (April 30). <http://en.wikipedia.org/wiki/Qualitative_psychological_research>.

_____.2011h. "Data Mining" (April 30). <http://en.wikipedia.org/wiki/Data_mining>.

_____.2011i. "Statistics" (April 30). <http://en.wikipedia.org/wiki/Statistics>.

_____.2011j. "Descriptive Statistics" (April 30). <http://en.wikipedia.org/wiki/Descriptive_statistics>.

_____.2011k. "Exploratory Data Analysis" (April 30). <http://en.wikipedia.org/wiki/Exploratory_data_analysis>.

_____.2011l. "Predictive Analytics" (April 30). <http://en.wikipedia.org/wiki/Predictive_analytics>.

_____.2011m. "List of Cogntiive Biases" (May 01). <http://en.wikipedia.org/wiki/List_of_cognitive_biases>.

_____.2011o. "Bounded Rationality" (May 01). <http://en.wikipedia.org/wiki/Bounded_rationality>.

_____.2011p. "Attribute Substitution" (May 01). <http://en.wikipedia.org/wiki/Attribute_substitution>.

_____.2011q. "Attribution Theory" (May 01). <http://en.wikipedia.org/wiki/Attribution_theory>.

_____.2011r. "Cognitive Dissoance" (May 01). <http://en.wikipedia.org/wiki/Cogntiive_dissonance>.

_____.2011s. "Heuristic" (January 27). <http://en.wikipedia.org/wiki/Heuristic>.

_____.2011t. "Adaptive Bias" (May 01). <http://en.wikipedia.org/wiki/Adaptive_bias>.

_____.2011u. "Misuse of Statistics" (May 01). <http://en.wikipedia.org/wiki/Misuse_of_statistics>.

_____.2011v. "Phenomenology" (May 01). <http://en.wikipedia.org/wiki/Phenomenology>.

_____.2011w. "Positivism" (May 01). <http://en.wikipedia.org/wiki/Positivism>.

_____.2011x. "Text Analytics" (May 01). <http://en.wikipedia.org/wiki/Text_analytics>.

_____.2011y. "Evolution" (May 01). <http://en.wikipedia.org/wiki/Evolution>.

_____.2011z. "Evolutionary Algorithm" (May 01). <http://en.wikipedia.org/wiki/Evolutionary_algorithm>.

_____.2011aa. "Critical Theory" (May 01). <http://en.wikipedia.org/wiki/Critical_theory>.

_____.2011bb. "Constructivist Epistemology" (May 01). <http://en.wikipedia.org/wiki/Constructivist_epistemology>.

_____.2011cc. "Postpositivism" (May 01). <http://en.wikipedia.org/wiki/Postpositivism>.

_____.2011dd. "Statistical Hypothesis Testing" (May 01). <http://en.wikipedia.org/wiki/Statistical_hypothesis_testing>.

_____.2011ee. "Action Research" (May 02). <http://en.wikipedia.org/wiki/Action_research>.

_____.2011ff. "Hermeneutics" (May 03). <http://en.wikipedia.org/wiki/Hermeneutics>.

_____.2011gg. "Paradigm" (May 03). <http://en.wikipedia.org/wiki/Paradigm>.

_____.2011hh. "Thomas Kuhn" (May 03). <http://en.wikipedia.org/wiki/Thomas_Kuhn>.

_____.2011ii. "Cognitive Bias" (May 03). <http://en.wikipedia.org/wiki/Cognitive_bias>.

_____.2011jj. "Scientific Modeling" (May 05). <http://en.wikipedia.org/wiki/

Scientific_modeling>.

______.2011kk. “Mathematical Modeling” (May 05). <http://en.wikipedia.org/wiki/ Mathematical_modeling>.

______.2011ll. “Bayesian Probability” (May 05). <http://en.wikipedia.org/wiki/ Bayesian_probability>.

______.2011mm. “Social Statistics” (May 05). <http://en.wikipedia.org/wiki/ Social_statistics>.

______.2011nn. “Analytics” (May 06). <http://en.wikipedia.org/wiki/Analytics>.

______.2011oo. “Interpretation (Logic)” (April 13). <http://en.wikipedia.org/ wiki/ Interpretation_(Logic)>.

______.2011pp. “Unsolved Problems in Linguistics” (February 20). <http://en.wikipedia.org/wiki/Unsolved_problems_in_linguistics>.

______.2011qq. “Portfolio Analytics” (May 06). <http://en.wikipedia.org/wiki/ Portfolio_analytics>.

______.2011rr. “Social Statistics” (May 07). <http://en.wikipedia.org/wiki/ Social_statistics>.

______.2011ss. “Survey Sampling” (May 07). <http://en.wikipedia.org/wiki/ Survey_sampling>.

______.2011tt. “Cluster Analysis” (May 07). <http://en.wikipedia.org/wiki/ Cluster_analysis>.

______.2011uu. “Factor Analysis” (May 07). <http://en.wikipedia.org/wiki/ Factor_analysis>.

______.2011vv. “Causality” (May 08). <http://en.wikipedia.org/wiki/Causality>.

______.2011ww. “Probability Interpretatons” (May 09). <http://en.wikipedia.org/ wiki/probability_interpretations>.

______.2011xx. “Mental Model” (May 10). <http://en.wikipedia.org/ wiki/Mental_model>.

______.2011yy. “Humanistic Psychology” (May 12). <http://en.wikipedia.org/ wiki/Humanistic_psychology>.

______.2011zz. “Hermeneutic Circle” (May 12). <http://en.wikipedia.org/ wiki/Hermeneutic_circle>.

______.2011aaa. “Genetic Algorithm” (May 12). <http://en.wikipedia.org/ wiki/Genetic_algorithm>.

______.2011bbb. “Genetic Regulatory Networks” (May 13). <http://en.wikipedia.org/ wiki/Genetic_regulatory_networks>.

______.2011ccc. “Phronesis” (May 14). <http://en.wikipedia.org/wiki/Phronesis>.

______.2011ddd. “Bent Flyvbjerg” (May 14). <http://en.wikipedia.org/wiki/Bent_ Flyvbjerg>.

______.2011eee. “Science Wars” (May 14). <http://en.wikipedia.org/wiki/Science_

wars>.

_____.2011fff. "Praxis Intervention" (May 14). <http://en.wikipedia.org/wiki/Praxis_intervention>.

_____.2011ggg. "Ethnography" (May 16). <http://en.wikipedia.org/wiki/Ethnography>.

_____.2011hhh. "Grounded Theory" (May 18). <http://en.wikipedia.org/wiki/Grounded_theory>.

_____.2008. "Posthumanism" (October 20). <http://en.wikipedia.org/wiki/Posthumanism>.

_____.2008a. "Transhumanism" (October 20). <http://en.wikipedia.org/wiki/Transhumanism>.

_____.2008b. "Lowest Common Denominator" (March 26). <http://en.wikipedia.org/ wiki/Least_common_denominator>.

_____.2008c. "Infinity" (November 11). <http://en.wikipedia.org/wiki/Infinity>.

_____.2008d. "Transfinite Number" (November 10). <http://en.wikipedia.org/wiki/Transfinite_ number>.

_____.2008e. "Quantum Mechanics" (November 26). <http://en.wikipedia.org/wiki/Quantum_mechanics>.

_____.2008f. "Copenhagen Interpretation" (November 26). <http://en.wikipedia.org/wiki/Copenhagen_interpretation >.

_____.2008g. "Many Worlds Interpretation" (November 26). <http://en.wikipedia.org/wiki/Many_Worlds_Interpretation>.

_____.2008h. "Hyperspace (Science Fiction)" (November 23). <http://en.wikipedia.org/wiki/Hyperspace_(science_fiction)>.

_____.2008i. "Time Travel" (November 26). <http://en.wikipedia.org/ wiki/Time_travel#Time_dilation>.

_____.2008j. "Wormhole" (November 25). <http://en.wikipedia.org/wiki/Wormhole>.

_____.2007. "Linguistics" (March 04). <http://en.wikipedia.org/wiki/Linguistics>.

Wikiversity (WKV). 2008. "Why 10 Dimensions" (November 23). <ttp://en.wikiversity.org/wiki/Why_10_dimensions>

Wittgenstein, Ludwig.1953. *Philosophical Investigations*. G. E. M. Anscombe and R. Rhees, eds. G.E.M. Anscombe, tr. Oxford: Blackwell.

Wolpert, D. H.; & W. G. Macready. 1995. *No Free Lunch Theorems for Optimisation*. NM: Santa Fe Institute.

INDEX

•A•

•D•

•E•

•F•

•G•

•H•

•I•

•J•

•K•

•L•

•M•

•N•

•O•

•P•

•Q•

•R•

•S•

•U•

•V•

•W•

•X•

•Y•

•Z•